# DNA Sequencing II: Optimizing Preparation and Cleanup

# Jones and Bartlett Upper Level Titles in Bioscience

# DNA Sequencing II: Optimizing Preparation and Cleanup

*Edited By*
Jan Kieleczawa, Ph.D.
*Wyeth Research, Cambridge, Massachusetts*

JONES AND BARTLETT PUBLISHERS
*Sudbury, Massachusetts*
BOSTON     TORONTO     LONDON     SINGAPORE

**World Headquarters**
Jones and Bartlett Publishers
40 Tall Pine Drive
Sudbury, MA 01776
978-443-5000
info@jbpub.com
www.jbpub.com

Jones and Bartlett Publishers Canada
6339 Ormindale Way
Mississauga, Ontario L5V 1J2
CANADA

Jones and Bartlett Publishers International
Barb House, Barb Mews
London W6 7PA
UK

Jones and Bartlett's books and products are available through most bookstores and
online booksellers. To contact Jones and Bartlett Publishers directly, call 800-832-0034,
fax 978-443-8000, or visit our website www.jbpub.com. Substantial discounts on bulk
quantities of Jones and Bartlett's publications are available to corporations, professional
associations, and other qualified organizations. For details and specific discount
information, contact the special sales department at Jones and Bartlett via the above
contact information or send an email to specialsales@jbpub.com.

**Production Credits**
Chief Executive Officer: Clayton Jones
Chief Operating Officer: Don W. Jones, Jr.
President, Higher Education and Professional Publishing: Robert W. Holland, Jr.
V.P., Design and Production: Anne Spencer
V.P., Sales and Marketing: William Kane
V.P., Manufacturing and Inventory Control: Therese Connell
Executive Editor, Science: Stephen L. Weaver
Managing Editor, Science: Dean W. DeChambeau
Editorial Assistant, Science: Molly Steinbach
Senior Production Editor: Louis C. Bruno, Jr.
Marketing Manager: Andrea DeFronzo
Text Design: Louis C. Bruno, Jr.
Cover Design: Anne Spencer
Composition: SNP Best-set Typesetter, Ltd., Hong Kong
Printing and Binding: Malloy
Cover Printing: Malloy

**Library of Congress Cataloging-in-Publication Data**
DNA sequencing II : optimizing preparation and cleanup / edited by Jan Kieleczawa
     p. ; cm.
  Includes bibliographical references and index.
  ISBN 0-7637-3383-0 (alk. paper)
  1. DNA—analysis.   2. Nucleotide sequence—Methodology.   I. Kieleczawa, Jan.
  [DNLM:  1.  Sequence Analysis, DNA—methods.   2.  DNA—isolation &
     purification. QU 450 D6297 2006]
  QP624.D1749 2006
  572.8′633—dc22                                                      2005052099
  6048

Printed in the United States of America
10  09  08  07  06      10  9  8  7  6  5  4  3  2  1

# Brief Table of Contents

# Contents

# Contributors

**Deven Atnoor**
Biological Technologies Department, Wyeth Research, Cambridge, Massachusetts
*Chapter 17*

**Katarzyna Bajson**
Biological Technologies Department, Wyeth Research, Cambridge, Massachusetts
*Chapter 12*

**Evgeniya V. Belogubova**
Group of Molecular Diagnostics, N.N. Petrov Institute of Oncology, St. Petersburg, Russia
*Chapter 6*

**Konstantin G. Buslov**
Group of Molecular Diagnostics, N.N. Petrov Institute of Oncology, St. Petersburg, Russia
*Chapter 6*

**Russ Carmical**
Seqwright, Inc., Houston
*Chapter 17*

**Piotr Chomczynski**
Molecular Research Center, Inc., Cincinnati
*Chapter 16*

**Junping Chen**
Plant Stress and Germplasm Development Laboratory, USDA-ARS, Lubbock, Texas
*Chapter 4*

**Christopher M. Clee**
The Wellcome Trust Sanger Institute, The Wellcome Trust Genome Campus, Hinxton, Cambridgeshire, United Kingdom
*Chapter 10*

**Rita De Gasperi**
Department of Psychiatry, The Mount Sinai School of Medicine of New York University, and Psychiatry Research, Bronx Veterans Administration Medical Center, Bronx
*Chapter 2*

**Mitchel J. Doktycz**
Life Sciences Division, Oak Ridge National Laboratory, Oak Ridge, Tennessee
*Chapter 5*

**Markryan Dwyer**
Biological Technologies Department, Wyeth Research, Cambridge, Massachusetts
*Chapter 7*

**Gregory A. Elder**
Department of Psychiatry, The Mount Sinai School of Medicine of New York University, and Psychiatry Research, Bronx Veterans Administration Medical Center, Bronx
*Chapter 2*

**Miguel A. Gama Sosa**
Department of Psychiatry, The Mount Sinai School of Medicine of New York University, and Psychiatry Research, Bronx Veterans Administration Medical Center, Bronx
*Chapter 2*

**Richard A. Gibbs**
Department of Molecular and Human Genetics, Baylor College of Medicine, Houston
*Foreword*

**Erik Gustafson**
Agencourt Bioscience Corporation, Beverly, Massachusetts
*Chapter 9*

**Kaido P. Hanson**
Group of Molecular Diagnostics, N.N. Petrov
Institute of Oncology, St. Petersburg, Russia
*Chapter 6*

**Yoshihide Hayashizaki**
The DNABook Team, Japan Science and
Technology Agency, Kanagawa; Genome
Exploration Research Group and the Core
Group of the Genome Network Project,
RIKEN Yokohama Institute, Kanagawa; and
Genome Science Laboratory, RIKEN Wako
Main Campus, Saitama, Japan
*Chapter 15*

**Andrew Hill**
Biological Technologies Department, Wyeth
Research, Cambridge, Massachusetts
*Chapter 17*

**Peter R. Hoyt**
Department of Biochemistry and Molecular
Biology, Oklahoma State University,
Stillwater
*Chapter 5*

**Evgeny N. Imyanitov**
Group of Molecular Diagnostics, N.N. Petrov
Institute of Oncology, St. Petersburg, Russia
*Chapter 6*

**Darryl L. Irwin**
Institute for Molecular Biosciences, and
Australian Genome Research Facility, The
University of Queensland, Brisbane,
Australia
*Chapter 11*

**Jan Kieleczawa**
Biological Technologies Department, Wyeth
Research, Cambridge, Massachusetts
*Chapters 1, 7, 12, 14, 17*

**Midori Kobayashi**
The DNABook Team, Japan Science and
Technology Agency, Kanagawa, Japan
*Chapter 15*

**Sergei A. Kozyavkin**
Fidelity Systems, Inc., Gaithersburg,
Maryland
*Chapter 13*

**Ekatherina Sh. Kuligina**
Group of Molecular Diagnostics, N.N. Petrov
Institute of Oncology, St. Petersburg, Russia
*Chapter 6*

**Roger S. Lasken**
Center for Genomic Sciences, Allegheny-
Singer Research Institute, Pittsburgh
*Chapter 8*

**Tony Li**
Biological Technologies Department, Wyeth
Research, Cambridge, Massachusetts
*Chapter 1*

**Fei Lu**
Seqwright, Inc., Houston
*Chapter 17*

**Karol Mackey**
Molecular Research Center, Inc., Cincinnati
*Chapter 16*

**Kevin McKernan**
Agencourt Bioscience Corporation, Beverly,
Massachusetts
*Chapter 9*

**Keith R. Mitchelson**
Institute for Molecular Biosciences, The
University of Queensland, Brisbane,
Australia
*Chapter 11*

**Andrey R. Pavlov**
Fidelity Systems, Inc., Gaithersburg,
Maryland
*Chapter 13*

**Nadejda V. Pavlova**
Fidelity Systems, Inc., Gaithersburg,
Maryland
*Chapter 13*

**Ken Paynter**
Seqwright, Inc., Houston
*Chapter 17*

**Lloydia Reynolds**
Biological Technologies Department, Wyeth
Research, Cambridge, Massachusetts
*Chapter 7*

**Jane Rogers**
The Wellcome Trust Sanger Institute, The
Wellcome Trust Genome Campus, Hinxton,
Cambridgeshire, United Kingdom
*Chapter 10*

**Richard Sheldon**
Biological Technologies Department, Wyeth
Research, Cambridge, Massachusetts
*Chapter 7*

**Alexei I. Slesarev**
Fidelity Systems, Inc., Gaithersburg,
Maryland
*Chapter 13*

**Evgeny N. Suspitsin**
Group of Molecular Diagnostics, N.N. Petrov
Institute of Oncology, St. Petersburg, Russia
*Chapter 6*

**Lois Tack**
PerkinElmer LAS, Downers Grove, Illinois
*Chapter 5*

**Alexandr V. Togo**
Group of Molecular Diagnostics, N.N. Petrov
Institute of Oncology, St. Petersburg, Russia
*Chapter 6*

**Gary E. Truett**
Department of Nutrition, University of
Tennessee, Knoxville
*Chapter 3*

**Anthony P. West**
The Wellcome Trust Sanger Institute, The
Wellcome Trust Genome Campus, Hinxton,
Cambridgeshire, United Kingdom
*Chapter 10*

**William W. Wilfinger**
Molecular Research Center, Inc., Cincinnati
*Chapter 16*

**Paul Wu**
Biological Technologies Department, Wyeth
Research, Cambridge, Massachusetts
*Chapters 1, 14, 17*

**Zhanguo Xin**
Plant Stress and Germplasm Development
Laboratory, USDA-ARS, Lubbock, Texas
*Chapter 4*

# Preface

This is the second volume for a developing Jones and Bartlett Publishers series on methods in DNA sequencing technology. Similar to the first volume, *DNA Sequencing: Optimizing the Process and Analysis* (2005), *DNA Sequencing II: Optimizing Preparation and Cleanup* is devoted to techniques related to the analysis of DNA. Its primary goal is to assemble both overview and experimental articles into a single volume focused on various methods for extraction, cleanup, quantitation, and analysis of DNA.

This volume has four, distinct sections. The first and biggest theme includes eight chapters devoted to purification of DNA (and to a lesser degree RNA) from a variety of sources, each serving a different purpose. The choice of method varies when preparing difficult DNAs (Chapter 1), bacterial artificial chromosomes (BACs) (Chapter 2), DNA from animal (Chapter 3) or plants tissues (Chapter 4), DNAs for microarrays (Chapter 5), isolating DNA from preserved sources (Chapter 6), or amplifying whole genomes (Chapter 8). At least where DNA sequencing is concerned, TempliPhi methods to produce template DNAs are as good as other well-established, "classical" protocols for isolation of DNA (Chapter 7).

The second theme is the cleanup of DNA fragments used in specific applications. Chapter 9 describes the uses of magnetic beads for a variety of applications. Complementary to this topic is Chapter 10, which describes setup and operation of a high-throughput pipeline in one of the premiere DNA sequencing centers, Sanger Institute. Chapter 11 compares five protocols for cleanup of polymerase chain reaction (PCR) fragments suitable both for low- and high-throughput applications. Chapter 12 compares twenty different protocols, in four distinct categories, for cleaning of sequencing reactions before loading onto capillary electrophoresis instruments.

The third theme contains two chapters on short- (Chapter 14) and long-term (Chapter 15) storage of DNA under a variety of conditions.

Finally, the last section contains chapters on numerous ways to quantify DNA and RNA (Chapter 16) in which an interesting new parameter (Purity Quotient-PQ) is suggested as a measure of purity of nucleic acids. Chapter 13 describes new DNA polymerases with a range of unusual properties that may find broad applications in many molecular biology techniques. This volume ends with Chapter 17, which briefly touches upon many tools, mainly software programs that may be found in a typical modern biology laboratory.

The scope and content of articles presented in this volume are guided by the knowledge and experience of the contributing authors. Any omission of other methods, protocols, and analysis tools is unintentional. On the other hand, inclusion of any method, protocol, or software tool does not constitute their advertisement and represents the experiences and preferences of the contributing authors. As always, the reader is strongly encouraged to apply and test any method that is suitable for a particular environment and application. Although quite comprehensive, this by no means is an exhaustive or complete assembly of articles on DNA technologies. Notably, we did not include information devoted to extraction of DNA from hazardous and forensic sources. Future volumes will expand on the knowledge in this current text and will encompass many other methodologies (e.g., single nucleotide polymorphism [SNP] analysis, genotyping, high-content screening, etc.) and tools (e.g., primer design, automated fragment assembly and analysis, or gene annotation, etc.). Of course, your suggestions and recommendations would be greatly appreciated.

All of the figures in this volume have been reproduced in black and white. Some, for which color is critical, have been duplicated in color. Such figures are cited in the text as, for example, "Figure 9-11; color Plate 4." Figure legends also indicate if an image appears as a color plate. All color plates are gathered together in an insert near the middle of the book.

## Acknowledgments

The production of any book, but a scientific one in particular, is a complex task and involves many people and groups of people. I like to thank everyone involved with this book for pulling together and making sure that all inquires were answered and deadlines met. Without your dedication and contributed chapters this book would be still in the "dream stage."

In particular, I thank the DNA sequencing staff and management at Biological Technologies Department, Wyeth Research in Cambridge for all the help and encouragement during the course of research and preparation of manuscripts; Stephen Weaver of Jones and Bartlett, Publishers, the science editor of this volume, for superb editorial guidance; the entire editorial and production staff at Jones & Bartlett, including Lou Bruno, Dean DeChambeau, Rebecca Seastrong, Molly Steinbach, and Anne Spencer; and, lastly, my truly supporting wife Carla and children Alex, Kasia, and Michael.

Jan Kieleczawa
Wyeth Research
Cambridge, Massachusetts
November 2005

# Foreword

## DNA Sequencing Is a Technology and a Passion

Twenty five years ago, as a graduate student, I was fascinated by the sight of a DNA sequence ladder exposed on a celluloid film. The image and information on the film were simple and elegant and provided a major contrast to the vague behaviors of cells on plates or in drug or radiation experiments. It was immediately apparent that it would be much easier to do great science on something as stable and qualitatively predictable as primary DNA sequence than to study other biological processes that were so difficult to control. This was a contrast between a simple digital pattern and the complex analog behaviors of living systems. The sight of the primary DNA sequence data changed my view of biology, and I was convinced that there was in fact a way to access the underpinning of biological processes that we had been otherwise limited to studying only at a very high phenomenological level.

This somewhat romantic view was in direct contrast to the harsh reality of actually generating DNA sequence—which required large amounts of radioactivity and other toxic chemicals and a fair knowledge of chemistry. In the beginning, the generation of DNA sequence was tough. Plenty of others shared the DNA bug, however, and the early 1980s were a period of enthusiastic DNA cloning and manipulation. After fears about DNA cloning as a major biological hazard were left behind, the act of gene cloning to simply access the true primary structure of the molecule, to make proteins, and measure expression became a standard. In those heady days before polymerase chain reaction (PCR) and the common use of fluorescence, there was little effort to generate sequence data unless there was a compelling experimental reason to do so. Nevertheless a community of individuals who were intrigued by this technology slowly grew. In general, this community was not regarded in exactly the same way as were mainstream biologists—instead the sequencers were seen as a rather technical lot, with practical—almost "engineering style"—interests.

These biologists who drifted off to study DNA through DNA sequencing activities were quite "edgy" and not part of the mainstream of biology at all. In fact, they were craftsmen, and their craft was to produce beautiful—and reproducible—gels of DNA sequence ladders. Not everyone could do it, and it was a fine compliment to be known in the field as someone with "good hands."

In the mid-1980s molecular biology underwent major changes as PCR was discovered and developed and four-color fluorescent DNA sequencing emerged. Shortly afterward, as we began to get an understanding of the nature of mutations occurring in human cells, a bold proposal emerged to sequence the entire human genome. This idea took seed and gave rise to the Human Genome Project (HGP), which has since been completed, celebrated, and described elsewhere. Since that time, an astounding number of genomes have been completed, and today there is truly a staggering amount of data in the public databases. Even the most imaginative and optimistic individuals are a little surprised by the sheer volume of DNA sequence, now available.

A surprising aspect of the first five years of the thirteen year HGP history was a period of discontinuity between the efforts of the earliest DNA sequence craftsmen and the shaping of the larger program. In the early 1990s, DNA sequencing was thought of as cumbersome and the objective view was that none of the methods at the time would scale properly to allow a 3 billion–base genome to be completed. Therefore, it was concluded, new methods that would be radically different would have to be invented so that the job could be done. That view turned out to be entirely wrong, and the project was eventually completed using only incremental developments of the 1985 technologies. As a consequence, a whole new generation of scientists who began DNA sequencing for the HGP were introduced to the craft of DNA sequencing. Some of them fell in love with sequencing, just like others did a decade before, while others simply enjoyed a more transient relationship.

This contrast between the ongoing efforts of DNA sequencers in the early 1980s and the new wave of sequencing enthusiasts who came to the technology for the HGP is relevant to this book. This wonderful collection of chapters is brought to you mainly by the veterans of the earlier period, where so many of the basic technologies were first developed and tested, and by a few who came later but fell in love with the technology. Each is a pioneer of today's DNA sequencing methods. The issues that are important to them are reflected in different chapter titles—and are essential knowledge that you can really use if you want to generate beautiful DNA sequences of your own.

Many of the chapters deal with methods for getting good DNA with which to perform the sequence reactions. This focus is in the right place as the adage "garbage in—garbage out" was never more true than in this technology! Other chapters focus on a mix of different aspects of the sequencing process. Yet others look at complete sequencing pipelines and at the software tools that are needed. Throughout is the theme of enabling the reader to be fully equipped to tackle all kinds of sequencing demands.

At the time of writing, the sequencing of mitochondria from a wooly mammoth has just been announced. If anyone has any doubt that there remains a need for craftsmanship in DNA sequencing, they can search online to find a kit for wooly mammoth mitochondrial DNA sequencing. It is not there, of course, because it is just another example of how the interesting problems are not standard ones, and once the problems get more varied then you need real skills to solve them. Congratulations, Jan, on a fine book.

Richard Gibbs
Director, BCM–Human Genome Sequencing Center and
Wofford Cain Professor
Department of Molecular and Human Genetics
Baylor College of Medicine
Houston

# Preparation of Difficult DNA Templates Using Seven Different Commercial Methods

Jan Kieleczawa, Tony Li, and Paul Wu
*Wyeth Research, Cambridge, MA*

From the deoxyribonucleic acid (DNA) sequencing point of view, the most critical factor for obtaining good data is the quality and the second important factor is, to some degree, the quantity of the DNA preparation. Over the last 25 years, significant effort has been devoted to the development of efficient and convenient methods yielding sufficient amounts of high quality plasmid DNA preparations for the variety of molecular biology procedures. Virtually all of these methods are variations of the original alkaline lysis protocol developed by Birnboim (3, 4). The basic steps of this protocol include the disruption of bacterial walls using sodium dodecyl sulfide (SDS) at a high-pH solution, and then neutralization and separation of all cell debris from the lysate containing the plasmid DNA of interest. The separation can be accomplished using number of techniques, the most common of which are centrifugation, filtration (sometimes combinations of the two), or binding to magnetic beads (10 and references therein). Extensive review of various modern plasmid preparation methods is described by Flick (9 and references therein). It is worth noting that the quality of DNA preparations using most of commercial kits is sufficient to remove majority of proteins and, therefore, alleviates the need for time-consuming and hazardous-waste generating steps of phenol/chloroform extractions.

Currently, three are more then ten commercial suppliers of the plasmid DNA kits (9) that can be used both in manual and automated modes. Often, suppliers try to distinguish themselves from others by emphasizing yield, quality, or the speed of the protocol, and sometimes they will provide limited data comparing their product to supplier A, B, or C. This practice, though potentially useful, could be misleading as the

*DNA Sequencing II: Optimizing Preparation and Cleanup*
Edited by Jan Kieleczawa
©2006 Jones and Bartlett Publishers

data can be presented in a way that suits particular objective of the supplier. To our best knowledge, there are few independent data comparing the variety of commercial kits on the basis of quality, quantity, and readiness for downstream applications that may include, but are not limited to, DNA sequencing, restriction analysis, transformation, etc. One such attempt is described by Ma et al. (15); however, their study is potentially flawed because the culture growth conditions were not optimal for some of the preparation methods, so the data may be unfavorably skewed, especially when measuring the yield of DNA preparation. On the other hand, such studies can be valuable when comparing different DNA preparation using the same growth conditions.

This chapter examines whether the type of DNA preparation kit affects the quality of the DNA sequencing in series of thirteen difficult templates. The templates used in this study are described in more detail in an earlier study (compared to the list in Kieleczawa's study, one additional (88% GC-rich template) was added for the current study) (12). In short, these templates belong to four categories:

1. GC-rich; 6 templates with 63–90% GC content
2. Those templates with various short repeats (3 templates)
3. Those with an Alu-repeat (1 template)
4. Those with ribonucleic acid interference (RNAi)-plasmid hairpins (3 templates)

Each of these templates was prepared using six different commercial liquid-culture-based kits and the TempliPhi-based method (6, 14).

## Material and Methods

The following commercial plasmid purification kits were used in this study:

1. Amersham's TempliPhi™ DNA Sequencing Template Amplification Kit, catalog #25-6400-50 (Amersham Biosciences, Inc., Piscataway, NJ, USA; Amersham is currently part of GE Healthcare).
2. EdgeBiosystems' Plasmid 96 MiniPrep kit catalog #21914 (EdgeBiosystems, Inc., Gaithersburg, MD, USA).
3. Eppendorf's Perfectprep® Plasmid Mini kit, catalog #0032005454 (Brinkmann, Inc, Westbury, NY, USA).
4. Marligen's Rapid Plasmid Miniprep system®, catalog #11453-016 (Marilgen Biosciences, Inc., Ijamsville, MD, USA).
5. Promega's Wizard® Plus SV Miniprep DNA Purification System, catalog #A1330 (Promega Corporation, Madison, WI, USA).

6. Qiagen's QIAprep® Spin Miniprep Kit, catalog #27104 (Qiagen, Inc, Valencia, CA, USA).
7. Sigma's GeneElute™ Miniprep kit, catalog #PLN-350 (Sigma-Aldrich, St. Louis, MO, USA).

This chapter refers to these kits as Amersham, EdgeBiosystems, Eppendorf, Marligen, Promega, Sigma, or Qiagen kits, and the data are grouped separately for the culture-based kits (kits #2–7) and TempliPhi-based kit (kit #1).

The plasmid DNAs of original 12 difficult templates were transfected into MaxEfficiency® DH10B chemical competent cells according to manufacturer's protocol (Invitrogen, Carlsbad, CA, USA) and grown on SB (superbroth) agar plates with an appropriate selection marker. A single colony was then inoculated into 10 mL of Terrific Broth medium and the culture was grown overnight at 37°C as recommended by standard molecular biology protocols (2, 16). One milliliter of a liquid culture was then used for plasmid preparation using the above kits (#2–7) following the protocols suggested by manufacturers, except for EdgeBiosystems kit (see below). Whenever there was a choice between spin and vacuum protocols, the spin option was used. The following modification to the EdgeBiosystems protocol was used: after the centrifugation step, the pellet was resuspended in 100 μL of TE-RNAse solution (10 mM Tris/1 mM EDTA, pH 8.0 + 0.1 mg/mL of RNAse). Then, 100 μL of lysis buffer was added and mixed by inversion several times. After a 5-minute incubation at room temperature, 100 μL of neutralization buffer was added and the tube was spun down for 5 minutes at 14,000 rpm in an Eppendorf 5415c centrifuge. The supernatant was transferred to a new tube containing 210 μL of 2-propanol. This mixture was then centrifuged for 10 minutes at room temperature at the same speed as above. The pellet was washed with 0.25 mL of 70% ethanol and air dried for 15 minutes before being resuspended in 100 μL of TEsl (10 mM Tris/0.01 mM EDTA, pH 8.0).

The DNAs were checked on a 1% agarose gel (catalog #15510-027; Invitrogen Inc.) using procedures described earlier (2, 16). For assessment of quality and quantity of each DNA, the low DNA mass ladder markers from Invitrogen were used (catalog #10068-013; Invitrogen Inc.).

All fluorescent dye-terminators were purchased from Applied Biosystems (Foster City, CA, USA). Betaine was from Sigma-Aldrich. The set of seven sequence enhancer Rx reagents was from Invitrogen; Single-stranded binding protein (SSB) was from Promega Corporation. All other reagents were of highest available purity. If used, other reagents were purchased from sources as described earlier (11).

The TempliPhi-driven amplification reactions of difficult template DNAs were carried out as described in earlier studies (6, 9, 14). In addition, some template DNAs were amplified in the presence of SSB to take advantage of the DNA-unwinding properties of this protein (5).

## DNA Sequencing

Each DNA was sequenced using four different protocols that are referred to in the tables as described below.

### Protocol #1

Protocol #1 corresponds to a standard ABI-like protocol described in reference 1. Briefly, in a 0.2 mL tube mix x µL of DNA (final amount ~ 200 ng) with 1 µL of 5 µM primer and y µL of TEsl (10 mM Tris/0.01 mM EDTA, pH 8.0) to a final volume of 7 µL. Add 3 µL of 1.5 diluted BigDye™ V3.0 terminator mix (0.8 mL of undiluted Taq mix + 0.4 mL of ABI's 5 × Dye-terminator buffer = 4 × final Taq dilution) and start cycle sequencing protocol (5 sec at 50°C/2 min at 60°C/10 sec at 96°C) × 40.

### Protocol #2

In a 0.2 mL tube mix x µL of DNA (final amount ~ 200 ng) with 1 µL of 5 µM primer and y µL of TEsl (10 mM Tris/0.01 mM EDTA, pH 8.0) to a final volume of 7 µL. Heat-denature at 98°C for 5 minutes (11). Place the tube(s) on ice, add Taq dye-terminator mix as above, and cycle as in Protocol #1.

### Protocols #3 and #4

These two protocols varied between different templates and represent the two best sequencing protocols as described earlier (12). They always included the denaturation step, the same cycling regime as in Protocols #1 and #2, and the presence of an additive or a different dye-terminator mix (dGTP V3.0 or BigDye V3.1).

For the DNAs amplified with TempliPhi method, 1 µL of a product was used for sequencing using three different protocols (equivalent to the above Protocols #1–3).

The sequenced reactions were purified as described (13) and run on an ABI3100 DNA genetic analyzer (purchased from Applied Biosystems Inc.) equipped with a 50-cm capillary and using procedure recommended by the manufacturer. The data were analyzed in terms of read lengths using phred $Q \geq 20$ criteria (8) with each data point being the average of four sequencing reactions.

**Table 1-1.** **Performance of six plasmid preparation methods.**

| DNA Preparation Method | DNA Sequencing Method | | | | | | | |
|---|---|---|---|---|---|---|---|---|
| | Method #1 | | Method #2 | | Method #3 | | Method #4 | |
| | *N* | % | *N* | % | *N* | % | *N* | % |
| EdgeBiosystems | 7 | 58.3 | 4 | 33.3 | 4 | 33.3 | 5 | 38.5 |
| Eppendorf | 0 | 0 | 0 | 0 | 0 | 0 | 0 | 0 |
| Marlin | 2 | 16.7 | 0 | 0 | 0 | 0 | 1 | 7.7 |
| Promega | 0 | 0 | 0 | 0 | 0 | 0 | 0 | 0 |
| Sigma | 0 | 0 | 0 | 0 | 5 | 41.7 | 4 | 30.8 |
| Qiagen | 3 | 25.0 | 8 | 67.3 | 3 | 25.0 | 3 | 23.0 |

Only the longest $Q \geq 20$ read lengths are included. $N$ = number of times the method gave the best results.

## Results and Discussion

Table 1-1 shows the data arranged according to the longest reads with phred $Q \geq 20$ values even if the difference is just a few bases. Despite the relatively small sample size, two methods were the most consistent across all sequencing protocols: EdgeBiosystems and Qiagen. This was especially evident for Protocols #1 and #2, where 60% to 68% of the time using EdgeBiosystems or Qiagen DNA preparation methods resulted in superior data. Once other reagents were included in the sequencing protocol, more even distribution is observed across different preparation methods.

As the differences of just a few bases may be accidental, we recalculated the data from Table 1-1 to include data that are within 10% of the longest reads (Table 1-2). Again, the EdgeBiosystems and Qiagen methods seem to be the most consistent regardless of the sequencing protocol used. The presence of various sequencing additives (Protocols #3 and #4) minimizes differences between all DNA preparation methods. Table 1-3 shows the numerical example of two DNAs prepared using all six methods. The DNA #1 is around 70% GC-rich template and DNA #2 is the shRNA plasmid. As is evident from Table 1-3 and especially for Protocols #1 and #2, the difference in read lengths can be two- to threefold depending on the plasmid preparation method used. Figures 1-1 and 1-2 show chromatograms for two DNAs prepared with two different methods.

We cannot explain these differences easily as the composition of each kit is proprietary and the observation from the agarose gel images are not sufficiently detailed to draw necessary conclusions.

**Table 1-2.** Performance of six plasmid preparation methods.

| DNA Preparation Method | DNA Sequencing Method | | | | | | | |
|---|---|---|---|---|---|---|---|---|
| | Method #1 | | Method #2 | | Method #3 | | Method #4 | |
| | N | % | N | % | N | % | N | % |
| EdgeBiosystems | 11 | 35.5 | 9 | 19.7 | 10 | 18.9 | 11 | 18.7 |
| Eppendorf | 2 | 6.5 | 6 | 13.0 | 6 | 11.3 | 7 | 11.9 |
| Marlin | 6 | 19.3 | 7 | 15.2 | 9 | 17.0 | 10 | 15.3 |
| Promega | 3 | 9.7 | 6 | 13.0 | 7 | 13.2 | 10 | 16.9 |
| Sigma | 2 | 6.4 | 8 | 17.4 | 10 | 18.9 | 12 | 20.3 |
| Qiagen | 7 | 22.6 | 10 | 21.7 | 14 | 20.7 | 10 | 16.9 |

The $Q \geq 20$ read length for any method was counted if it was within 10% of the best performing method. $N$ = number of times the method gave the best results.

The idea to use TempliPhi-derived DNA to sequence difficult templates arose from the mechanism of action of this enzyme (6, 9, 14). We were hoping that rosette-like structures with many "unfinished" strands would facilitate the sequencing, especially through fragments of templates that form strong secondary structures. The SSB facilitated the unwinding and separation of DNA strands, which should have further improved the amplification process. Although the SSB enhanced several-fold the amount of DNA products (2- to 3-fold increase in the presence of 1 µg of SSB in 20 µL reaction, and up to 10-fold increase in the presence of 5 µg of SSB under the same conditions), we did not observe any improvements in the read length for four tested templates. Table 1-4 shows the comparison of $Q \geq 20$ reads for data obtained using either Qiagen or TempliPhi methods. In addition to the twelve templates used in reference 12, we added additional difficult templates in this part of the work (please also see Chapter 7, where the comparison between "classical" DNA purification methods and TempliPhi is described for non-difficult templates). Based on the data presented in the Table 1-4, we did not observe any improvements in read lengths when using TempliPhi-generated templates. In fact, when heat denaturation step was part of a standard sequencing protocol, the read lengths were generally longer with the same templates prepared by classical methods. It is worth noting that recently GE Healthcare has been promoting a new TempliPhi-based sequencing tool for reading through some classes of difficult templates (7). It remains to be seen if this tool applied to the set of templates presented in this work and using modified sequencing protocols will offer any advantages.

**Table 1-3.** The Q ≥ 20 read length for two different plasmid DNAs.

| DNA Preparation Method | DNA Sequencing Method | | | | | | | |
| --- | --- | --- | --- | --- | --- | --- | --- | --- |
| | Method #1 | | Method #2 | | Method #3 | | Method #4 | |
| | DNA1 | DNA2 | DNA1 | DNA2 | DNA1 | DNA2 | DNA1 | DNA2 |
| EdgeBiosystems | 355 | 564 | 496 | 572 | 666 | 628 | 609 | 588 |
| Eppendorf | 300 | 215 | 490 | 374 | 656 | 512 | 612 | 428 |
| Marlin | 391 | 407 | 424 | 516 | 668 | 585 | 607 | 543 |
| Promega | 242 | 341 | 416 | 551 | 669 | 611 | 613 | 580 |
| Sigma | 311 | 285 | 432 | 481 | 677 | 598 | 664 | 510 |
| Qiagen | 218 | 405 | 292 | 579 | 665 | 599 | 619 | 572 |

The DNA #1 and DNA #2 correspond to DNA #1 and DNA #12 in Table 1-4, respectively.

(a)

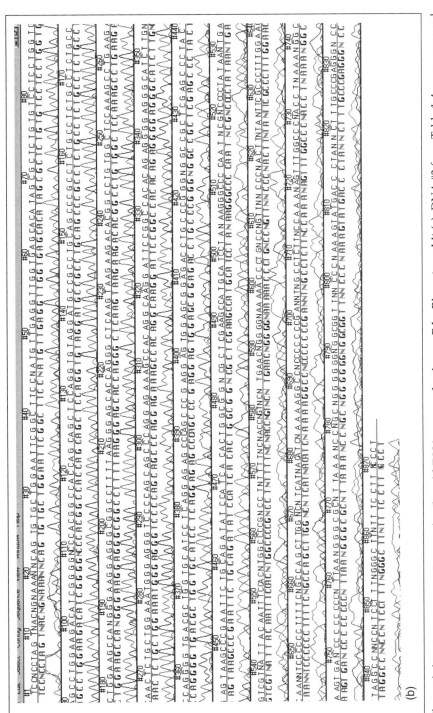

**Figure 1-1.** Chromatogram image of a plasmid DNA prepared using an EdgeBiosystems kit (a). DNA #2 from Table 1-4 was sequenced using standard sequencing protocol. Q ≥ 20 = 509 bases. (b) Chromatogram image of a plasmid DNA prepared using Promega kit. DNA #2 from Table 1-4 was sequenced using standard sequencing protocol. Q ≥ 20 = 378 bases.

(a)

10

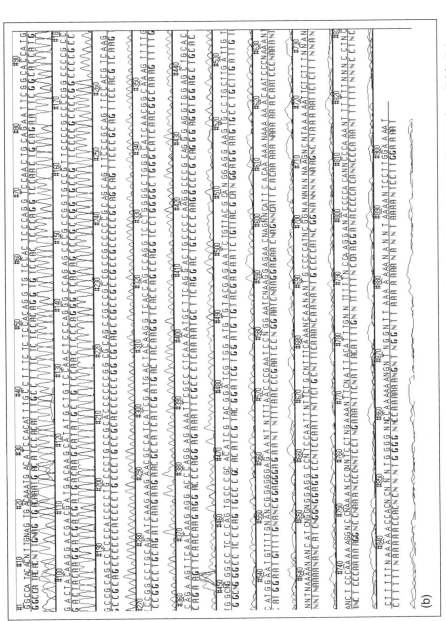

**Figure 1-2.** Chromatogram image of a plasmid DNA prepared using Qiagen kit (a). DNA #6 from Table 1-4 was sequenced using modified sequencing protocol. $Q \geq 20 = 591$ bases. (b) Chromatogram image of a plasmid DNA prepared using Eppendorf kit. DNA #6 from Table 1-4 was sequenced using modified sequencing protocol. $Q \geq 20 = 452$ bases.

**Table 1-4.** Comparison of read lengths between DNAs prepared using either the TempliPhi or Qiagen methods.

| DNA # | Characteristics | Preparation Method | DNA Sequencing Method | | |
|---|---|---|---|---|---|
| | | | **#1** | **#2** | **#3** |
| 1 | CCT/TTTCCC | TempliPhi | 340 | 515 | 594 |
| | | Qiagen | 375 | 495 | 647 |
| 2 | TCC/GCC | TempliPhi | 635 | 640 | 626 |
| | | Qiagen | 599 | 659 | 642 |
| 3 | CTT/CCCT/CCTT | TempliPhi | 0 | 0 | 0 |
| | | Marligen | 337 | 458 | 454 |
| 4 | Alu-repeats | TempliPhi | 128 | 147 | 292 |
| | | Qiagen | 214 | 240 | 259 |
| 5 | 90% GC | TempliPhi | 111 | 278 | 285 |
| | | Marligen | 0 | 292 | 327 |
| 6 | 88% GC | TempliPhi | 143 | 165 | 181 |
| | | Qiagen | 313 | 502 | 587 |
| 7 | 75% GC | TempliPhi | 148 | 610 | 567 |
| | | Qiagen | 565 | 642 | 600 |
| 8 | 63% GC | TempliPhi | 524 | 591 | 596 |
| | | Qiagen | 556 | 575 | 596 |
| 9 | 63% GC + strong hairpin | TempliPhi | <100 | <100 | <100 |
| | | Qiagen | <100 | 358 | 100 |
| 10 | 54% GC | TempliPhi | 634 | 663 | 634 |
| | | Qiagen | 640 | 671 | 643 |
| 11 | shRNA hairpin 1 | TempliPhi | 497 | 487 | 501 |
| | | Qiagen | 609 | 663 | 653 |
| 12 | shRNA hairpin 2 | TempliPhi | 355 | 421 | 571 |
| | | Qiagen | 270 | 343 | 526 |

DNAs #3 and #5 were prepared using Marligen protocol. The data are expressed in terms of read length using $Q \geq 20$ values.

## Conclusions

We have sequenced a number of difficult templates prepared using several commercially available kits to find out if there are any significant differences between them. Two kits (EdgeBiosystems' Plasmid 96 MiniPrep and Qiagen's QIAprep Spin Miniprep) seem to be most consistent and produced the longest reads under most tested experimental conditions. DNAs prepared using remaining kits result in somewhat shorter (though sometimes significantly) reads, especially when a standard DNA sequencing protocol was used. Contrary to our expectations, we did not observe any improvements in read lengths when the difficult templates were prepared using the TempliPhi protocol. In fact, for two templates either we did not get any data or the reads were two- to threefold shorter compared to reads using the classical method for preparation of DNAs. In the future, we will expand this type of work to encompass a wider variety of difficult templates. Another approach would be to collect more templates belonging to the same class and to establish a potential link (if any) between plasmid preparation method and the class of templates.

## References

1. *ABI PRISM® BigDye™ Terminator v3.1 Cycle Sequencing Kit.* 2002. Protocol. Part number 4337035 Rev. A. Foster City, CA: Applied Biosystems.
2. Ausubel, F.M., Brent, R., Kingston, R.E., Moore, D.D., Seidman, J.D., Smith, J.A., and Struhl, K. (eds). 1998. *Current Protocols in Molecular Biology.* New York: John Wiley & Sons.
3. Birnboim, H.C., and Doly, J. 1979. A rapid alkaline extraction procedure for screening recombinant plasmid DNA. *Nucleic Acids Res* 7: 1513–1523.
4. Birnboim, H.C. 1983. A rapid alkaline extraction method for the isolation of plasmid DNA. *Methods Enzymol* 100: 243–255.
5. Chase, J.W., and Williams, K.R. 1986. Single-stranded DNA binding proteins required for DNA replication. *Ann Rev Biochem* 55: 103–136.
6. Dean, R.B., Nelson, J.R., Giesler, T.L., and Lasken, R.S. 2001. Rapid amplification of plasmid and phage DNA using phi29 DNA polymerase and multiply-primed rolling circle amplification. *Genome Res* 11: 1095–1099.
7. GE Healthcare. 2004. *New Tool for Resolving Difficult to Sequence Templates* (brochure). Chalfont, St Giles, UK: GE Healthcare.
8. Ewing, B., Hillier, L., Wendl, M.C., and Green, P. 1998. Base-calling of automated sequencer traces using phred. I. Accuracy assessment. *Genome Res* 8: 175–185.
9. Flick, P.K. 2005. Plasmid preparation methods for DNA sequencing. In: Kieleczawa, J., ed. *DNA Sequencing: Optimizing the Process and Analysis.* Sudbury, MA: Jones and Bartlett, 99–115.
10. Hawkins, T.L., O'Connor-Morin, T., Roy, A., and Santillan, C. 1994. DNA purification and isolation using a solid-phase. *Nucleic Acids Res* 22: 4543–4544.
11. Kieleczawa, J. 2005. Controlled heat-denaturation of DNA plasmids. In: Kieleczawa, J., ed. *DNA Sequencing: Optimizing the Process and Analysis.* Sudbury, MA: Jones and Bartlett, 1–10.
12. Kieleczawa, J. 2005. Sequencing of difficult templates. In: Kieleczawa, J., ed. *DNA Sequencing: Optimizing the Process and Analysis.* Sudbury, MA: Jones and Bartlett, 27–34.

13. Kieleczawa, J. 2005. Simple modifications of the standard DNA sequencing protocol allow for sequencing through siRNA hairpins and other repeats. *J Biomol Tech.* 16: 220–223.

14. Nelson, J.R., Cai, Y.C., Giesler, T.L., et al. 2002. TempliPhi™: Phi29 DNA polymerase-based rolling circle amplification of templates for DNA sequencing. *BioTechniques* 32(suppl): S44–S47.

15. Ma, P., Bacio, W., and Heiner, C. 2003. *Template preparation for DNA sequencing,* Abstract #P-038. In: Savannah, GA: XVth Genome Sequencing and Analysis Conference.

16. Sambrook, J., and Russell, D.W. 2001. *Molecular Cloning,* 3rd ed. Cold Spring Harbor, NY: Cold Spring Harbor Laboratory Press.

# 2

# Isolation and Use of Bacterial and P1 Bacteriophage-Derived Artificial Chromosomes

**Miguel A. Gama Sosa,**[1,2] **Rita De Gasperi,**[1,2] **and Gregory A. Elder**[1,2]

*[1]Department of Psychiatry, The Mount Sinai School of Medicine of New York University, NY, and [2]Psychiatry Research, Bronx Veterans Administration Medical Center, NY*

Bacterial artificial chromosomes (BACs) and P1 bacteriophage-derived artificial chromosomes (PACs) are low copy number circular vectors (1–2 copies/cell) capable of maintaining and stably propagating large inserts of cloned DNA (50–350 kB) in *Escherichia coli* (31, 37, 71). They provide a unique resource for the physical characterization of large blocks of genomic DNA (66). Because of their high genetic stability, they have been used to map and sequence eukaryotic genomes as diverse as those of human, mouse, rat, rice, etc. In addition, they constitute powerful tools for the functional discovery of new genes by complementation of mutations, the identification of key regulatory genomic sequences crucial for proper gene expression, for the creation of authentic models of human genetic diseases in transgenic animals, and for the genetic engineering of plants (9, 20, 27).

One central parameter for the successful use of BACs and PACs in automated DNA sequencing, transfections in cultured cells, and for preparing transgenic animals via pronuclear injection is the quality of the DNA preparations. Methods for the preparation of such large DNAs generally call for growing the recombinant bacteria in media such as LB or NZCYM (68) with the appropriate selection antibiotic until late log-stationary phase (A660 = 1.2–5.0) depending on the subsequent methodology. Although several protocols suggest growing bacteria at 37°C, we recommend growing the cultures at 30°C because certain sequences such as tandem and

*DNA Sequencing II: Optimizing Preparation and Cleanup*
Edited by Jan Kieleczawa
©2006 Jones and Bartlett Publishers

inverted repeats may adopt unusual DNA structures at the higher temperature (slipped DNAs and hairpins, respectively) and be deleted by the bacterial host (18). The bacteria are then harvested by centrifugation and resuspended in a hypotonic buffer usually containing 10 mM EDTA and RNase A. After alkaline lysis with 0.2 N NaOH, 1% SDS, the BAC/PAC DNA is separated from the bacterial genomic DNA and protein by Hirt precipitation with cold concentrated potassium acetate (3, 4). After centrifugation, the BAC/PAC DNA in the supernatant can be concentrated by ethanol or isopropanol precipitation, resuspended, and further purified by several methods depending on subsequent needs. It should be stressed that throughout the BAC/PAC DNA isolation procedures, manipulations leading to DNA shearing and nicking should be minimized by reducing pipetting after cell lysis, mixing reagents gently, never vortexing, and using sterile nuclease-free glassware and plasticware (Figure 2-1).

## DNA Sequencing

When attempting to sequence large templates such as BACs and PACs, DNA quality becomes crucial for obtaining the highest quality and longest sequencing read lengths. One simple protocol that can be utilized is to purify the BAC/PAC DNA from a Hirt supernatant with sequential phenol and chloroform/isoamyl alcohol extractions (24:1, v/v), followed by ethanol precipitation. However, more consistent results are obtained with the use of commercial isolation systems that employ anion exchange column chromatography (Table 2-1). These methodologies employ silica beads of defined size with positively charged diethylaminoethyl (DEAE) groups. Purification of BAC/PAC DNA is based on the interaction between the negatively charged phosphates of the DNA backbone and positively charged DEAE groups on the surface of the resin. Large-construct DNA remains tightly bound to the DEAE groups over a wide range of salt concentrations. Many impurities (such as RNA, protein, carbohydrates, and small metabolites) can be eliminated by washing with medium salt buffers, while the large-construct DNAs remain bound until eluted with a high-salt buffer.

High quality BAC/PAC DNA can also be prepared by cesium chloride (CsCl)-ethidium bromide isopyknic centrifugation by adjusting the density of the solution to 1.59 g/mL with solid CsCl (1, 10). However, this methodology, although yielding DNA of the highest quality, becomes limiting and time consuming when isolating multiple samples. Other purification methods have included utilizing solid phase supercationic polyelectrolytes to remove contaminating proteins (replacing the phenol and chloroform extractions). In this method, the polymer chains are initially extended in a high-energy state because of an overall net negative

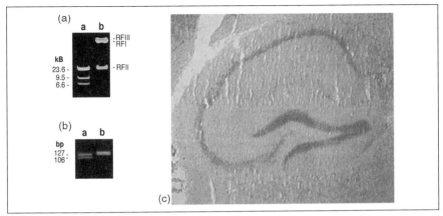

**Figure 2-1.** **Mutagenesis and expression of the human Presenilin 1 gene (*PSEN1*) in transgenic mice.** (a) Agarose gel (0.5%) electrophoresis of purified 204 kB PAC DNA harboring the entire human *PSEN1* gene with a mutation in exon 6 corresponding to the familial Alzheimer's disease (FAD) M146V mutant allele. Mutagenesis was performed via *recA*-mediated homologous recombination (30). The resulting *PSEN1* M146V PAC DNA was purified using a commercial ion exchange purification kit including the treatment with an ATP-dependent exonuclease to remove contaminating bacterial host DNA (Qiagen Inc.) as described (18) and used to prepare transgenic (tg) mice. a, Fragments of λ DNA digested with *Hind*III (0.5 μg); b, supercoiled (RFI), linear (RFII) and nicked (RFIII) forms of the PAC DNA. (b) Genotyping of transgenic animals using *Bsp*HI restriction analysis of PCR amplified genomic DNA. a, normal human *PSEN1* tg; b, mutant *PSEN1* M146V tg. (c) Expression of the human *PSEN1* M1y6V allele in the hippocampus of a transgenic (tg) mouse. In situ hybridization analyses using a digoxigenin-labeled cRNA probe homologous to 3′ untranslated human *PSEN1* sequences (70). Hybridization was visualized with Fab fragments from sheep anti-digoxigenin antibody conjugated with alkaline phosphatase in the presence of nitro blue tetrazolium chloride (NBT) and 5-bromo-4-chloro-3-indolyl-phosphate toluidine salt (BCIP). Note the high levels of expression of the mutant M146V presenilin in the hippocampal dentate gyrus, CA1, CA2, and CA3 regions as observed for the endogenous *PSEN1* (15).

charge. When introduced to protein solutions, the charges are neutralized and the polymer chains collapse to a more favorable energy state, whereas the BAC/PAC DNAs in solution are unreacted (33). Excellent detailed protocols for the isolation of BAC/PAC DNA for sequencing can also be found on the World Wide Web (Table 2-1).

## Transfection into Cultured Cells

Lipid-based DNA delivery systems have proven reliable and highly efficient methods for transfecting mammalian cells. Most preparations

**Table 2-1. Electronic links for BAC/PAC resources.**

**Detailed protocols for the isolation, sequencing and use of BAC/PAC DNAs:**
http://www.med.umich.edu/tamc/BACDNA.html
http://www.genome.clemson.edu/groups/bac/protocols/
   protocols2new.html
http://www.genome.wustl.edu/tools/protocols/
http://www.genome.ou.edu
http://www.cnb.uam.es/~montoliu
http://www.bacpac.chori.org
http://www.protocol-online.org/prot/Molecular_Biology/Sequencing/
   Genomic_Sequencing/
http://www.genome.ou.edu/proto.html
http://www.hbz7.tamu.edu/homelinks/tool/protocol.htm
http://www.biovisa.net/protocol/list_by_inst.php3?id=14
http://www.mcri.edu.au/Downloads/cell_gene_therapy/
   PACBAC_protocols.rtf

**Commercial BAC/PAC DNA isolation systems and reagents:**
http://www.ligochem.com/downloads/procipitatea.htm
http://www.clontech.com/archive/OCT99UPD/NucleoBond.shtml
http://www.fronine.com.au/plck1.asp
http://www.marligen.com/products/highPurityPlasmid.htm
http://www.nestgrp.com
http://www1.qiagen.com/Products/Plasmid/
   QIAGENPlasmidPurificationSystem/QIAGENLargeConstructKit.aspx

**Commercial BAC modification kit**
http://www.genebridges.com

The reader is encouraged to visit these sites to learn more about specifics of each method. We separated these links to point out detailed protocols for the isolation, sequencing and use of BAC/PAC DNAs as well as for commercial BAC/PAC DNA isolation systems and reagents.

consist of a cationic polymer or amphophile, sometimes in combination with a neutrally charged fusogenic lipid. Cationic lipids form complexes with DNA through interactions with the negatively charged phosphate backbone resulting in micelles of multilamellar structure, which enclose the DNA molecules. Fusion of the lipid-DNA complex (lipoplex) with the cell membrane results in the entry of the lipoplex into the cytoplasm possibly via an endosomal pathway, with subsequent DNA release and transport into the nucleus. In a systematic study using atomic force microscopy to determine the parameters affecting BAC transfection efficiency into cultured cells, it was found that BAC DNAs prepared using commercial DNA purification kits (anionic-exchange chromatography) showed variable contamination with debris (53). This debris was attributed to dissociation of the resin from the purification columns (53), and/or to trace amounts of contaminating protein (18), which co-eluted with the BAC DNA (18). When lipid-DNA complexes were prepared with these BAC DNA preparations large aggregates that could not enter the cell were produced. However, this debris could be easily removed by a gentle extraction with buffer-saturated phenol:chloroform followed by ethanol precipitation (18, 53). BAC DNA prepared by CsCl isopyknic centrifugation also showed excellent transfection efficiencies. This study concluded that DNA purity and concentration are the most important parameters in the formation of transfection efficient lipoplex complexes. Cell transfection of large artificial chromosomes was also shown to be improved by DNA compaction (52). This method relies on the ability of a small fragment (Z2) of the polydactyl zinc finger protein RIP60 to bind and multimerize on DNA to induce DNA condensation by DNA looping.

Lipofectamine has also been utilized to transfect BAC DNA into stably transformed hamster cell lines such as CHO and DUKX (29). In this study, a BAC transgene containing the murine *Cdc6* gene was found to be expressed at levels that correlated with BAC copy number. In addition, the *Cdc6* gene was regulated normally during the cell cycle and the transfected protein bound to chromatin was degraded during apoptosis and associated with the spindle apparatus during mitosis (29). BAC DNA has also been stably introduced into murine fibroblasts using electroporation. Stable functional complementation of adenosine phosphoribosyltransferase (*aprt*) deficiency was achieved using BAC DNA containing a human *aprt* gene that had been purified using a commercial anion exchange column followed by linearization with *Not I* (28). A transfection method of artificial chromosomes that relies on the use of cationic amphophiles and high frequency ultrasound has been reported to produce transfection efficiencies of up to 53% in COS cells (64). In this method, cultured cells are preincubated with liposomes consisting of the cationic lipid SAINT-2 (N-methyl-4[dioleyl]methylpyridiniumchloride) and the phospholipid dioleylphosphatidylethanolamine (molar ratio 1:1) followed by ultrasound treatment. Interestingly, no detectable delivery occurred when cells were treated alone with either ultrasound or liposomes. It is believed that integration of the lipids into the liposome creates unstable membrane domains that are prone to ultrasound-induced pore formation.

High efficiency BAC DNA delivery into mammalian cells has also been achieved using a complex of psoralen-inactivated adenovirus carrier, and a polycationic DNA condensation agent such as polylysine or polyethyleneimine (PEI) (2). BAC/PEI/adenovirus complexes were reported to give approximately tenfold greater levels of transient gene delivery than the polylysine/adenovirus system and 2 to 3 log higher levels than commercial cationic lipid systems when tested in primary fibroblasts, HeLa and A549 cells. As expected, the transfection efficiency in HeLa cells was inversely proportional to the BAC size: 170 kB, 7.4%; 130 kB, 11.0%; 75 kB, 7.7%; and 7 kB, 39%.

## Animal Transgenesis

Transgenic animals are generated to study gene function in development and disease, to create in vivo models of human genetic disease, and to produce recombinant proteins of biological interest. It is known that host sequences surrounding the transgene integration site can modify transgene expression, leading to ectopic, weak, or sometimes undetectable expression (chromosomal position effects). However, the inclusion of all regulatory elements that are associated with a given transcription unit in a transgenic construct usually insures optimal levels and spatiotemporal patterns of expression that are free of positional effects. Because of their large cloning capacity, large construct DNAs such as BACs and PACs allow inclusion of all the structural and regulatory sequences of an average gene (<100 kB), making them ideal to overcome chromosomal position effects in transgenic animals. Transgenesis using larger loci requires the more tedious use of yeast artificial chromosomes (cloning capacity >2 MB) (20, 21, 54).

Several methods, including some already mentioned above, have been used to purify BAC/PAC DNA for transgenesis. These include gel filtration through Sepharose 4B-CL (Amersham Biosciences, Uppsala, Sweden) (77), CsCl–ethidium bromide isopyknic centrifugation (1, 10), and silica-based anion exchange chromatography from commercial sources (Table 2-1). Gel filtration through Sepharose 4B-CL allows BAC/PAC DNA to be prepared with minimal manipulation, ensuring that it will remain essentially intact and without any significant amount of shearing (77). However, while simple, this method also results in copurification of host genomic DNA. By contrast, CsCl–ethidium bromide isopyknic centrifugation yields supercoiled BAC/PAC DNA of the highest quality, which is free of host chromosomal DNA (1, 10). After ultracentrifugation, ethidium bromide is removed by gentle extractions with 0.1 M NaCl-saturated butanol and CsCl is subsequently removed by dialysis. Silica-based anion exchange chromatography also yields high quality BAC/PAC DNA for microinjection. Removal of contaminating chromosomal DNA with an adenosine 5′-triphosphate (ATP)-dependent exonuclease (Qiagen, Valencia, CA, USA) further decreases the risk of generating transgenic animals that may have also incorporated bacterial genomic DNA (Figure 2-1 and Table 2-1). As mentioned above, a gentle organic extraction of the purified BAC/PAC DNA with phenol chloroform removes any potential debris, such as leaked resin from the column and/or trace amounts of contaminating protein, which may interfere with the pronuclear microinjection process itself or with transgene integration (18, 53). Commercial anion exchange chromatographic systems (Table 2-1) are the easiest to use for this purpose because they do not require the special equipment or laboratory skills required for gel filtration chromatography or CsCl isopyknic centrifugation.

For transgenesis, most studies have microinjected linearized BAC/PAC clones or genomic restriction fragments purified by pulse-field gel electrophoresis (73) or Sepharose 4B-CL chromatography (77). Also, linearization of BAC clones at the unique loxP site of the pBeloBAC vector with a small oligonucleotide containing a loxP sequence and Cre recombinase has been reported (56). However, we as well as others have not found it necessary to either linearize BAC/PAC DNAs or to remove vector sequences (18, 72). The negative impact of vector sequences on transgene expression often noted for small constructs does not seem to occur with genes contained within large DNA constructs. This offers considerable advantages, because the size, complexity, and often lack of a complete and ordered nucleotide sequence make the selection of restriction sites difficult in BAC/PAC DNAs.

In preparing DNA for pronuclear microinjection it is recommended that the PAC/BAC DNA be kept at a concentration of $1 \mu g/mL$ in injection buffer containing $10 mM$ Tris-HCl, pH, 7.5, $0.5 mM$ EDTA, $30 nM$ spermine, $70 nM$ spermidine, and $0.1 M$ NaCl (13, 18–21). The high ionic strength of this buffer ($0.1 M$ NaCl) appears to stabilize large construct DNA molecules in solution (20, 55). The addition of polyamines as compacting agents promotes formation of DNA-polyamine complexes that prevent shearing upon handling and microinjection of large construct DNAs (20, 55). Prepared under these conditions, BAC/PAC DNA samples are stable for several months at 4°C. It must be stressed that accurate determination of DNA concentration in the microinjection preparation is very important. This preferably should be done by comparing the ethidium bromide–ultraviolet (UV)–induced fluorescence of the BAC/PAC DNA with those of known DNA standards such as bacteriophage lambda DNA (Figure 2-1A). In order to prevent nicking, for long-term storage of highly purified BAC/PAC DNA, it is recommended that the DNA samples be stored as ethanol precipitates at low temperature (−20 to −80°C).

## Plant Transgenesis and Transformation

Because plant somatic cells are totipotent, plants can be regenerated from single cells grown in vitro. Therefore, genetic modifications can be made in somatic cells that will develop into adult plants of single-cell origin. BAC/PAC DNA can be introduced physically into protoplasts by electroporation, thereby avoiding the plant cell wall. It can also be introduced across intact cell walls by single cell microinjection and biolistic bombardment of plant cells with DNA-coated gold particles. This latter technique has been used for genetic complementation and gene identification by positional cloning in potatoes (16) and *Arabidopsis* (34), and also for in vivo recombination with genomic DNA in tobacco (9). Alternatively, BAC DNA can be introduced biologically via *Agrobacterium tumefaciens*-mediated DNA transfer (17, 25). BAC vectors have also been engineered for transformation of large DNA inserts into plant genomic DNA. Binary bacterial artificial chromosomes (BIBAC) have been designed that replicate in both *E. coli* and *A. tumefaciens*, and have all the features required for transferring large inserts of DNA into the plant genome (17, 25, 74).

## Retrofitting of BAC and PAC DNAs

During the past five years, an upsurge has occurred in the development of techniques for performing genetic manipulations in large construct DNAs (60). These methods have allowed their use in areas as diverse as preparing animal models of human disease (19, 42, 50), tagging in vivo expression of genes (23, 26), identifying genes within particular chromosomal regions (35, 67), introducing selectable markers (32, 44, 48), and merging BACs with overlapping complementary sequence to obtain complete genetic units (38), or merging of a BAC with a high content of centromeric alphoid sequences with another BAC/PAC containing the genetic information to produce human artificial chromosomes (47). BACs have also been maintained as episomes in human cells by retrofitting with the Epstein-Barr virus (EBV) latent replication origin *oriP* and the viral *EBNA-1* gene (11, 44). Retrofitting has also benefited studies involving the creation of deletion mutants for physical mapping and for functional analyses in which some or all aspects of the native gene expression needs to be preserved (5–8, 75).

Because of their large size, BAC and PAC retrofitting in *E. coli* is based on homologous recombination (recombinogenic engineering or recombineering) mediated by the *E. coli* (*RecA*), lambdoid Rac prophage (*RecE/RecT*), lambda bacteriophage (Redα/Redβ, *exo*, and *bet*) or the bacteriophage P1 (*Cre-loxP*) recombination systems. The main function of *RecA* in DNA metabolism is to promote DNA strand exchange coupled to ATP hydrolysis (12). *RecA*-mediated recombination between circular molecules requires homologous regions of 1 kB or longer. The first successful genetic modification of a BAC clone was accomplished via *RecA*-mediated homologous recombination (77). This consisted of introducing an IRES-

LacZ tag gene into a 131 kB BAC containing the murine zinc finger *RU49* gene. This system uses a recombination plasmid (shuttle vector) with a temperature-sensitive replication initiation protein (*RepA^{ts}*) gene for the identification of cointegrants, the *RecA* gene and the tetracycline-resistance gene (*Tet^R*) for positive selection combined with a counter selection strategy using fusaric acid. The shuttle vector containing the IRES-LacZ tag was trans formed into BAC-containing cells and transformants were selected for *Tet^R* at the permissive temperature of 30°C. Co-integrants were subsequently selected at the non-permissive temperature of 43°C, and double homologous regions resolved first by growing the *Tet^R* clones without Tet selection and then by negative selection with fusaric acid yielding either the original or the IRES-LacZ tag retrofitted BACs.

A similar system was developed for retrofitting PAC clones (*Kan^R*) spanning the human β-globin locus (30). This system utilizes a temperature-sensitive shuttle vector for replication harboring the *recA* gene, a chloramphenicol-resistance marker (*Cm^R*) for positive selection, and the *rpsL+* allele for counter selection with streptomycin. We have recently used this system for the mutagenesis of a PAC harboring the entire human presenilin 1 gene (*PSEN1*) for the expression of the *PSEN1* M146V mutant allele in transgenic mice (Gama Sosa et al., unpublished data; Figure 2-1). A third system for RecA-mediated homologous recombination in BAC/PAC clones also uses a temperature-sensitive shuttle vector for replication with chloramphenicol and kanamycin resistance genes for positive selection and the levansucrase gene (*sacB*) of *Bacillus amyloliquefaciens* for negative selection (39). The RecA function is provided in *trans* from a separate plasmid.

Homologous recombination between a linearized plasmid shuttle vector or a DNA fragment containing homologous target regions and artificial chromosomes has also been done in bacterial strains expressing either the functionally equivalent Rac prophage RecE/RecT or the bacteriophage lambda Redα/Redβ protein pairs (ET recombination and lambda-mediated recombination) (60, 61, 79, 81). RecE and Redα are both 5′–3′exonucleases and RecT and Redβ are annealing proteins. In order to catalyze the homologous recombination, a double-stranded break is initiated by a functional interaction between RecE and RecT or between Redα and Redβ in regions of homology. RecE or Redα then degrade the DNA from the double-stranded breaks, creating 3′-OH single-stranded overhangs. Subsequently, RecT or Redβ binds to the homologous single-stranded regions forming protein-nucleic acid complexes, which, by strand invasion, will be recombined. Because linear DNA is unstable in bacteria due to the *E. coli* RecBCD exonuclease activity, the recombination is assisted with the lambda-encoded *Gam* protein, which inhibits *RecBCD* activity. Alternatively, one can use RecBCD deficient bacterial strains. Similar recombination efficiencies as those found with *RecE/RecT* have been observed with Redα/Redβ. With these systems, a linear targeting DNA carrying a region as short as 35 to 65 nucleotides of homology can be recombined into a circular DNA molecule (60). More efficient recombination occurs when the homologous regions flank the shuttle cassette, although the *RecE/RecT* system has been shown to also recombine internal homologous regions in linear DNA molecules.

Since the initial modification of the mouse *Hoxa* genetic complex in a P1 clone (*neo* insertion upstream of the *Hoxa3* proximal promoter and a 6.2 kB deletion of a region between the *Hoxa3* and *Hoxa4* genes) (60, 79), this technology has been extensively used to engineer large intact DNA constructs (57, 62, 63, 65). The flexibility and reliability of this methodology has been used in various contexts, including precise insertions, deletions, and point mutations, sometimes in combination with counter-selection and site-specific recombination steps (58, 59, 61–63, 65). More recently, a commercial counter-selection BAC modification system, which uses both Red and ET recombination, has been created (Gene Bridges; Table 2-1). Also, Red-mediated homologous recombination has been used to combine overlapping bacterial artificial chromosomes to assemble a full-length tyrosinase-related protein-1 (*Tyrp-1*) gene (80).

Another homologous recombination system makes use of *E. coli* strains (derivatives of DH10B) expressing the *gam* and the *red* recombination genes *exo* and *bet*. The *exo* protein is also a 5′–3′ exonuclease that creates single-stranded overhangs on introduced linear DNA, and *bet* is a single stranded DNA binding protein that protects these overhangs and promotes annealing in the subsequent homologous recombination. The utility of this recombination system in BAC retrofitting was illustrated by targeting the 3′ end of the murine neuron-specific enolase gene (*Eno2*) in a BAC with *cre*, which allowed the generation of transgenic mice expressing Cre in all mature neurons (41). This recombination system has also been used to generate subtle mutations in BAC DNAs using a two-step "hit and fix" method with short denatured PCR fragments (78). The "hit" step consists in generating a 6- to 20-nucleotide change around the base where the mutation has to be generated. The subsequent "fix" consists of reverting these altered nucleotides to the original sequence and simultaneously introducing the desired subtle alteration. Because several nucleotides are changed in each step, PCR primers specific for such alterations can be designed.

The *cre-loxP* recombination system has also been used for introduction of genetic markers (36, 44) and for the development of in vivo site-specific recombination in plant and fungal genomes via *loxP* recombination (9). This system takes advantage of the presence of *loxP* sites in the vector backbone of all BACs and PACs and of the unique specificity of cre recombinase for the 34 bp *loxP* sequence. For retrofitting, prior to the insertion of the DNA into the BAC at the *loxP* site it is necessary to remove or inactivate any high copy origin of replication in the shuttle vector that would make the BAC/PAC multicopy. This can be done by removing the DNA region containing the origin of replication by restriction endonuclease digestion, and recircularization by ligation, followed by its in vitro recombination with purified BAC DNA with *cre* recombinase. The recombined products are then transfected into *E. coli*. This in vitro method has been used to introduce green fluorescent protein (GFP) and *neo* markers into BACs (36). Another approach to retrofitting with *cre-loxP* is to use a shuttle retrofitting vector with the R6K gamma origin of replication, which can only replicate in *E. coli* strains expressing the *pir* gene product, the π protein (22, 32). Once the plasmid is retrofitted into the BAC, the gamma origin is not functional in the *pir* negative *E. coli* DH10B host. More efficient *cre*-mediated retrofitting of BACs has been achieved in vivo by expressing *cre* protein in the *E. coli* host carrying the BAC. One approach uses a shuttle plasmid that in addition to the desired marker of interest harbors a genetic cassette with the tac promoter driving the expression of the fusion gene *GST-loxP-cre* (76). Once transfected into the BAC containing bacteria, expression of the *GST-loxP-cre* results in a functional cre protein that will promote recombination between the *loxP* sites in the retrofitting vector and the BAC. This will result in integration of the shuttle vector into the BAC at the *loxP* site and in the interruption of the *GST-loxP-cre* fusion gene leading to cre inactivation. Alternatively, *cre* can be expressed in *trans* from replication incompetent plasmids cotransfected with the retrofitting shuttle vector.

In vivo *cre*-mediated recombination between two *lox P* sites has been used for the generation of human artificial chromosomes in which a BAC containing >70 kB of centromeric α-satellite sequences, telomeric sequences, a *lox P* site and a selectable marker (*neo*) were recombined with another BAC/PAC to generate a hybrid human artificial chromosome (24, 40, 46, 47, 49). It has also been found that the telomeric sequences are not essential for the de novo chromosome formation (14).

## Future Directions

The ability to clone and sequence large construct DNAs has revolutionized our understanding of eukaryotic genomic organization including the presence and relative positions of genetic regulatory regions. This information has allowed these constructs to be used to prepare transgenic organisms in which the transgene expression mimics the levels and cell-specificity of the gene in its host. The progressive simplification of the methods for isolation, manipulation, and modification (retrofitting) of BAC and PAC DNAs have allowed their use in areas as diverse as DNA sequencing, gene mapping, and transgenesis.

One area in which the use of these large constructs is still in its infancy is gene therapy. Genetic defects have been rescued in several model systems with PAC or BAC transgenes (35, 42, 51, 67, 69), including some in which rescue was not possible with conventional cDNAs (13). For example, transgenic rescue of the galactocerebrosidase-deficiency in the twitcher mouse (a model of human Krabbe's disease) with a human galactocerebrosidase cDNA was only partially successful with the mice living a maximum of 66 days (45). However, the central nervous system (CNS) disease was fully rescued in transgenic twitcher animals with a BAC transgene containing the entire human galactocerebrosidase gene (13). Methods for utilizing the advantages of BACs and PACs for gene therapy in humans remain to be developed. Nevertheless, recently in mice, luciferase expression has been induced in muscle by injection of naked BAC DNA followed by in vivo electroporation (43). Luciferase expression was also seen in liver after injection of naked BAC DNA into the tail vein (43). Thus, despite their large size, BACs and PACs can likely be utilized in many of the same settings where gene therapy with smaller plasmid DNAs have been contemplated, but with the expectation that they will exhibit the tighter cell type specificity and expression levels expected of large construct DNA.

## Conclusion

Large construct DNAs such as BACs and PACs are powerful molecular tools for genomic sequencing, gene mapping and complementation, characterization of regulatory regions, assembly of entire transcription units or genes, and transgenesis. They also offer considerable future promise as gene therapy vectors in situations where tightness of expression is critical, which leads us to agree with others that, "the full potential of large construct DNAs has not yet been realized" (27).

### Acknowledgments

This work was supported by research grants from the Hunter's Hope Foundation, the Multiple Sclerosis Foundation, and by grants AG20139 and AG05138 from the National Institutes on Aging. We would like to thank Dr. Christine Magin-Lachman for reviewing this manuscript and for her helpful comments.

## References _____

1. Antoch, M.P., Song, E.J., Chang, A.M., et al. 1997. Functional identification of the mouse circadian clock gene by transgenic BAC rescue. *Cell* 89: 655–667.

2. Baker, A., and Cotten, M. 1997. Delivery of bacterial artificial chromosomes into mammalian cells with psoralen-inactivated adenovirus carrier. *Nucleic Acids Res* 25: 1950–1956.

3. Birnboim, H.C. 1983. A rapid alkaline extraction method for the isolation of plasmid DNA. *Methods Enzymol* 100: 243–255.

4. Birnboim, H.C., and Doly, J. 1979. A rapid alkaline extraction procedure for screening recombinant plasmid DNA. *Nucleic Acids Res* 7: 1513–1523.

5. Casanova, E., Lemberger, T., Fehsenfeld, S., et al. 2002. Rapid localization of a gene within BACs and PACs. *Biotechniques* 32: 240–242.

6. Chatterjee, P.K., and Coren, J.S. 1997. Isolating large nested deletions in bacterial and P1 artificial chromosomes by in vivo P1 packaging of products of Cre-catalysed recombination between the endogenous and a transposed loxP site. *Nucleic Acids Res* 25: 2205–2212.

7. Chatterjee, P.K., Shakes, L.A., Srivastava, D.K., et al. 2004. Mutually exclusive recombination of wild-type and mutant loxP sites in vivo facilitates transposon-mediated deletions from both ends of genomic DNA in PACs. *Nucleic Acids Res* 32: 5668–5676.

8. Chatterjee, P.K., Yarnall, D.P., Haneline, S.A., et al. 1999. Direct sequencing of bacterial and P1 artificial chromosome-nested deletions for identifying position-specific single-nucleotide polymorphisms. *Proc Natl Acad Sci U S A* 96: 13276–13281.

9. Choi, S., Begum, D., Koshinsky, H., et al. 2000. A new approach for the identification and cloning of genes: the pBACwich system using Cre/lox site-specific recombination. *Nucleic Acids Res* 28: E19.

10. Chrast, R., Scott, H.S., and Antonarakis, S.E. 1999. Linearization and purification of BAC DNA for the development of transgenic mice. *Transgenic Res* 8: 147–150.

11. Coren, J.S., and Sternberg, N. 2001. Construction of a PAC vector system for the propagation of genomic DNA in bacterial and mammalian cells and subsequent generation of nested deletions in individual library members. *Gene* 264: 11–18.

12. Cox, M.M. 2003. The bacterial RecA protein as a motor protein. *Annu Rev Microbiol* 57: 551–577.

13. De Gasperi, R., Friedrich, V.L., Perez, G.M., et al. 2004. Transgenic rescue of Krabbe disease in the twitcher mouse. *Gene Ther* 11: 1188–1194.

14. Ebersole, T.A., Ross, A., Clark, E., et al. 2000. Mammalian artificial chromosome formation from circular alphoid input DNA does not require telomere repeats. *Hum Mol Genet* 9: 1623–1631.

15. Elder, G.A., Tezapsidis, N., Carter, J., et al. 1996. Identification and neuron specific expression of the S182/presenilin I protein in human and rodent brains. *J Neurosci Res* 45: 308–320.

16. Ercolano, M.R., Ballvora, A., Paal, J.U.R., et al. 2004. Functional complementation analysis in potato via biolistic transformation with BAC large DNA fragments. *Mol Breed* 13: 15–22.

17. Frary, A., and Hamilton, C.M. 2001. Efficiency and stability of high molecular weight DNA transformation: an analysis in tomato. *Transgenic Res* 10: 121–132.

18. Gama Sosa, M., De Gasperi, R., Wen, P.H., et al. 2002. BAC and PAC DNA for the generation of transgenic animals. *BioTechniques* 33: 51–53.

19. Gama Sosa, M.A., Friedrich, V.L. Jr., DeGasperi, R., et al. 2003. Human midsized neurofilament subunit induces motor neuron disease in transgenic mice. *Exp Neurol* 184: 408–419.

20. Giraldo, P., and Montoliu, L. 2001. Size matters: use of YACs, BACs and PACs in transgenic animals. *Transgenic Res* 10: 83–103.

21. Giraldo, P., and Montoliu, L. 2002. Artificial chromosome transgenesis in pigmentary research. *Pigment Cell Res* 15: 258–264.

22. Gong, S., Yang, X.W., Li, C., et al. 2002. Highly efficient modification of bacterial artificial chromosomes (BACs) using novel shuttle vectors containing the R6Kgamma origin of replication. *Genome Res* 12: 1992–1998.

23. Gong, S., Zheng, C., Doughty, M.L., et al. 2003. A gene expression atlas of the central nervous system based on bacterial artificial chromosomes. *Nature* 425: 917–925.

24. Grimes, B.R., Schindelhauer, D., McGill, N.I., et al. 2001. Stable gene expression from a mammalian artificial chromosome. *EMBO Rep* 2: 910–914.

25. Hamilton, C.M., Frary, A., Lewis, C., et al. 1996. Stable transfer of intact high molecular weight DNA into plant chromosomes. *Proc Natl Acad Sci U S A* 93: 9975–9979.

26. Heintz, N. 2000. Analysis of mammalian central nervous system gene expression and function using bacterial artificial chromosome-mediated transgenesis. *Hum Mol Genet* 9: 937–943.

27. Heintz, N. 2001. BAC to the future: the use of bac transgenic mice for neuroscience research. *Nat Rev Neurosci* 2: 861–870.

28. Hejna, J.A., Johnstone, P.L., Kohler, S.L., et al. 1998. Functional complementation by electroporation of human BACs into mammalian fibroblast cells. *Nucleic Acids Res* 26: 1124–1125.

29. Illenye, S., and Heintz, N.H. 2004. Functional analysis of bacterial artificial chromosomes in mammalian cells: mouse Cdc6 is associated with the mitotic spindle apparatus. *Genomics* 83: 66–75.

30. Imam, A.M., Patrinos, G.P., de Krom, M., et al. 2000. Modification of human beta-globin locus PAC clones by homologous recombination in *Escherichia coli*. *Nucleic Acids Res* 28: E65.

31. Ioannou, P.A., Amemiya, C.T., Garnes, J., et al. 1994. A new bacteriophage P1-derived vector for the propagation of large human DNA fragments. *Nat Genet* 6: 84–89.

32. Kaname, T., and Huxley, C. 2001. Simple and efficient vectors for retrofitting BACs and PACs with mammalian neoR and EGFP marker genes. *Gene* 266: 147–153.

33. Kelley, J.M., Field, C.E., Craven, M.B., et al. 1999. High throughput direct end sequencing of BAC clones. *Nucleic Acids Res* 27: 1539–1546.

34. Kemp, A., Parker, J., and Grierson, C. 2001. Biolistic transformation of Arabidopsis root hairs: a novel technique to facilitate map-based cloning. *Plant J* 27: 367–371.

35. Kibar, Z., Gauthier, S., Lee, S.H., et al. 2003. Rescue of the neural tube defect of loop-tail mice by a BAC clone containing the Ltap gene. *Genomics* 82: 397–400.

36. Kim, S.Y., Horrigan, S.K., Altenhofen, J.L., et al. 1998. Modification of bacterial artificial chromosome clones using Cre recombinase: introduction of selectable markers for expression in eukaryotic cells. *Genome Res* 8: 404–412.

37. Kim, U.J., Birren, B.W., Slepak, T., et al. 1996. Construction and characterization of a human bacterial artificial chromosome library. *Genomics* 34: 213–218.

38. Kotzamanis, G., and Huxley, C. 2004. Recombining overlapping BACs into a single larger BAC. *BMC Biotechnol* 4: 1.

39. Lalioti, M., and Heath, J. 2001. A new method for generating point mutations in bacterial artificial chromosomes by homologous recombination in *Escherichia coli*. *Nucleic Acids Res* 29: E14.

40. Larin, Z., and Mejia, J.E. 2002. Advances in human artificial chromosome technology. *Trends Genet* 18: 313–319.

41. Lee, E.C., Yu, D., Martinez de Velasco, J., et al. 2001. A highly efficient *Escherichia coli*-based chromosome engineering system adapted for recombinogenic targeting and subcloning of BAC DNA. *Genomics* 73: 56–65.

42. Liao, J., Kochilas, L., Nowotschin, S., et al. 2004. Full spectrum of malformations in velo-cardio-facial syndrome/DiGeorge syndrome mouse models by altering Tbx1 dosage. *Hum Mol Genet* 13: 1577–1585.

43. Magin-Lachmann, C., Kotzamanis, G., D'Aiuto, L., et al. 2004. In vitro and in vivo delivery of intact BAC DNA: comparison of different methods. *J Gene Med* 6: 195–209.

44. Magin-Lachmann, C., Kotzamanis, G., D'Aiuto, L., et al. 2003. Retrofitting BACs with G418 resistance, luciferase, and oriP and EBNA-1—new vectors for in vitro and in vivo delivery. *BMC Biotechnol* 3: 2.

45. Matsumoto, A., Vanier, M.T., Oya, Y., et al. 1997. Transgenic introduction of human galactosylceramidase into twitcher mouse: Significant phenotype improvement with a minimal expression. *Developmental Brain Dysfunction* 10: 142–154.

46. Mejia, J.E., Alazami, A., Willmott, A., et al. 2002. Efficiency of de novo centromere formation in human artificial chromosomes. *Genomics* 79: 297–304.

47. Mejia, J.E., and Larin, Z. 2000. The assembly of large BACs by in vivo recombination. *Genomics* 70: 165–170.

48. Mejia, J.E., and Monaco, A.P. 1997. Retrofitting vectors for *Escherichia coli*-based artificial chromosomes (PACs and BACs) with markers for transfection studies. *Genome Res* 7: 179–186.

49. Mejia, J.E., Willmott, A., Levy, E., et al. 2001. Functional complementation of a genetic deficiency with human artificial chromosomes. *Am J Hum Genet* 69: 315–326.

50. Merscher, S., Funke, B., Epstein, J.A., et al. 2001. TBX1 is responsible for cardiovascular defects in velo-cardio-facial/DiGeorge syndrome. *Cell* 104: 619–629.

51. Michalkiewicz, M., Michalkiewicz, T., Ettinger, R.A., et al. 2004. Transgenic rescue demonstrates involvement of the Ian5 gene in T cell development in the rat. *Physiol Genomics* 19: 228–232.

52. Montigny, W.J., Houchens, C.R., Illenye, S., et al. 2001. Condensation by DNA looping facilitates transfer of large DNA molecules into mammalian cells. *Nucleic Acids Res* 29: 1982–1988.
53. Montigny, W.J., Phelps, S.F., Illenye, S., et al. 2003. Parameters influencing high-efficiency transfection of bacterial artificial chromosomes into cultured mammalian cells. *Biotechniques* 35: 796–807.
54. Montoliu, L. 2002. Gene transfer strategies in animal transgenesis. *Cloning Stem Cells* 4: 39–46.
55. Montoliu, L., Bock, C.T., Schutz, G., et al. 1995. Visualization of large DNA molecules by electron microscopy with polyamines: application to the analysis of yeast endogenous and artificial chromosomes. *J Mol Biol* 246: 486–492.
56. Mullins, L.J., Kotelevtseva, N., Boyd, A.C., et al. 1997. Efficient Cre-lox linearisation of BACs: applications to physical mapping and generation of transgenic animals. *Nucleic Acids Res* 25: 2539–2540.
57. Muyrers, J.P., Zhang, Y., Benes, V., et al. 2000. Point mutation of bacterial artificial chromosomes by ET recombination. *EMBO Rep* 1: 239–243.
58. Muyrers, J.P., Zhang, Y., Buchholz, F., et al. 2000. RecE/RecT and Redalpha/Redbeta initiate double-stranded break repair by specifically interacting with their respective partners. *Genes Dev* 14: 1971–1982.
59. Muyrers, J.P., Zhang, Y., and Stewart, A.F. 2000. ET-cloning: think recombination first. *Genet Eng (NY)* 22: 77–98.
60. Muyrers, J.P., Zhang, Y., and Stewart, A.F. 2001. Techniques: recombinogenic engineering—new options for cloning and manipulating DNA. *Trends Biochem Sci* 26: 325–331.
61. Muyrers, J.P., Zhang, Y., Testa, G., et al. 1999. Rapid modification of bacterial artificial chromosomes by ET-recombination. *Nucleic Acids Res* 27: 1555–1557.
62. Narayanan, K., Williamson, R., Zhang, Y., et al. 1999. Efficient and precise engineering of a 200 kb beta-globin human/bacterial artificial chromosome in *E. coli* DH10B using an inducible homologous recombination system. *Gene Ther* 6: 442–447.
63. Nefedov, M., Williamson, R., and Ioannou, P.A. 2000. Insertion of disease-causing mutations in BACs by homologous recombination in *Escherichia coli*. *Nucleic Acids Res* 28: E79.
64. Oberle, V., de Jong, G., Drayer, J.I., et al. 2004. Efficient transfer of chromosome-based DNA constructs into mammalian cells. *Biochim Biophys Acta* 1676: 223–230.
65. Orford, M., Nefedov, M., Vadolas, J., et al. 2000. Engineering EGFP reporter constructs into a 200 kb human beta-globin BAC clone using GET Recombination. *Nucleic Acids Res* 28: E84.
66. Osoegawa, K., Tateno, M., Woon, P.Y., et al. 2000. Bacterial artificial chromosome libraries for mouse sequencing and functional analysis. *Genome Res* 10: 116–128.
67. Probst, F.J., Fridell, R.A., Raphael, Y., et al. 1998. Correction of deafness in shaker-2 mice by an unconventional myosin in a BAC transgene. *Science* 280: 1444–1447.

68. Sambrook, J., and Russell, D. 2001. *Molecular Cloning, A Laboratory Manual,* 3rd ed. Cold Spring Harbor, NY: Cold Spring Harbor Laboratory Press.
69. Sarsero, J.P., Li, L., Holloway, T.P., et al. 2004. Human BAC-mediated rescue of the Friedreich ataxia knockout mutation in transgenic mice. *Mamm Genome* 15: 370–382.
70. Schaeren-Wiemers, N., and Gerfin-Moser, A. 1993. A single protocol to detect transcripts of various types and expression levels in neural tissue and cultured cells: in situ hybridization using digoxigenin-labelled cRNA probes. *Histochemistry* 100: 431–440.
71. Shizuya, H., Birren, B., Kim, U.J., et al. 1992. Cloning and stable maintenance of 300-kilobase-pair fragments of human DNA in *Escherichia coli* using an F-factor-based vector. *Proc Natl Acad Sci U S A* 89: 8794–8797.
72. Sinn, P.L., Davis, D.R., and Sigmund, C.D. 1999. Highly regulated cell type-restricted expression of human renin in mice containing 140- or 160-kilobase pair P1 phage artificial chromosome transgenes. *J Biol Chem* 274: 35785–35793.
73. Strong, S.J., Ohta, Y., Litman, G.W., et al. 1997. Marked improvement of PAC and BAC cloning is achieved using electroelution of pulsed-field gel-separated partial digests of genomic DNA. *Nucleic Acids Res* 25: 3959–3961.
74. Takken, F.L., Van Wijk, R., Michielse, C.B., et al. 2004. A one-step method to convert vectors into binary vectors suited for *Agrobacterium*-mediated transformation. *Curr Genet* 45: 242–248.
75. Wade-Martins, R., White, R.E., Kimura, H., et al. 2000. Stable correction of a genetic deficiency in human cells by an episome carrying a 115 kb genomic transgene. *Nat Biotechnol* 18: 1311–1314.
76. Wang, Z., Engler, P., Longacre, A., et al. 2001. An efficient method for high-fidelity BAC/PAC retrofitting with a selectable marker for mammalian cell transfection. *Genome Res* 11: 137–142.
77. Yang, X.W., Model, P., and Heintz, N. 1997. Homologous recombination based modification in *Escherichia coli* and germline transmission in transgenic mice of a bacterial artificial chromosome. *Nat Biotechnol* 15: 859–865.
78. Yang, Y.P., and Sharan, S.K. 2003. A simple two-step, "hit and fix" method to generate subtle mutations in BACs using short denatured PCR fragments. *Nucleic Acids Res* 31: E80.
79. Zhang, Y., Buchholz, F., Muyrers, J.P., et al. 1998. A new logic for DNA engineering using recombination in *Escherichia coli*. *Nat Genet* 20: 123–128.
80. Zhang, Y., Muyrers, J.P., Rientjes, J., et al. 2003. Phage annealing proteins promote oligonucleotide-directed mutagenesis in *Escherichia coli* and mouse ES cells. *BMC Mol Biol* 4: 1.
81. Zhang, Y., Muyrers, J.P., Testa, G., et al. 2000. DNA cloning by homologous recombination in *Escherichia coli*. *Nat Biotechnol* 18: 1314–1317.

# 3

# Preparation of Genomic DNA from Animal Tissues

## Gary E. Truett
*Department of Nutrition, The University of Tennessee, Knoxville, TN*

Preparation of genomic DNA from tissues and cell cultures can either be: (a) laborious and time-consuming, requiring multiple steps, long enzymatic digestion times, detergents and organic solvents; or (b) rapid and inexpensive, involving fairly innocuous reagents. The choice depends on the sample type and quality and purity of the DNA desired. While pure DNA was once a necessity for manipulation of genomic DNA, the remarkable specificity of the polymerase chain reaction (PCR) has made DNA purity less critical. In fact, "purification" sometimes leads to the co-isolation of PCR inhibitors that interfere with DNA amplification (4, 22). There are two main methods for rapid and inexpensive preparation of DNA for PCR, alkaline lysis and proteinase K digestion, with many variations, and also several alternatives for isolation of high molecular weight DNA.

## Alkaline Lysis Protocols

Alkaline lysis was initially developed to isolate plasmid DNA from bacteria under the principle that an alkaline solution denatures bacterial DNA, while plasmid DNA remained double-stranded (3). With the introduction of PCR, alkaline lysis was adapted to the preparation of PCR—quality genomic DNA from animal cells and tissues. Samples are incubated briefly in NaOH (27, 28) or KOH (12, 19), and then neutralized with Tris buffer. These reagents are inexpensive, so that the main cost in alkaline lysis protocols is labor and plasticware. The time required for alkaline protocols is also brief; depending on the sample type, DNA preparation might require only five to thirty minutes with sample processing limited to two pipetting steps. These advantages make alkaline lysis protocols particularly useful for high-throughput applications.

A typical example of an alkaline lysis protocol is that of Rudbeck and Dissing (27), who described alkaline lysis of forensic samples, including whole blood, semen or buccal swabs. There are minor processing differences among various types of tissue samples, but the essential protocol is to incubate 5 μL of sample or a dry buccal swab in 20 μL of 0.2 M NaOH or KOH, either at room temperature, or at 75°C for 5 to 30 minutes, and then add 180 μL of 0.04 M Tris-HCl, pH 7.5. Five microliters of lysate are recommended for each 50 μL PCR amplification. Under these conditions, the small alkaline lysis volume, 20 μL, limits the amount of tissue that is efficiently lyzed. Alkaline lysis of larger tissue amounts could be achieved by increasing the volumes of the alkaline lysis and the Tris buffer, or by increasing the alkaline lysis volume and the concentration of the Tris buffer.

A striking feature of alkaline lysis protocols is that the neutralizing buffer is usually in the pH range of 7.5 to 8.5, which is the pH range in which DNA is usually stored and in which many reactions are performed. Because the neutralizing buffer is already near the final target pH, its starting pH limits its buffering capacity; a better choice would be a neutralizing buffer from the opposite end of the pH spectrum from that of the alkaline lysis reagent. By using Tris HCl as the neutralizing buffer, which has a pH of 4.5 when dissolved in water with no pH adjustment, the buffering capacity is improved. This allows a larger volume of alkaline lysis reagent to be used, and a greater amount of tissue to be lysed in a minimal volume. We used this principle to design the hotSHOT (from hot Sodium HydrOxide and Tris) alkaline lysis protocol (28).

The hotSHOT method (Table 3-1) was designed for high-throughput DNA preparation in a format consistent with resources available in genetics laboratories, principally to prepare DNA for PCR genotyping. Two reagents, 25 mM NaOH/0.2 mM EDTA (the EDTA is optional) and 40 mM Tris HCl (*not Tris base!*), are prepared by dissolving the reagents in water without adjusting their pH value. This makes an alkaline lysis buffer with a pH of 12 and an acidic neutralizing reagent with a pH of 4.5, reagents that are stable at room temperature. (A novice error that leads to failure of the protocol is the use of Tris base, which is often labeled by its distributors as simply Tris, to make the neutralizing buffer. The use of Tris base to make the neutralizing buffer will prevent neutralization and can cause PCR amplification to fail.) Small tissue samples are collected in either 96-well plates or PCR tubes, and 75 μL of 25 mM NaOH/0.2 mM EDTA are added to the tubes. The tubes are heated to 95°C for 5 to 30 minutes, depending on the tissue. Some tissues do not completely solubilize, but unlysed tissue does not interfere with PCR. After the incubation 75 μL of 40 mM Tris, pH 4.5 are added to the tubes to generate

**Table 3-1.** HotSHOT alkaline lysis protocol for preparation of PCR quality DNA.

**Reagents**

**Alkaline lysis reagent (25 mM NaOH/0.2 mM EDTA)**
1 g NaOH
0.074 g Na$_2$ EDTA
1 liter H$_2$O
Do not adjust the pH, it will be about 12.
Store at room temperature.

**Tris HCl buffer (40 mM Tris HCl)**
6.3 g Tris HCl (*not Tris base!*)
1 liter H$_2$O
Do not adjust the pH, it will be about 5.
Store at room temperature.

**Abbreviated Protocol**

1. Collect a small tissue sample in a PCR tube or 96-well plate.
2. Add 75 µL alkaline lysis reagent.
3. Heat to 95°C for 5 to 30 minutes.
4. Add 75 µL 40 mM Tris HCl, pH 4.5 to neutralize.
5. The lysate is ready for PCR.

PCR-quality DNA in a solution of 20 mM Tris/0.1 mM EDTA with a pH of about 8. This lysate can make up as much as 50% of the PCR volume (Figure 3-1). In fact, the product of the hotSHOT lysis from a single neonatal rat toe, for example, can be diluted many times to provide sufficient DNA for thousands of genotyping reactions (Figure 3-2). A second common error in this protocol is the use of too much tissue in the alkaline lysis reaction. An ear punch, a single mouse toe, or a 2 mm section of mouse tail provides plenty of DNA for many PCR amplifications. Too much tissue, such as 1 cM section of tail, is not efficiently solubilized by the alkaline reagent and can lead to PCR failure.

Despite alkaline lysis DNA being unpurified, samples perform remarkably well in PCR amplification. For example, we compared the performance of DNA prepared by the hotSHOT method to DNA prepared by a traditional proteinase K digestion, phenol/chlorofrom extraction and alcohol precipitation, we found that the hotSHOT DNA produced more of the specific PCR product and less nonspecific PCR product (Figure 3-3) (28). Furthermore, hotSHOT DNA had a lower PCR failure rate than DNA prepared by phenol-chloroform extraction of proteinase K digests

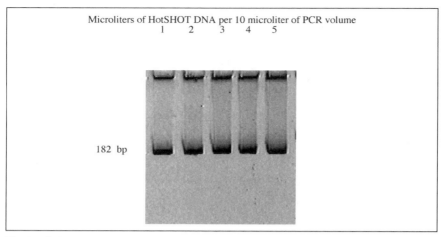

**Figure 3-1. Effect of hotSHOT lysate volume on PCR amplification.** To determine whether the volume of hotSHOT DNA influenced the quality of the PCR product, 1 to 5 μL of hotSHOT DNA were amplified in 10 μL PCR volumes. Forward and reverse primers were combined at 200 nM each with PCR buffer, 3 mM MgCl$_2$, 0.2 mM each dATP, dCTP, dGTP and dTTP, 0.12 U Taq DNA polymerase (Promega, Madison, WI, USA), in a 10 μL volume. Reactions were covered with 10 μL mineral oil and amplified in a PTC-100 thermal cycler (MJ Research, Watertown, MA, USA) for 40 cycles at 95°C for 40 seconds and 60°C for 45 seconds. Following amplification, 10 μL loading dye were added to each well (4 μL 60% sucrose and 1 mM cresol red plus 6 μL water (15). Twelve microliters of the reaction were loaded on a 12% vertical polyacrylamide mini-gel (8 × 10 cm) and electrophoresed at 165 V for 45 min in 1 × TBE buffer (89 mM Tris/89 mM boric acid/2 mM EDTA). Gels were stained for about 2 min in 1 × TBE containing 1 μg/mL ethidium bromide and imaged under UV light using an Ambis Imaging System (Scanalytics, Billerica, MA, USA). Amounts of hotSHOT DNA ranging from 10% to 50% of the PCR volume gave equivalent PCR results. This suggests that PCR is insensitive to impurities in the hotSHOT DNA preparation. The larger volumes are easier to pipette accurately with multichannel pipettes.

(28). Several possibilities may contribute to the better performance of hotSHOT DNA. First, isolation of DNA by phenol-chloroform extraction of proteinase K digests can co-isolate PCR inhibitors (4, 22). Also, alkaline lysis might reduce the complexity of the nucleic acid pool in a sample because highly basic solutions hydrolyze RNA. Finally, Bourke et al. (4) demonstrated that treating DNA samples with NaOH eliminates PCR inhibitors and improves amplification. Forensic DNA samples were prepared by phenol-chloroform extraction followed by Microcon-100 column purification. DNA samples that failed to amplify were loaded on Microcon-100 columns and washed with 0.4 N NaOH followed by centrifugation for three washes. After the third wash, DNA was neutralized

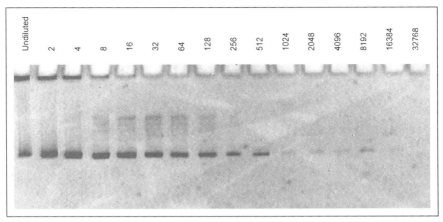

**Figure 3-2.** **Effect of hotSHOT lysate dilution on PCR.** The volume of the hotSHOT solution is 150 μL, which is sufficient for about 130 PCR assays if 1 μL volumes are used. For high throughput genome scans, it could be desirable to dilute the DNA so that larger volumes can be pipetted and larger numbers of assays can be performed. To estimate the effect of dilution of hotSHOT DNA, we prepared the hotSHOT DNA from a single toe of a 5-day-old mouse. The concentration of DNA in the undiluted sample was 60 ng/μL, as measured with fluorometric assay that uses Hoescht dye (6). The hotSHOT DNA was serially diluted with equal volumes of 10 mM Tris/0.1 mM EDTA, pH 8. PCR reactions were prepared with 1 μL of DNA in 10 μL PCR volumes, as described in Figure 3-1. The amount of PCR product decreased with dilution, but a specific band was obtained even at a 512-fold dilution. These results indicate that substantial dilution of hotSHOT DNA could be made. All reagents mentioned in described protocols were of highest available purity.

by washing with 400 mL of 10 mM Tris, pH 7.5, and finally eluted in 10 mM Tris, pH 7.5. This procedure caused 28 samples that previously failed to amplify to become amplifiable, suggested that PCR inhibitors were either destroyed or washed away from the DNA by NaOH washes.

We have also adapted the hotSHOT protocol for use with fecal samples of mice during eradication of *Helicobacter* infections (29). Because whole fecal pellets are the preferred sample size, we vortexed fecal pellets in 2 mL of alkaline lysis reagent for each fecal pellet in a 5 mL polypropylene tube, incubated the sample at 95°C in a water bath for 10 minutes, and neutralized the solution with 2 mL of 40 mM Tris-HCl, pH 4.5. Using this method, we were able to detect DNA from fecal bacteria and also mouse genomic DNA present in the fecal sample. This makes it possible to perform completely noninvasive genotyping, if necessary.

The main advantages of alkaline lysis methods are that they are inexpensive, rapid, free of organic solvents, and they eliminate PCR inhibitors.

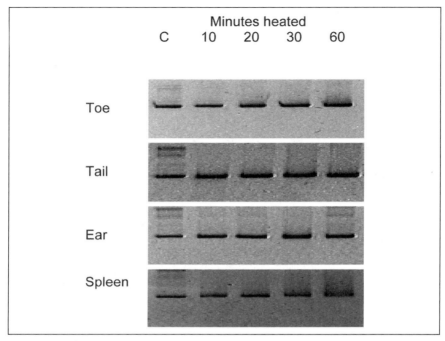

**Figure 3-3.** Effect of heating time and tissue on hotSHOT DNA amplification. The performance of hotSHOT DNA in PCR was tested by amplifying a 182 bp PCR product and compared to the performance of DNA prepared by a conventional proteinase K digestion, followed by phenol:chloroform extraction and ethanol precipitation as described later in this chapter. The PCR product from the conventional extraction is in the first lane (C). The variation in soluble DNA among tissues and times had little influence on the amplification of the PCR product. These results suggest that the amount of heating time is not critical to the outcome and may be chosen for the convenience of the investigator. We use primarily neonatal toes and ear punches, and chose a 30 minute heating time. Note also that the control DNA, labeled C, produced more nonspecific PCR products than the hotSHOT DNA. This was confirmed in another experiment that compared hotSHOT DNA to DNA prepared by other methods and found that hotSHOT produced the cleanest PCR products (data not shown). Reproduced from Truett, G.E., Heeger, P., Mynatt, R.L., et al. 2000. Preparation of PCR-quality mouse genomic DNA with hot sodium hydroxide and tris (hotSHOT). *Biotechniques* 29: 52–54 with permission of Eaton Associates.

One disadvantage is that hotSHOT DNA is sheared and cannot be used for Southern blots, although PCR products of at least 1.9 kb can be amplified (28). Another disadvantage is that the DNA may degrade faster unless it is stored frozen. If samples are needed for archiving, it is more efficient and probably more reliable to store larger or multiple tissue samples that can be processed again, if needed.

## Proteinase K Digestion

Another popular alternative for rapid preparation of unpurified PCR-quality DNA is simple proteinase K digestion, followed by heat inactivation of proteinase K. Proteinase K is a protease isolated from the saprophytic fungus, *Tritirachium album*; it is stable in wide pH and temperature range, but denatures above 65°C. Originally, proteinase K was employed to remove histones and other proteins that bound so tightly to DNA that they resisted extraction with phenol alone (14). By lysing cell nuclei with detergents and digesting the lysate with proteinase K, protein could be dislodged from DNA, and then easily removed with phenol extraction. Since the introduction of PCR, it has become apparent that purification of DNA away from cell lysates is not essential. Many tissues can be digested with proteinase K with little processing to generate PCR-quality lysates, without the use of organic solvent to remove protein and lipids.

Proteinase K digestion of tissue samples collected from mice and rats has been widely used for genotyping. For example, Malumbres et al. (22) demonstrated that DNA for PCR can be prepared from mouse tissues collected when transgenic mice are marked by toe-clipping for identification, by digesting the toe with proteinase K in 200 µL of GNT-K buffer (50 mM KCl, 1.5 mM $MgCl_2$, 10 mM Tris-HCl, pH 8.5, 0.01% gelatin, 0.45% Nonidet P-40, 0.45% Tween-20 and 100 µg/mL proteinase K) at 55°C for 2 hours and then denaturing the proteinase K by heating to 95°C for 15 minutes. This produces high-molecular weight DNA that is suitable for PCR amplification of PCR products up to 4.5 kb or greater.

Proteinase K digestion is also useful for small tissue samples. McClive and Sinclair (23) used a slight modification to prepare DNA from the yolk sac of mouse embryos for sex identification. A small piece of yolk sac is placed in 50 µL digestion buffer (50 mM KCl, 10 mM Tris-HCl (pH 8.3), 0.1 mg/mL gelatin, 0.45% NP40, 0.45% Tween-20, 200 µg/mL proteinase K) for 15 minutes at 55°C, then heated to 95°C for 10 minutes to denature the proteinase K. One microliter of lysate is used in a 20 µL PCR volume. The smaller, thinner tissue sample presumably aids in rapid digestion to reduce incubation time from 2 hours to 15 minutes. Katoh et al. (17) have also used proteinase K digestion to prepare small tissue samples for PCR. They placed 8-cell embryos in 20 µL of 1× PCR buffer with 40 µg/mL proteinase K, covered the buffer with paraffin oil to inhibit evaporation, incubated the samples at 55°C for 60 minutes, then 95°C for 5 minutes, and used 3.5 to 5 µL, 17.5% to 25% of the lysate volume, in PCR. This approach was developed for genetic quality control of frozen embryos. Its simplicity should make it useful for other very small samples. Use of the entire lysate volume might make this approach sufficiently sensitive to achieve amplification from fewer cells. A commercial proteinase K based reagent,

Quantilyse (Hamilton Thorn Biosciences, Beverly, MA, USA), with a proprietary buffer composition, has been shown to allow amplification of DNA from single cells (26). Single lymphocytes were digested with $10\,\mu L$ Quantilyse for 30 minutes at 50°C and then for 95°C for 10 minutes. The entire sample was amplified using quantitative real-time PCR to detect DNA from single cells.

The main advantages of simple proteinase K digestion and denaturation are its low cost, minimal processing (addition of a single reagent to the sample), and rapid processing time. The main limit to processing time is probably the tissue size, as larger samples require more time to be digested with proteinase K. A significant advantage of proteinase K digestion methods over alkaline lysis is that they generate high molecular weight DNA. However, high molecular weight DNA is not usually required for PCR and most DNA manipulations that require high molecular weight DNA might perform better with purified DNA. PCR and real time PCR seems to be the exceptions because isolation of high molecular weight DNA from proteinase K-digested samples can also isolate PCR inhibitors.

## Isolation of High Molecular Weight DNA from Proteinase K Digests

Purification of high molecular weight DNA from proteinase K digests can be performed by several approaches. In the past, phenol/chlorofrom extraction was the most popular method. Small tissue samples are usually digested in buffer that includes proteinase K and a detergent to enhance its activity, such as 10 mM Tris pH 8, 1 mM EDTA, 1% SDS, 250 to $750\,\mu g/mL$ proteinase K for 4 hours or overnight at 55° to 65°C. This may be performed in a water bath, but it generally proceeds better if gentle mixing can be provided in a hybridization oven or similar device. For larger tissues, such as spleen and liver, it is preferable to first mash the tissue in 10 mM Tris, 1 mM EDTA, pH 8 with a glass dounce tube, or by mashing smaller amounts of tissue between frosted glass slides. The tissue is then added to a centrifuge tube, to which sufficient 10% Triton X-100 is added to reach a concentration of 1% Triton X-100, which lyses the plasma membrane but not the cell membrane. The cell nuclei are pelleted by centrifugation, and then resuspended in 10 mM Tris, pH 8, 1 mM EDTA with vortexing. Once the pellet is suspended, sufficient 10% SDS is added to reach 1% SDS, which ruptures the nuclear membranes. Proteinase K is added to 250 to $750\,\mu g/mL$ and the tube is incubated overnight at 55° to 65°C, with mixing, if possible. It is critical that the sample digest until it forms a freely moving solution with no semisolid regions.

Once the sample is completely digested, it can be next extracted with phenol and chloroform. First, one volume of phenol, pH 8, is added to the sample and mixed gently without vortexing. An option at this point is to add Dow Corning high vacuum grease (Fisher Scientific, Hampton, NH, USA; catalog number 14-635-5D) to the tube. When the sample is centrifuged, vacuum grease migrates to the interface between the lower organic phase and upper aqueous phase, allowing a subsequent extraction to be performed in the same tube and improving the separation of interfacial proteins into the phenol phase. Vacuum grease can be loaded into a 10 mL syringe with the barrel cut off, or it can be loaded into 12.5 mL Eppendorf repeator pipette for easy dispensing when the volumes are small. After centrifugation, the upper aqueous phase must be carefully transferred to a clean tube if no vacuum grease is used. If vacuum grease is used, one volume of chloroform is added for another extraction. If vacuum grease is not used, it is usually necessary to perform a second extraction with $\frac{1}{2}$ volume phenol and $\frac{1}{2}$ volume chloroform, and a third extraction with 1 volume chloroform. After all of the organic extractions are completed, the aqueous phase is transferred to a clean tube, and 10 M ammonium acetate to a final concentration of 2.5 M ammonium acetate are added to the sample. One volume of isopropanol is added to precipitate the DNA. The DNA can be centrifuged if there is a small amount, or it can be spun out on a pipette tip, drained against the side of the tube, and transferred to a clean tube. The DNA is then washed with 70% ethanol, and either centrifuged or picked up and drained against the side of the tube. It is important to drain as much ethanol as possible before dissolving the DNA in 10 mM Tris, 1 mM EDTA, pH 8 (T10E1 buffer). Depending on the amount of DNA isolated, it might take a few hours or a day to completely dissolve the DNA.

This approach to isolating high molecular weight DNA produces DNA that can be cut with restriction enzymes, used for Southern blots, or used for cloning projects. However, it usually requires at least one day of labor-intensive processing and can take as long as three days to complete. Furthermore, it is unnecessary and sometimes even counterproductive if the DNA is only to be used for PCR, because it can isolate PCR inhibitors.

## Isolation of DNA from Proteinase K Digests without Organic Solvents

Because the organic solvents phenol and chloroform are toxic and must be discarded as organic waste, many investigators prefer not to use them. An alternative to organic extraction of DNA involves precipitation of proteins with a high salt concentration, followed by precipitation of DNA with alcohol, based on Miller et al. (24). The basic protocol begins with proteinase K digestion in 10 mM Tris, 1 mM EDTA, 0.01 M NaCl, 1% SDS, pH 8. After digestion is complete, 6 M NaCl is added to a final concentration of 1.5 M NaCl and the samples are mixed by inversion, causing proteins to precipitate. The proteins are centrifuged out and the supernatant is transferred to a new tube. Two volumes of ethanol are added to precipitate the DNA. DNA can either be centrifuged, if there is a small amount, or the DNA fibers can be spun out on a pipette tip, drained at the side of the tube, and transferred to a third tube. The DNA is dissolved in 10 mM Tris, 1 mM EDTA, pH 8, for several hours or overnight. When large amounts of DNA are isolated, it is useful to dissolve the DNA with gentle mixing on a nutator.

The advantages of the protein salting out protocol are that high molecular weight DNA can be prepared with no organic solvents, at a quality that is useful for restriction enzyme digestion and Southern blotting. The DNA is also stable enough to be stored in a refrigerator for long periods without substantial degradation (G. Truett; data not shown). Resins are also used to isolate DNA after proteinase K digestion to produce PCR-quality DNA that is free of inhibitors (5). After proteinase K digestion, Chelex 100 is added to 20% of the volume. The samples are incubated for 30 minutes at 55°C and 6 M NaCl is added to a final concentration of 1.5 M to precipitate proteins. After centrifugation, the supernatant is transferred to a clean tube and 2 volumes of ethanol are added to precipitate DNA. The DNA can either be centrifuged or spun out with a pipette tip and dissolved in 10 mM Tris, 1 mM EDTA, pH 8. The primary advantages of the Chelex-100 resin technique is that it produces PCR quality DNA free of inhibitors and isolated from other tissue components, so that it stores well with refrigeration. However, it does increase the processing time over alkaline lysis and proteinase K lysate methods.

## Isolation of Genomic DNA with Guanidinium Salts

Guanidinium salts are widely used for the isolation of RNA because they rapidly denature proteins, including RNase, but they are also used for isolation of DNA. The use of guanidinium salts for the rapid isolation of DNA is the basis of some commercial DNA isolation methods, such as DNAzol (Molecular Research Center, Inc., Cincinnati, OH, USA), which is a combination of guanidinium salts and detergents from which DNA can be selectively precipitated with ethanol (10). Several other commercial methods for isolation of DNA and RNA from the same sample are based on Chomczynski's method using guanidinium thiocyanate, phenol, and chloroform to partition RNA and DNA into different phases, so that each can be precipitated separately (9). Guanidinium salts have also been used to isolate sperm DNA by a relatively simple method (16). Sperm were washed with phosphate-buffered saline and pelleted by centrifugation. They were lyzed in 6 M guanidinium, 30 mM sodium citrate, pH 7.0, 0.5% sarkosyl, 200 µg/mL proteinase K, and 0.3 M β-mercaptoethanol, and then incubated at 55°C for 3 to 4 hours. Two volumes of isopropanol were added and the tube was inverted until DNA precipitated. This DNA was suitable for restriction digestion and Southern blotting. This and other guanidinium salt methods are more involved than the DNAzol method, and may have no significant advantage over this proprietary reagent.

## Isolation of Fetal DNA from Maternal Blood

DNA can be prepared from whole blood, blood cells, or buffy coat cells by virtually any of the protocols described above for tissues. An emerging area of research has developed over the detection of fetal DNA in maternal blood. It has been long recognized that fetal DNA is present in maternal blood, suggesting that maternal blood might be useful for fetal diagnostics. The logical expectation that fetal DNA would be found mainly in the cellular fraction of maternal blood might have delayed progress in this area. Instead, it appears that most fetal DNA is actually found in the cell-free fraction of maternal blood, the plasma, in fragmented form (7, 20, 21). Fetal DNA becomes detectable in maternal blood by five weeks of gestation, increasing to parturition and rapidly clearing after parturition (2). The amount of fetal blood in maternal plasma might also correlate with fetal disorders, such that measuring the quantity may be as important clinically as testing for specific DNA sequences.

The detection of fetal DNA in maternal blood poses special problems. The two foremost challenges are that fetal DNA needs to be concentrated for optimal sensitivity and that care must be taken to limit the introduction of more maternal DNA during isolation. The requirement of concentration of DNA eliminates shortcut protocols, such as alkaline lysis and proteinase K digestion. When multiple DNA isolation procedures are compared, there is much less variation in total fetal DNA isolated by the different procedures than there is in the total maternal DNA that is isolated (8, 13); this suggests that the blood processing protocols cause rupture of maternal cells. Some have suggested that the addition of formaldehyde to the blood collection tube substantially reduces the amount of maternal DNA in the plasma fraction, presumably by stabilizing the cell membranes and limiting hemolysis (1, 13); however, recently this has been challenged (11). The primary need in this rapidly developing area is the development of standardized methods for collection and processing of maternal blood to minimize maternal cell rupture. Once these challenges are overcome, an optimal DNA isolation and concentration protocol can be developed.

## Preparation of Genomic DNA for In Silico Hybridization

The arrival of the polymerase chain reaction (PCR) technology for the iso-lation of specific sequences revolutionized molecular biology and pro-moted the development of rapid, simple, and inexpensive technologies for DNA preparation. As DNA sequence detection by biosensors and microarrays becomes increasingly important for the detection for specific pathogens, species sources, and genetic variants in humans and other animals, in silica preparation and detection of DNA will become neces-sary. The most efficient DNA detection methods will also be independent of DNA amplification. Direct detection of DNA without labeling the DNA would greatly simplify this process. Mir and Katakis (25) have proposed the use of competitive displacement of labeled probes as a solution to this critical function. Li et al. (18) have proposed another strategy in which DNA hybridization alters impedance of a microelectrode. Once practical solutions to direct DNA detection are effectively solved, these methods will need to prepare DNA from small sample sizes, including cell lysis, removal of DNA binding proteins, reducing DNA strand length, and denaturing double-stranded DNA through either enzymatic, chemical, mechanical, ultrasonic, or electrical approaches.

## References

1. Benachi, A., Yamgnane, A., Olivi, M., et al. 2005. Impact of formaldehyde on the in vitro proportion of fetal DNA in maternal plasma and serum. *Clin Chem* 51: 242–244.
2. Birch, L., English, C.A., O'Donoghue, K., et al. 2005. Accurate and robust quantification of circulating fetal and total DNA in maternal plasma from 5 to 41 weeks of gestation. *Clin Chem* 51: 312–320.
3. Birnboim, H.C., and Doly, J. 1979. A rapid alkaline extraction procedure for screening recombinant plasmid DNA. *Nucleic Acids Res* 7: 1513–1523.
4. Bourke, M.T., Scherczinger, C.A., Ladd, C., and Lee, H.C. 1999. NaOH treatment to neutralize inhibitors of Taq polymerase. *J Forensic Sci* 44: 1046–1050.
5. Burkhart, C.A., Norris, M.D., and Haber, M. 2002. A simple method for the isolation of genomic DNA from mouse tail free of real-time PCR inhibitors. *J Biochem Biophys Methods* 52: 145–149.
6. Cesarone, C.F., Bolognesi, C., and Santi, L. 1979. Improved microfluoro-metric DNA determination in biological material using 33258 Hoechst. *Anal Biochem* 100: 188–197.
7. Chan, K.C., Zhang, J., Hui, A.B., et al. 2004. Size distributions of maternal and fetal DNA in maternal plasma. *Clin Chem* 50: 88–92.

8. Chiu, R.W., Poon, L.L., Lau, T.K., et al. 2001. Effects of blood-processing protocols on fetal and total DNA quantification in maternal plasma. *Clin Chem* 47: 1607–1613.

9. Chomczynski, P. 1993. A reagent for the single-step simultaneous isolation of RNA, DNA and proteins from cell and tissue samples. *Biotechniques* 15: 532–537.

10. Chomczynski, P., Mackey, K., Drews, R., and Wilfinger, W. 1997. DNAzol: a reagent for the rapid isolation of genomic DNA. *Biotechniques* 22: 550–553.

11. Chung, G.T., Chiu, R.W., Chan, K.C., et al. 2005. Lack of dramatic enrichment of fetal DNA in maternal plasma by formaldehyde treatment. *Clin Chem* 51: 655–658.

12. Cui, X.F., Li, H.H., Goradia, T.M., et al. 1989. Single-sperm typing: determination of genetic distance between the G gamma-globin and parathyroid hormone loci by using the polymerase chain reaction and allele-specific oligomers. *Proc Natl Acad Sci U S A* 86: 9389–9393.

13. Dhallan, R., Au, W.C., Mattagajasingh, S., et al. 2004. Methods to increase the percentage of free fetal DNA recovered from the maternal circulation. *JAMA* 291: 1114–1119.

14. Hilz, H., Wiegers, U., and Adamietz, P. 1975. Stimulation of proteinase K action by denaturing agents: application to the isolation of nucleic acids and the degradation of 'masked' proteins. *Eur J Biochem* 56: 103–108.

15. Hodges, E., Boddy, S.M., Thomas, S., and Smith, J.L. 1997. Modification of IgH PCR clonal analysis by the addition of sucrose and cresol red directly to PCR reaction mixes. *Mol Pathol* 50: 164–166.

16. Hossain, A.M., Rizk, B., Behzadian, A., and Thorneycroft, I.H. 1997. Modified guanidinium thiocyanate method for human sperm DNA isolation. *Mol Hum Reprod* 3: 953–956.

17. Katoh, H., Oda, K., Hioki, K., and Muguruma, K. 2003. A genetic quality testing system for early stage embryos in the mouse. *Exp Anim* 52: 397–400.

18. Li, C.M., Sun, C.Q., Song, S., et al. 2005. Impedance labelless detection-based polypyrrole DNA biosensor. *Front Biosci* 10: 180–186.

19. Lin, Z., and Floros, J. 2000. Protocol for genomic DNA preparation from fresh or frozen serum for PCR amplification. *Biotechniques* 29: 460–466.

20. Lo, Y.M., Corbetta, N., Chamberlain, P.F., et al. 1997. Presence of fetal DNA in maternal plasma and serum. *Lancet* 350: 485–487.

21. Lo, Y.M., Tein, M.S., Lau, T.K., et al. 1998. Quantitative analysis of fetal DNA in maternal plasma and serum: implications for noninvasive prenatal diagnosis. *Am J Hum Genet* 62: 768–775.

22. Malumbres, M., Mangues, R., Ferrer, N., et al. 1997. Isolation of high molecular weight DNA for reliable genotyping of transgenic mice. *Biotechniques* 22: 1114–1119.

23. McClive, P.J., and Sinclair, A.H. 2001. Rapid DNA extraction and PCR-sexing of mouse embryos. *Mol Reprod Dev* 60: 225–226.

24. Miller, S.A., Dykes, D.D., and Polesky, H.F. 1988. A simple salting out procedure for extracting DNA from human nucleated cells. *Nucleic Acids Res* 16: 1215.

25. Mir, M., and Katakis, I. 2005. Towards a fast-responding, label-free electrochemical DNA biosensor. *Anal Bioanal Chem* 381: 1033–1035.

26. Pierce, K.E., Rice, J.E., Sanchez, J.A., and Wangh, L.J. 2002. QuantiLyse: reliable DNA amplification from single cells. *Biotechniques* 32: 1106–1111.

27. Rudbeck, L., and Dissing, J. 1998. Rapid, simple alkaline extraction of human genomic DNA from whole blood, buccal epithelial cells, semen and forensic stains for PCR. *Biotechniques* 25: 588–592.

28. Truett, G.E., Heeger, P., Mynatt, R.L., et al. 2000. Preparation of PCR-quality mouse genomic DNA with hot sodium hydroxide and tris (hotSHOT). *Biotechniques* 29: 52–54.

29. Truett, G.E., Walker, J.A., and Baker, D.G. 2000. Eradication of infection with *Helicobacter* spp. by use of neonatal transfer. *Comp Med* 50: 444–451.

# 4 Extraction of Genomic DNA from Plant Tissues

Zhanguo Xin and Junping Chen
*Plant Stress and Germplasm Development Laboratory, United States Department of Agriculture—Agriculture Research Service, Lubbock, TX, USA*

Many molecular biology studies require the preparation of genomic DNA. The isolation of genomic DNA from bacteria, yeast, and animal cells has become routine. However, extraction of genomic DNA from plant tissues is hampered by two major factors. (1) Plant cells are enclosed in strong cell walls that must be disrupted before DNA can be released. (2) Plant tissues usually contain high levels of carbohydrates, phenolics, and other phytochemicals that may inhibit the downstream manipulations of the genomic DNA, such as restriction digestion and PCR amplification.

Traditionally, methods for extracting genomic DNA from plant tissues include grinding of plant cells in a liquid nitrogen-chilled mortar. The ground plant tissue is then lysed with an ionic detergent, and treated with protease to release genomic DNA, which can be separated from other plant cell ingredients by cesium chloride density gradient centrifugation. This traditional method yields highly pure genomic DNA that is essentially free of inhibitory compounds. However, the method requires days to prepare the DNA and involves the handling of high concentration of ethidium bromide, a strong mutagen that may cause cancer (6, 13, 15). For this reason, this method currently is not often used and, thus, it is not discussed further in this chapter. Interested readers can review the studies of Reynolds and Williams (12) and Richards et al. (13) for details about this method.

In this chapter, a commonly used cetyltrimethylammonium bromide (CTAB) method, commercial kits, and a simple high throughput method to isolate genomic DNA from plant tissues are described. The principles of these methods and their relative strengths and weaknesses are dis-

***DNA Sequencing II: Optimizing Preparation and Cleanup***
Edited by Jan Kieleczawa
©2006 Jones and Bartlett Publishers

cussed to help readers to choose or modify a method for their respective research purpose. Emphasis is given to a simple high-throughput (HTP) method to prepare genomic DNA for PCR amplification.

## Plant Materials and Reagents

### Plant Materials

Arabidopsis and tobacco were grown in a growth chamber at 21°C with 100 µmoL quanta/m² constant light in a commercial potting mix Sunshine No.1 (Sun Gro Horticulture Inc., Bellevue, WA, USA). Sorghum (*Sorghum bicolor*) and cotton (*Gossypium hirsutum*) were grown in a field on site. Moss (*Tortula ruralis*) tissues were taken from cultures grown from spores on sterilized MS Murashiege-Skoog medium (9). Pine needles were collected from an ornamental Afghanistan pine tree (*Pinus eldarica*) growing on site.

### Chemicals

CTAB (H9151, cetyltrimethylammonium bromide), PVP (PVP-40, polyvinylpyrrolidone), bovine serum albumin (BSA; A4503), Tween®20 (P1379, polyoxythylene-sorbitan monolaurate), Tris-HCl (catalog # T3253), and Tris-Cl (catalog # T1503) were purchased from Sigma (St. Louis, MO, USA). Hot-Star Taq DNA polymerase was purchased from Qiagen, Inc. (Valencia, CA, USA). All other reagents were of highest purity available.

### Solutions and Buffers for the CTAB Method

### CTAB Extraction Buffer

The CTAB extraction buffer consists of: 2% (w/v) CTAB, 100 mM Tris-Cl, pH 8.0, 20 mM EDTA, pH 8.0, 1.4 M NaCl, and 0.2% (v/v) β-mercaptoethanol (add just before use).

### CTAB Precipitation Buffer

The CTAB precipitation buffer is 1% (w/v) CTAB, 50 mM Tris-Cl, pH 8.0, 10 mM EDTA, pH 8.0.

### High-Salt TE

The high-salt TE is 10 mM Tris-Cl, pH 8.0, 1 mM EDTA, pH 8.0, and 1 M NaCl.

### Solutions and Buffers for the High-Throughput Method

#### HTP Solution A

The HTP solution A is 100 mM NaOH and 2% Tween 20 (prepared daily/fresh from 10 M NaOH and 20% Tween 20 stock solutions).

#### HTP Solution B

The HTP solution B is 100 mM Tris-HCl (catalog # T-3253), Sigma, 2 mM EDTA. It is unnecessary to adjust the pH. Alternatively, HTP solution B can be made with 100 mM Tris-Cl (catalog # T1503, Sigma; pH 7.0), and 2 mM EDTA.

### Solutions and Buffers for the Commercial Kits

These were usually provided with kits.

## PCR Conditions

The PCR conditions described below are used to amplify the genomic DNA prepared with the simple high throughput method. The total PCR reaction volume is 20 μL containing 2 μL 10× Qiagen PCR buffer, 1 μL DNA template, 1.5 mM MgCl$_2$, 0.2 mM each dATP, dCTP, dGTP, dTTP, 0.25 μM each forward and reverse primer, 0.1% BSA (w/v), 1% PVP (w/v), and 0.5 U units of Qiagen's Hot-Star Taq DNA polymerase. Amplifications were carried out with MJR PTC-200 thermocyclers (purchased from MJ Research, Waltham, MA, USA). The initial step of 95°C for 15 minutes was followed by 40 cycles of 94° for 15 seconds, 56° for 15 seconds, and 72°C for 1 minute 30 seconds, and 1 cycle of extension for 10 minutes at 72°C. PCR products were analyzed by electrophoresis on 4% agarose gels in 0.5× TBE according to standard molecular biology protocols (14). This PCR condition works well for purer genomic DNA prepared with CTAB and commercial kits.

## CTAB Method

### Principle

This method is widely used to isolate genomic DNA from plants. Ground plant tissue is lysed in the presence of the non-ionic detergent CTAB, which forms soluble complex with nucleic acids at NaCl concentrations >0.6 M (4). The complex becomes insoluble when NaCl concentration is

**Table 4-1. A basic protocol of the CTAB method for extracting DNA from plant tissue.**

| Step | Description |
|------|-------------|
| 1 | Ground ~100 mg fresh tissue in liquid nitrogen chilled 1.5 mL centrifuge tube with a disposable pestle. (Keep tissue frozen until CTAB extraction buffer is added.) Add 0.5 mL CTAB extraction buffer preheated to 60°C and mix well; ensure that all the tissue is mixed with extraction buffer. Incubate the tube at 60°C for 30 minutes with occasional mixing. |
| 2 | Add 0.5 mL chloroform/isoamyl alcohol (24 : 1) and mix well. Spin at 8000× $g$ for 5 minutes. Transfer the aqueous phase to a new tube. |
| 3 | Add 1.33× CTAB precipitation buffer. Precipitate nucleic acid-CTAB complex with low centrifugation force at 500× $g$ for 5 minutes. |
| 4 | Wash the pellet two times with 80% ethanol. Resuspend the pellet in 200 μL high salt TE. |
| | *(Optional RNase digestion to remove RNA. Add 10 μL of 10 mg/mL RNase A, then incubate at 37°C for 30 minutes, followed by standard phenol/chloroform and chloroform extraction.)* |
| 5 | Add 0.6 volume of isopropanol or 2.5 volume of ethanol, and incubate at room temperature for 5 minutes. Precipitate DNA at 8000× $g$ for 5 minutes. Wash two times with 80% ethanol. |
| 6 | Air-dry the tube and resuspend DNA in 20 μL TE. |

lowered below 0.5 M. Polysaccharides, phenolic compounds, and other enzyme-inhibiting contaminants remain soluble under these conditions; thus, they can be efficiently removed with the supernatant (9, 10). The CTAB-nucleic acid complex is redissolved in high salt TE (10 mM Tris-Cl, 1 mM EDTA, pH 8, and 1 M NaCl) and precipitated with ethanol or iso-propanol. A basic protocol for this method is presented in Table 4-1.

## *Procedure*

A small amount of fresh tissue (<100 mg) is grounded in a 1.5 mL centrifuge tube with a disposable plastic pestle (Table 4-1). The centrifuge tube is occasionally chilled in liquid nitrogen to keep the plant tissue frozen until the addition of CTAB extraction buffer. To release the genomic DNA from the plant tissue, 0.5 mL of CTAB extraction buffer pre-heated to 60°C is added to the centrifuge tube. It is critical to keep the tissue frozen before the addition of the extraction buffer and to resuspend the tissue into extraction buffer as quickly as possible to avoid degradation of DNA by DNase released from plant tissue. The tube is briefly and gently mixed to resuspend the ground tissue into the extraction buffer in the shortest possible time. The tube is then incubated at 60°C for additional 30 minutes with occasional mixing. The mixture is extracted once with chloroform/isoamyl alcohol (24:1) to remove carbohydrates and other plant debris. The aqueous phase is transferred to a new tube and the salt concentration is adjusted to 0.5 M by adding 1.33× CTAB precipitation buffer. The CTAB-nucleic acid complex is precipitated by centrifugation at low speed (~500× $g$) for five minutes: increasing the centrifugation speed or time is not recommended as the pellet may become too compact and difficult to redissolve. After washing the pellet with 80% ethanol to remove excess of CTAB detergent, the pellet is resuspended in 200 μL high salt TE. The nucleic acid preparation at this point contains both DNA and RNA and a standard RNase digestion can remove the RNA. The DNA is precipitated with 2.5 volumes of ethanol or 0.6 volumes of isopropanol. The pellet is washed twice with 80% ethanol to remove the residual CTAB and salt; it is critical to remove most of ethanol after each wash before air-drying the pellet. The air-dried DNA pellet is dissolved in 20 μL water or TE (10 mM Tris-Cl, 1 mM EDTA, pH 8.0).

The protocol presented in Table 4-1 is a basic CTAB protocol for preparing genomic DNA from a small amount of plant tissue but it easily could be configured to extract DNA from a large amount of plant tissues. The ratio of plant tissue to CTAB extraction buffer should be maintained at 1 g : 4 mL (based on fresh weight) (12, 13). Dried plant tissue can also be used, however, the amount of CTAB precipitation buffer needs to be adjusted to a final NaCl concentration <0.5 M in the precipitation step.

Many variations of this protocol have been developed to prepare genomic DNA from different plant species and tissue types (5, 7, 11, 12). For most plant tissues, step 3 of the CTAB-nucleic acid complex can be omitted and the genomic DNA can be precipitated by ethanol or isopropanol (3). Elimination of this step significantly speeds up the DNA extraction with little compromise of the DNA quality. Furthermore, the DNA yield is usually higher with this modified method. For tissues containing high levels of phenolic compounds, 1% (w/v) soluble PVP can be added to CTAB extraction buffer (11–13). The method has also been amended for high throughput process of isolation genomic DNA from plant tissues (5).

## Advantages of the CTAB Method

The method is versatile and it is used successfully to extract DNA from different plant species. Many versions and adaptations have been developed to suit different needs. The above brief discussion is just a guideline for optimization of the method for a particular plant tissue. It is recommended that each laboratory or user validate the protocol for their individual or particular application.

### Disadvantages of the CTAB Method

The quality of DNA obtained with the CTAB method is usually not very high but adequate for most applications. Sometimes the presence of trace amount of CTAB affects the downstream manipulations (10, 11); however, the CTAB is very soluble in ethanol and the precipitation of DNA with ethanol (instead of isopropanol), and two washings of DNA pellet with 80% ethanol are usually sufficient to remove the trace amount of CTAB.

## Commercial Kits for DNA Extraction From Plants

### Principle

Many commercial kits are available for isolating genomic DNA from plant tissues and they easily can be found through an Internet search. A partial list of the commercial kits for DNA extraction from plant tissues is listed in Table 4-2. Kits come with individual preparations in single or 96-well plate formats for high throughput applications. Most of the popular kits to isolate genomic DNA are based on the reversible binding of DNA to silica particles in the presence of the chaotropic salts like NaI, NaClO$_4$, or guanidium hydrochloride (1). The protocols usually begin with the disruption of plant tissue in buffer containing high concentrations of chaotropic salt. After removing the plant debris by filtration or centrifugation, the crude extract is loaded on a column packed with silica particles. The column is washed with buffered 80% ethanol to remove other plant tissue ingredients and the remaining chaotropic salt. The genomic DNA can be recovered by elution of the column with DNase-free water or TE. The whole procedure can be completed within an hour and the DNA isolated is pure and clean if proper procedure is followed.

### Advantages and Disadvantages of Commercial Kits

Commercial kits are usually reliable and the quality of genomic DNA obtained with commercial kits is generally high. If the recommended procedure is followed, good quality DNA can be obtained on the first try. The major disadvantage of commercial kits is cost, which ranges from $2 to $3 for processing one small sample (<100 mg) of fresh plant tissue. The

**Table 4-2.** Partial list of the commercial plant DNA extraction kits.

| Kit Name | Vendor | Price/Prep ($) | Internet Web Site |
|---|---|---|---|
| NucleoSpin Plant | BD Biosciences | 2.42 | http://www.bdbiosciences.com |
| DNA Extraction, Plants | Cartagen | 2.00 | http://www.cartagen.com |
| DNA Extraction | Microzone | 1.30 | http://www.microzone.co.uk |
| Genomic DNA Extraction | UBI | 1.86 | http://www.ubi.ca |
| Nucleon PhytoPure | GE Health | 3.20 | http://www5.amershambiosciences.com |
| DNeasy Plant Mini | Qiagen | 3.00 | http://www1.qiagen.com |
| GenSpin | Fisher Scientific | 2.80 | https://www1.fishersci.com |
| UltraClean™ Plant DNA | Mo Bio Labs | 2.00 | http://www.mobio.com |
| Wizard® Magnetic 96 DNA Plant System | Promega | 1.90 | http://www.promega.com |
| GenEluteä | Sigma | 2.00 | http://www.sigmaaldrich.com |

high cost of using commercial kits to extract DNA from plant tissues is often prohibitory for studies that require the processing of large numbers of samples. For laboratories with limited funds, other methods should be explored as warranted by a particular research goal.

## A Simple HTP DNA Extraction Method Suitable for PCR

Map-based cloning and marker-assisted breeding require the genotyping of thousands of individual progenies. Preparation of large numbers of DNA samples is often the limiting step for high-throughput genotyping program. We have devised a simple, low cost, high throughput method to prepare genomic DNA for PCR amplification (14, 16). With this method, one person can process up to two thousand plant samples in a single day starting from the collection of plant tissues, preparing genomic DNA, and then setting up the PCR reactions. Moreover, this method is versatile and can be used to prepare genomic DNA from a wide range of plant species including PCR recalcitrant tissues, such as pine needles (2).

### *Principle*

The tissue maceration step for DNA extraction is not only time consuming, but also a major source of sample cross-contamination. This step can be eliminated by incubating plant tissue in 96-well plates, at high temperatures under alkaline conditions, and in the presence of detergent that is compatible with PCR amplification. The alkaline conditions and the inclusion of detergent help to weaken plant cell walls and disrupt cellular membranes so that genomic DNA can be released into the extraction buffer. The DNA extract is neutralized with an acidic buffer to make the DNA preparation suitable for PCR amplification. We showed that the nonionic detergent Tween 20 has little effect on the efficiency or the specificity of PCR amplification at concentrations below 2% (14, 16). The genomic DNA preparation can be directly used for PCR amplification without the need for further clean up step. In some recalcitrant plant tissues, the crude DNA preparation contains high levels of compounds that can completely inhibit the PCR amplification. We found that the inclusion of 0.1% BSA and 1% soluble PVP in the PCR reaction mixture can alleviate the inhibition and enable the successful PCR amplification of the crude DNA preparations from many plant species (14, 16). Elimination of tissue maceration and the modification of PCR amplification conditions make it possible to establish a truly high throughput protocol to prepare genomic DNA from plant tissues.

## *Procedure*

Figure 4-1 outlines the protocol of the simple HTP method to prepare genomic DNA from plant tissues. A small piece of plant tissue (~30 mm$^2$) is inserted into a well of a 96-well plate with tweezers. The tweezers should be briefly rinsed in distilled water between samples to minimize carry-over contaminations. A 50 μL of HTP solution A was added to each well with a 12-channel pipette. The plate is sealed with thermosensitive sealing tape and incubated in a thermocycler for 10 minutes at 95°C. After the plate cools down to room temperature, aliquot of 50 μL of HTP neutralizing solution B is added to each well and approximately 0.5 to 1 μL

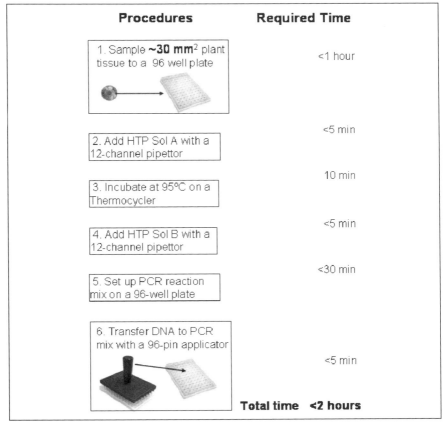

**Procedures**                **Required Time**

1. Sample **~30 mm$^2$** plant tissue to a 96 well plate          <1 hour

<5 min

2. Add HTP Sol A with a 12-channel pipettor

10 min

3. Incubate at 95°C on a Thermocycler

<5 min

4. Add HTP Sol B with a 12-channel pipettor

<30 min

5. Set up PCR reaction mix on a 96-well plate

6. Transfer DNA to PCR mix with a 96-pin applicator

<5 min

**Total time   <2 hours**

**Figure 4-1.** **A high-throughput method to prepare plant genomic DNA for PCR amplification.** DNA extraction and PCR amplification are carried out in 96-well format. The whole procedure from sampling plant tissues to setting up PCR amplification can be completed within two hours.

of this mixture can be transferred to PCR reaction mix with a 96-pin stainless steel applicator. The whole procedure from collecting plant tissue to setting up PCR reaction for one 96-well plate can be completed within 2 hours at a very low cost. For higher-throughput, multiple plates of plant tissues can be processed simultaneously. As stated earlier, one technician can process up to two thousand samples from collecting plant tissue to setting up the PCR reaction in a single working day without the aid of robotic technologies.

The most time-consuming step in this protocol is the collection of plant tissues. We have observed that a successful DNA extraction is possible with a wide range of sample sizes and tissue types. To speed up this step, we routinely collect tissue types that are the most convenient to sample, such as young leaves, cauline leaves, flower buds, small seedlings, etc. It is not necessary to freeze the plant tissue if they are processed within a few hours from a collection step. However, the tissue can also be stored at −70°C and processed at a later time.

A problem with this simple high-throughput method is that certain plant species contain a high level of compounds that are inhibitory to PCR amplification. This can be circumvented by inclusion of 0.1% BSA and 1% soluble PVP (mol wt 40,000) in the PCR reaction mix. Figure 4-2 shows the effect of adding BSA and PVP on the efficiency of PCR amplification in several plant species. The genomic DNA from Arabidopsis, tobacco, and sorghum prepared with this simple method can be directly amplified without the addition of BSA and PVP, although PCR efficiency can be enhanced by inclusion of these additives in the PCR reaction mix. However, the genomic DNA from cotton leaf and pine needles prepared with the simple method cannot be amplified without additives, and the addition of 0.1% BSA and 1% PVP in the PCR reaction mix allows the successful amplification of the genomic DNA. We routinely include 0.1% BSA and 1% PVP in the PCR reaction mix when the genomic DNA prepared with our simple HTP method is used as a template.

### Advantages of the Simple High-Throughput Method

The major advantage of the described method is the throughput and the low cost. As mentioned earlier, up to two thousand samples can be processed in a day by a single researcher. The estimated cost to process one plant sample for genomic DNA is less than $0.05. Another advantage is the very small amount of tissue needed for processing: a leaf disc of approximately 6 mm in diameter (<5 mg fresh weight) provides sufficient amount of genomic DNA for 100 PCR reactions. The DNA isolated with this simple HTP method is very stable. We have observed that the DNA can be stored at 4°C for more than a month and at −20°C for over three months without any apparent effect on the yield and specificity of PCR reactions.

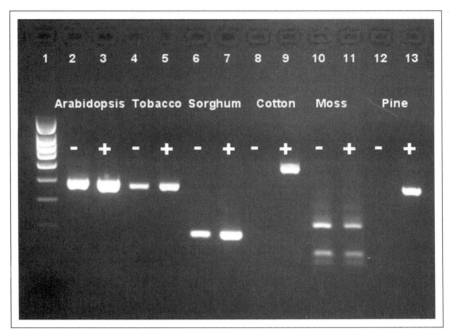

**Figure 4-2.** **Effects of BSA and PVP additives on PCR amplification of crude DNA preparations.** The crude DNA preparations from Arabidopsis, tobacco, sorghum, cotton, star mosses, and pine needles were amplified with species-specific primers. The Arabidopsis SSLP marker T4C21 was amplified with primers 5'-CGGCTTGATCTCCATTGATT-3' and 5'-TGGAGAGACCCATTTTGCAT-3'. The gene for green fluorescent protein was amplified from transgenic tobacco with primers 5'-ATCCCACTATCCTTCGCAAGAC-3' and 5'-GCGCTCTTGAA-GAAGTCGTG-3'. A drought-inducible EST (BG933199.1) from sorghum was amplified with primers 5'-GGCCATTTTTGGTAAGCAGA-3' and 5'-GTTGATTCGGCAGGTGAGTT-3'. The cellulose synthase A1 gene (*U58283.1*) from cotton was amplified with primers 5'-GGATCTGCACCCATCAATCT-3' and 5'-GCAAAGAGATGGGCTGAAAC-3'. The rehydrin gene family Tr288 (*AF275946*) from star moss was amplified with primers 5'-GCCCATGCC-GATAGCGTCCTTAGCC-3' and 5'-CGTCGGCATGGGCCCCAAC-3'. The trans-cinnamate-4-hydroxylase gene (*AF096998.1*) was amplified from pine DNA with primers 5'-TGTGGTGTCATCGCCGGATCT-3' and 5'-CGGAGGAAGAG CGGGTCGTC-3'. Symbols are (+) inclusion of 0.1% BSA and 1% PVP in PCR reaction mix; (−) no BSA or PVP in PCR reaction mix. *Source:* Adapted from Xin, Z., Velten, J.P., Oliver, M.J., and Burke, J.J. *Biotechniques* 34: 820–826. Used with permission.

### Disadvantages of the Simple HTP Method

The main disadvantage of this method is that the genomic DNA is mixed with other plant ingredients and the exact amount of DNA is unknown. Although the DNA can be amplified with PCR using modified conditions, it is not suitable for other applications that require purified genomic DNA. For applications that need larger amounts of purified genomic DNA the CTAB method or commercial kits are recommended.

## Conclusions

Many methods are available to isolate genomic DNA from plant tissues. Three basic protocols are presented in this chapter, and the principles and relative merits of each method are discussed. The CTAB method is the most flexible as it can be easily adapted to prepare genomic DNA from different amounts of plant tissues. Many variations of this method have evolved to isolate DNA from different plant species. The DNA quality obtained with CTAB method is fairly good and is applicable for most research purposes. A wide range of commercial kits is available for isolating highly purified genomic DNA from many plant tissues. Because of the high cost of commercial kits, other alternatives should be explored according to individual research objectives. A simple HTP method was presented that allows a single researcher to prepare genomic DNA from two thousand plant samples in a day. This method is particularly useful for molecular breeding, which usually requires the genotyping of thousands of individual samples (8).

## Disclaimer

Mention of trade names or commercial products in this article is solely for the purpose of providing specific information and does not imply recommendation or endorsement by the U.S. Department of Agriculture.

## *Acknowledgment*

The authors thank Lindsey Fox for excellent technical support.

## References _____

1. Boom, R., Sol, C.J.A., Salimans, M.M.M., et al. 1990. Rapid and simple method for purification of nucleic acids. *J Clin Microbiol* 28: 495–503.
2. Dong, J., and Dunstan, D.I. 1996. A reliable method for extraction of RNA from various conifer tissues. *Plant Cell Rep* 15: 516–521.
3. Doyle, J.J., and Doyle, J.L. 1987. A rapid isolation procedure for small quantities of fresh leaf tissue. *Phytochem Bull* 19: 11–15.
4. Dutta, S.K., Jones, A.S., and Stacey, M. 1953. The separation of desoxypentosenucleic acids and pentosenucleic acids. *Biochim Biophys Act* 10: 613–622.

5. Mace, E.S., Buhariwalla, H.K., and Crouch, J.H. 2003. A high-throughput DNA extraction protocol for tropical molecular breeding programs. *Plant Mol Biol Rep* 21: 459a–459h.

6. McCann, J., Choi, E., Yamasaki E., et al. 1975. Detection of carcinogens as mutagens in the salmonella/microsome test: assay of 300 chemicals. *Proc Natl Acad Sci U S A* 72: 5135–5139.

7. Michiels, A., Van den Ende, W., Tucker, M., et al. 2003. Extraction of high-quality genomic DNA from latex-containing plants. *Anal Biochem* 315: 85–89.

8. Morgante, M., and Salamini, F. 2003. From plant genomics to breeding practice. *Curr Opin Biotechnol* 14: 214–219.

9. Murashige, T., and Skoog, F. 1962. A revised medium for rapid growth and bioassays with tobacco tissue cultures. *Physiol Plant* 15: 473–497.

10. Murray, M.G., and Thompson, W.F. 1980. Rapid isolation of high molecular weight plant DNA. *Nucleic Acids Res* 8: 4321–4325.

11. Peist, R., Honsel, D., Twieling, G., et al. 2001. PCR inhibitors in plant DNA preparations. *Qiagen News* 3: 7–9.

12. Reynolds, M.M., and Williams, C.G. 2004. Extracting DNA from submerged pine wood. *Genome* 47: 994–997.

13. Richards, E., Reichardt, M., and Rogers, S. 1994. Preparation of genomic DNA from plant tissue. In: Ausubel, M.F., Brent, R., Kingston, R.E., et al., eds. *Current Protocols in Molecular Biology.* New York: John Wiley, 2.3.1–2.3.7.

14. Sambrook, J., and Russell, D.W. *Molecular Cloning: A Laboratory Manual*, 3rd ed. Cold Spring Harbor: Cold Spring Harbor Laboratory Press; 2001.

15. Singer, V., Lawlor, T., and Yue, S. 1999. Comparison of SYBR Green I nucleic acid gel stain mutagenicity and ethidium bromide mutagenicity in the *Salmonella*/mammalian microsome reverse mutation assay (Ames test). *Mutat Res* 439: 37–47.

16. Xin, Z., Velten, J.P., Oliver, M.J., and Burke, J.J. 2003. High-throughput DNA extraction method suitable for PCR. *Biotechniques* 34: 820–826.

# Preparation of cDNA Probes for DNA Microarrays

**Peter R. Hoyt,**[1] **Mitchel J. Doktycz, and Lois Tack**[2]

*Life Sciences Division, Oak Ridge National Laboratory, Oak Ridge, Tennessee*

Microarray core facilities are becoming commonplace both in academic and industrial settings. Many of these facilities use a common microarray format where PCR-amplified copy DNA (cDNA) clone inserts are subsequently spotted onto glass microscope slides. This "front-end" step for chip construction can involve significant time and effort compared to the actual chip hybridization experiment. Typical EST (expressed sequence tags) cDNA clone libraries may contain in excess of 10,000 unique clones and require considerable processing prior to spotting. Before the cDNA can be arrayed, the plasmid clones are grown in bacteria, plasmids isolated, inserts amplified, and amplified elements purified. Thorough quality control must be maintained throughout the process as well as precise sample tracking. Quality control may include gel electrophoresis, sample quantitation, and chemical purity measurements. Automated systems reduce the amount of labor involved in the process and also provide greater reproducibility when compared with manual procedures.

Automation of plasmid DNA isolation for PCR and sequencing has been reported using various systems (1, 4, 6, 10–12, 15, 16). Most fully automated systems are inflexible, such that customization or smaller quantity production is difficult. When studying organisms whose genomes are not sequenced, cDNA clones may be the only resource available. Often, only limited numbers of ESTs have been verified by sequencing, and the identities of the cDNAs annotated. For these projects,

[1]Current Address: Department of Biochemistry and Molecular Biology, Microarray Core Facility, Noble Research Center, Oklahoma State University, Stillwater, OK 74078.
[2]Current Address: PerkinElmer LAS, Downers Grove, Illinois 60515.

***DNA Sequencing II: Optimizing Preparation and Cleanup***
Edited by Jan Kieleczawa
©2006 Jones and Bartlett Publishers

laboratories may decide to produce their own probes and DNA arrays. To assist with these efforts, this chapter describes our realistic methodology for producing cDNA clones using semiautomated and automated instrumentation systems, and commercially available kits. A single flexible sample preparation workstation is discussed that can process samples at a controlled pace, yet is capable of high throughput when operated at full capacity. We used the Packard MultiPROBE II HT EX automated liquid handling instrument, integrated with the Gripper Integration Platform, a vacuum filtration system, and WinPREP® software as the central workstation for our customized processes. The advantages of this system include a versatile gripping arm that is capable of plate transfers and vacuum manifold assembly as well as the ability to address regions outside the workstation deck. This enables integration with accessories such as platform shakers, thermocyclers, plate readers, and stackers. Complete "walk-away" automation of sample processing was developed; however, semiautomated procedures involving minimal operator intervention (and realistic cost-benefits) are presented in this chapter. Salient evaluations and testing of steps leading to successful array printing are also included.

## Materials and Methods

### Reagents and Supplies

For testing, we used portions of the NIA 15K clone set #NIAH3000VU-11 (Incyte Genomics Inc., Palo Alto, CA, USA) (20), or purchased specific cDNA clones from Research Genetics (Huntsville, AL, USA). For plasmid isolations, the NIA 15K plates averaged cDNA inserts of 1.6 Kb, using pSPORT1 vector, in *E. coli* host strain DH10B. QIAprep 96 Turbo Miniprep Kits (catalog #27191), QIAquick 96 PCR purification kit (catalog #28181), and related reagents were purchased from Qiagen, Inc. (Valencia, CA, USA). PCR reagent kits were purchased from Perkin Elmer/Roche (Branchburg, NJ, USA) and Promega Corp. (Madison, WI, USA). Primers NIA1 (5′-GTTTTCCCAGTCACGACGTTG-3′) and NIA2 (5′-GAGCG-GATAACAATTTCACACAG-3′) were purchased from Integrated DNA Technologies, Inc. (Coralville, IA, USA). Superscript II Reverse Transcriptase was from Bethesda Research Laboratories (Bethesda, MD, USA).

Conductive, liquid-sensing pipette tips were purchased from Molecular BioProducts (San Diego, CA, USA). Chemicals and reagents for molecular biology or probe production were purchased from Sigma Chemical Co. (St. Louis, MO, USA). Plasticware was purchased from various manufacturers including standard 96-well microtiter assay plates (#3896; Corning Costar Co., Cambridge, MA, USA), deep well 96-well plates for growing bacteria (2 mL Uniplate #7701-5205; Whatman,

Florham Park, NJ, USA), 384-well polypropylene plates for spotting (Uniplate #7701-5101; Whatman), and black 96-well fluorescence reader plates (Dynatech microfluor® 1; Thermo Labsystems, Helsinki, Finland). Standard molecular biology protocols were followed when appropriate (16).

## Instrument Description/Configuration

The Packard Instruments MPII HT EX with Gripper Integration Platform and Nucleic Acid Extraction (NAE) Application Option is designed to fully automate plasmid DNA purification using a variety of commercially available kits and filter plates. The NAE template (13, 14) serves as a starting template for protocols requiring a vacuum manifold. The template directs a series of automated robotic sample processing steps including resuspension of bacterial cell pellets in 96-well, deep-well plates, timed incubations, manifold disassembly and reassembly, complex vacuum protocols, and final elution of purified plasmid DNA. An integrated DPC MicroMix5 shaker mediates resuspension, sample mixing, and filter plate blotting steps. Hardware can be purchased from Perkin-Elmer, or can be collected and modified in-house to create custom processes. For example, we used modified empty 1.2 mL collection tube racks as vertical positioning spacers during elution into flat-well microplates. Pipetting accuracy of these workstations can be adjusted by swapping larger or smaller syringes onto the pumps. For convenience when performing simple distributions of small liquid volumes, we also used a MultiPROBE II 4-Tip liquid handling robot equipped with smaller syringe pumps. Alternatively, hand-pipetting with multiple tip heads can be used to distribute very small volumes without having to switch syringes, or buy another workstation. Current information and upgrades to this instrument can be found on the Perkin Elmer Web site (http://las.perkinelmer.com/Catalog/ProductInfoPage.htm?ProductID=MPII-NAW).

## Plasmid Preparation

The cDNA clones used for probe production are obtained as frozen bacterial stocks in 96-well format. These stocks are thawed, and manually transferred into 96-well (2.0 mL) growth plates containing 1.5 mL of 2× YT bacterial media (Tryptone; 16.0 g/L, yeast extract; 10.0 g/L, NaCl; 5.0 g/L, 0.1 mM sodium citrate; 17.0 g/L, 1.0 mM $K_2HPO_4$; 36.0 mL/L, 1.0 M $KH_2PO_4$; 13.2 mL/L, 80% glycerol; 55.0 mL/L, 1.0 M $MgSO_4$; 0.4 mL/L, 2.0 M $(NH_4)_2SO_4$; 3.4 mL/L, pH 7.0) and the appropriate antibiotic (100 µg/mL ampicillin) using a 96-pin tool capable of 25 µL liquid transfers (#VP408; V&P Scientific, San Diego, CA, USA). Bacteria are grown 16 hours at 37°C with shaking (250 RPM). Subsequently, the bacterial cells are pelleted by centrifugation in a Sorvall RC5B with Sorvall SH3000 rotor (Kendro Laboratory Products, Asheville, NC, USA) at 5000× $g$ for 15 minutes, after which the plates are inverted and blotted to remove excess media. To begin the automated plasmid isolation, one or two 96-well plates of pelleted bacteria in 96-well format are placed on the MultiPROBE II's shaker module.

A vacuum manifold-defined plasmid isolation process was developed starting with a supplied WinPREP template. The QIAGEN QIAprep® 96 Turbo Miniprep Kit was chosen for developing automated nucleic acid plasmid DNA purification, and adapted to the template. A fully automated process was designed beginning with resuspension of bacterial cell pellets in the 96-well growth plate and ending with purified plasmid DNA samples in standard 96-well microplates. The samples are eluted using 100 μL of 10 mM Tris buffer, pH 8.0.

## *PCR Product Preparation*

PCR reactions used for microarray probe production contain 38 μL distilled water, to which is added 5 μL 10× reaction buffer, 4 μL 25 mM MgCl$_2$, 1 μL 10 mM dNTPs, 1 μL of a 10 μM solution of the appropriate 5′- and 3′-primers, 1 μL of plasmid DNA (~200 ng) and 1 unit Taq DNA polymerase. A different master mix is prepared containing the correct pair of oligonucleotide primers for the particular cloning vector used. The reactions were prepared robotically by aliquoting template DNA and the PCR master mix directly into 96-well thermocycler plates. Thermocycler plates were accurately positioned and secured during reagent addition by placing them inside a round-bottomed 96-well microtiter plate. After reactions were set up, the plates were manually removed from the deck and placed into the thermocycler instruments (GeneAmp 9600 thermocycler; Perkin Elmer/Roche). Amplification is performed using PCR conditions of 95°C for five minutes, followed by 40 cycles of 94°C for 30 seconds, 54°C for 30 seconds, and 73°C for 3.5 minutes, and finished with seven minutes at 72°C. Completed PCR reactions were returned to the robot deck for purification.

PCR products were purified from excess primers and nucleotides with a custom protocol for the MultiPROBE II workstation and the QIAGEN QIAquick™ 96 PCR purification kit using the supplied buffers. The concentration of the samples is estimated (see below), and the PCR purified products are then covered loosely and dried overnight at 50°C. The dried PCR products are resuspended in the appropriate volume of spotting buffer to prepare approximately 150 μg/mL solution. Spotting buffers were prepared in-house or purchased. In-house buffers included 3× SSC (standard saline citrate buffer: 150 mM NaCl, 15 mM NaCitrate, pH 7.0) buffer and 50% DMSO in water. These "chip-ready" probes in spotting buffer are stored frozen at –80°C, in 96-well format. For microarray construction the DNAs are redistributed into 384-well format using a redistribution protocol written on the MultiPROBE II.

## Gel Electrophoresis

Gel loading and electrophoresis of the plasmid DNA or PCR-amplified DNA are carried out on the deck of the MultiPROBE II. The DNA samples, stored in 96-well plates, are automatically loaded onto a 1.2% submerged agarose gel in TAE (40 mM Tris-acetate, 5 mM Na acetate, 1 mM EDTA, pH 7.8) buffer using the MultiPROBE II and an Owl Separation Systems (Portsmouth, NH, USA). Model D3-14 gel electrophoresis unit. The gel unit is modified by adding Lucite retainers 140 mm apart to the inside of the buffer chamber, to hold the gel plate precisely in the middle of the chamber during the loading process. A small (1.5 mm wide) spacer is placed between the gel tray and the wall of the buffer chamber to secure the gel. The combs are manually positioned at the top-left registration position during pouring. The left and top sides of the gel chamber are positioned onto the MultiPROBE II deck by pushing against standard plate adapter support tiles. The entire gel unit can be mapped on the deck using the WinPREP software labware definition library, and the well positions in the agarose gel are defined in a manner similar to other pieces of labware. To load samples, 15 μL of 1.5× DNA loading buffer (1× loading buffer is 8% glycerol, 0.01% bromphenol blue, 0.01% xylene cyanol) is added to each of two standard 96-well microplates. Then, 3 μL of purified plasmid, or 3 μL of PCR-amplified DNA are mixed into the loading buffer of each well. Without changing tips, the probes are immediately loaded into the wells of a submerged agarose gel. Electrophoresis is performed at 80 volts for 45 minutes.

## DNA Quantitation

A fluorescence-based quantitation assay is set-up using a four-tip Packard MultiPROBE II fitted with 500 μL syringe pumps for analysis and an HTS 7000 Plus bioassay plate reader (Perkin-Elmer). The DNA was diluted 1:10 with 10 mM Tris, and 50 μL was mixed with 50 μL of a 1:200 dilution of PicoGreen® (Molecular Probes, Eugene, OR, USA) reagent in a 96-well black flat-bottom Dynatech Microfluor® 1 plate (Thermo Labsystems). A standard concentration plot was generated under identical conditions with known DNA concentrations. Plates were read (fluorescence mode: Ex 485 nm; Em 535 nm) in the HTS 7000 plate reader. Measurements are taken using an instrument manual gain setting of "100," a lag time of 0 μsec, an integration time of 40 μsec, and three flashes of the light source. After linear regression of the standard concentration plot, the DNA concentration is calculated and added to a results file. DNAs with insufficient yields are flagged.

## Sample Tracking

Probes are tracked through the steps of plasmid and PCR preparation and purification, using the Microsoft Access™ database that is integrated with the WinPREP control software. Empty positions in plates can be used by flagging the well as "empty" throughout the clone processing. A clone "master" list is generated in Excel-compatible format that links the plate name, row, and column information to the clone identity, and can contain additional information. When sample annotation and tracking information is needed for spotting, those columns are selected and a new Clone-Name (CN) text-file is derived from positional information of the redistribution database. The CN text-file is input into CloneTracker™ (BioDiscovery, Inc., Marina Del Rey, CA, USA) software for tracking of the clone positions and gene information while spotting and analyzing microarrays. The resulting CloneTracker files are combined with set up information used by the VIRTEK SDDC-2 spotting instrument to track sample location after arraying.

## Microarray Production and Analysis

To "spot" DNA arrays, PCR products are thawed and gently vortexed prior to placement onto the deck of a VIRTEK Chip Writer (model SDDC-2; purchased from VIRTEK, currently part of BioRad Laboratories, Hercules, CA, USA) fitted with 4 Arrayit™ Stealth MicroSpotting pins (Telechem International Inc., Sunnyvale, CA, USA). Arrays are produced onto CMT-GAPS™ amino-silane coated microscope slides (catalog #2550, Corning Inc.). The spot pitch used was 187.5 µm. The arrays are dried, post-processed according to Hegde et al. (6), and stored in a desiccator until use.

Hybridization targets (labeled cRNA) are produced using Superscript II reverse transcriptase (#18064-022; Life Technologies, Rockville, MD, USA), 25 to 80 µg of total RNA and Cyanine 3-dUTP (#NEL578; Perkin Elmer) or Cyanine 5-dUTP (#NEL579; Perkin Elmer) fluorophores (6). Labeled cRNA is subsequently mixed with 10 µg of poly [dA] plus 10 µg of murine Cot-1 DNA (#18440-016; Life Technologies), dried, then resuspended in 20 µL of hybridization buffer containing 50% formamide, 4× SSC, and 0.1% SDS. The cRNA mixture is applied to the array, covered with a No. 1 (22 × 40 mm) cover glass (Corning), and placed in a fully humidified and sealed chamber at 42°C for a minimum of 16 hours. After hybridization, the arrays are washed once with 2× SSC, 0.1% SDS at 42°C for five minutes, twice in 0.2× SSC at room temperature for two minutes, and once in 0.1× SSC at room temperature for one minute. The microscope slide arrays are then placed into a slide holder and dried by centrifugation for 10 minutes at 500 rpm in a TJ-6 centrifuge using a 7-01 swinging bucket rotor (Beckman-Coulter, Palo Alto, CA, USA). Slides are imaged using a Packard BioSciences ScanArray 4000XL microarray confocal laser scanner (Perkin Elmer) at 5 or 10 micron resolution.

## Results

Probe preparation for DNA microarrays typically involves a number of sample preparation procedures as summarized in Figure 5-1. In general, these steps include plasmid preparation, PCR amplification, cleanup, and product characterization. The initial plasmid preparation steps are necessary only for preparation of a suitable template for subsequent PCR and for sample archiving. A single, small-scale preparation should provide sufficient template for hundreds of PCR amplifications, and an individual PCR reaction is sufficient for preparing spots for hundreds of microarrays. Although sample processing steps may only need to be conducted initially or perhaps on a yearly basis, these steps represent an initial bottleneck to the construction of a DNA microarray, especially considering the large number of reactions necessary to construct a high density microarray.

To automate the steps outlined in Figure 5-1, a MultiPROBE II HT EX workstation with Gripper option was utilized along with the supplied WinPREP software. An optional second instrument was used during DNA quantification setup and automated agarose gel loading, to avoid the task of switching syringes for pipetting smaller volumes. The labware setup used to automatically prepare plasmid DNA from 192 bacterial cultures using the MultiPROBE II with Gripper is shown in Figure 5-2. The positions of labware used for the solid phase extraction of 192 samples, and integration of a DPC shaking platform can be seen in Figure 5-2a. The Micromix5 shaker is important for resuspending cell pellets and mixing reagents during plasmid isolation. The Gripper arm was capable of moving a variety of plates, columns, and hardware, including disassembly and reassembly of the vacuum manifold as in Figure 5-2b.

The first step in the preparation of microarray cDNA probes is growing the bacterial clone stocks containing the plasmid DNA. When the bacterial stocks are provided in the same, 96-well format as the growth chambers, the inoculation of the growth chambers is simply a matter of using a 96-pin replicating tool capable of transferring 25 μL of stock culture (V&P Scientific). Additionally, the MultiPROBE is capable of efficiently performing replication of the stocks and inoculation of the growth chambers using a simple liquid transfer (1–5 μL). In our experience, a

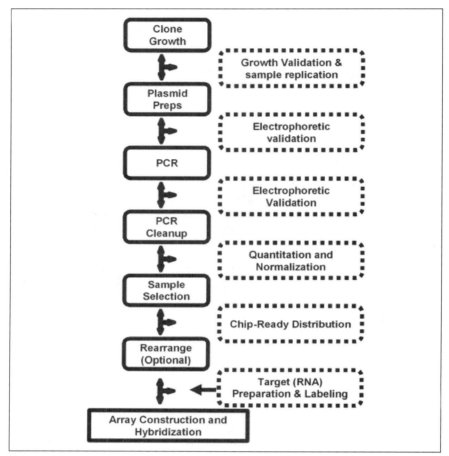

**Figure 5-1.** Flow-chart diagram of cDNA probe production strategy for automation and validation. Solid boxes represent direct sample flow and stages of probe automation leading to array production. Boxes with broken lines represent accessory processes that can be automated, leading to probe quality assurance and data production.

1.5 mL bacterial culture grown overnight in 2.0 mL wells of a 96-well plate can be expected to generate up to 10 µg of plasmid DNA (see below). This is sufficient for most user facilities to generate PCR products for many years (data not shown). There are many commercial kits available for high-throughput plasmid DNA purification, usually in 96-well format, and several of these kits are amenable to automation by a liquid handling robot (2, 5, 9, 11, 12, 16, 19). Our laboratory uses the QIAGEN QIAprep 96 Turbo miniprep system (10, 15), although most commercial plasmid purification kits would also be suitable. This system uses a vacuum man-

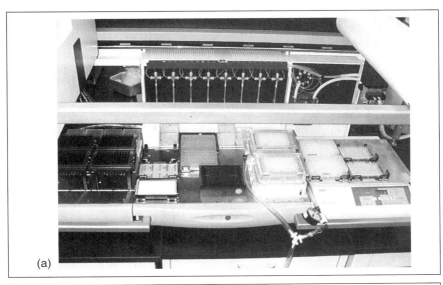

(a)

(b)

**Figure 5-2.** Images of the 8-tip MultiPROBE II EX HT Sample preparation workstation with the Gripper and nucleic acid extraction option (vacuum manifold and controls) used to prepare cDNA probes. (a) Deck layout for original plasmid preparation from bacterial pellets using Qiagen QIAprep 96 Turbo Miniprep kit. The integrated shaker can be seen at the lower right corner. A vacuum gauge was placed in line with the vacuum hoses to monitor pressure conditions during sample preparation. Racks of conductive tips, used for liquid level sensing, are seen on the left region of the deck. All buffer reservoirs are positioned along the rear edge of the deck. (b) Image showing the Gripper arm reassembling the vacuum manifold during probe processing. In this image, the lysed and precipitated samples have been filtered, the filter plate discarded (seen at the bottom right corner), and the filtrate directly loaded onto the binding columns. The columns with DNA are being transferred to the top of the vacuum manifold, in preparation for washing and eluting.

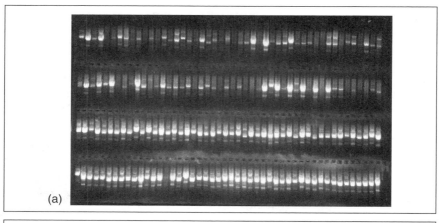

(a)

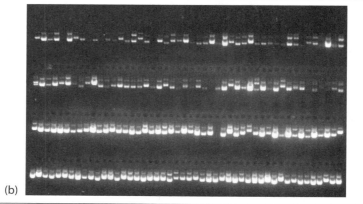

(b)

**Figure 5-3.** **Gel electrophoresis results of 192 plasmid preps isolated using fully automated processes.** Each lane represents 5 μL of 100 μL eluant stained with ethidium bromide. Clones are from plates H3002 and H3074 of the clone set of Tanaka et al. (20). (a) Plasmid preparation did not include a blotting step to remove residual EtOH. (b) Plasmid preparation used a blotting step and extended vacuum to dry prior to elution.

ifold to facilitate sample binding, washing, and elution from the purification columns. For fully automated procedures, this manifold must be disassembled, the DNA binding columns stored temporarily, and then reassembled with a collection plate on a vertical spacer for the final elution step. Using two manifolds, 192 samples can be processed simultaneously in less than two hours and several runs can be completed in a day (7).

When a vacuum manifold is used, care must be taken to dry the column beds and remove residual wash solution (containing salts and ethanol) from the tips of the column. Figure 5-3 compares plasmid DNA with contaminating wash solution to DNA isolated from properly dried columns. The 192 plasmid isolations performed in Figure 5-3a used minimal vacuum resulting in a slightly damp column bed, and most columns had droplets of wash solution clinging to the edges. The eluted plasmid DNA shows excessive streaking, and a denatured conformation is visible migrating faster than the covalently closed supercoiled DNA form.

In Figure 5-3b, these same clones were grown and re-isolated using extended vacuum times to dry the column beds, and the tips of the columns were robotically blotted prior to elution. The plasmid forms are predominately covalently closed, supercoiled DNA. The overall yield of plasmid DNA varied from clone to clone, and these intrinsic differences in plasmid amplification are consistently observed in the replicate experiments. Typical yields were 5 to 30 µg DNA in a final volume of 100 µL as determined by PicoGreen fluorescence analysis (see below). The $A_{260}/A_{280}$ ratios were measured to evaluate the quality of the plasmid preparations and the average ratio was approximately 1.7.

Plasmid DNA preparation can be omitted when preparing cDNA probes by amplifying the clone insert directly from colonies on plates, or from a small volume of the overnight culture (6). In this approach, the bacterial clones are cultured and a small aliquot is diluted and heated to lyse the bacteria and release the plasmid DNA. The goal for this approach is to reduce the costs associated with the plasmid preparation, and speed up the probe preparation process. The MultiPROBE workstation can be used to set up the appropriate dilutions and reactions; however, in our experience this method of amplification is less reliable (80% success, data not shown) than when using isolated plasmid as template. Figure 5-4 shows >92% success for 192 PCR reactions from the plasmid DNAs prepared in Figure 5-3b (7). Furthermore, the preparation of plasmid DNA provides an abundant and stable resource for storage of the clone library and can serve as a template for DNA sequencing if validation of a clone sequence is required.

The clone DNA resides as inserts within the prepared plasmids of the library, and must be amplified by PCR to remove most vector sequence elements and in order to obtain enough of the probe for spotting. PCR setup using the MultiPROBE workstation is a simple process of pipetting liquids to create a master mix, and then distributing the master mix to each well of a PCR plate (data not shown). For libraries where the inserts must be amplified using gene-specific primers, the WinPREP software can accept common spreadsheets as templates to direct the distribution of the primers to the correct well positions. We chose to perform the actual PCR reactions by manually removing the PCR plate from the workstation and transferring the plate to a thermocycler. The MultiPROBE Gripper is capable of transferring PCR plates into an "off-deck" thermocycler, and

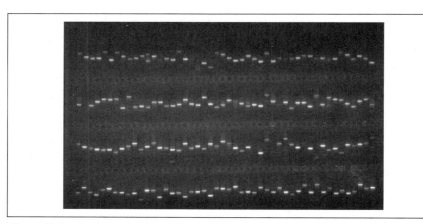

**Figure 5-4.** Gel electrophoresis results of 192 PCR products purified using the Qiagen QIAquick 96 nucleic acid purification kit. Clones used for PCR are the same as those in Figure 5-3.

the WinPREP software can control the operation of most automatic lids (using an RS232-port) to start PCR without user intervention. Manually removing the plate to a remote thermocycler enabled the workstation to continue operating in other tasks during the PCR process.

PCR cleanup is performed using the vacuum manifold and has similar reassembly requirements to plasmid isolation. Using the dual manifold setup 192 PCR products can be purified in 55 minutes. Figure 5-4 displays the electrophoresis results of 192 amplification reactions using the plasmid templates shown in Figure 5-3. Approximately 97% (187 of 192) samples were successfully amplified, although eight (4%) showed multiple bands. The number of reactions producing multiple bands is consistent with other reports (6). The purified products were recovered with insignificant loss of sample as determined by gel electrophoresis (data not shown).

The throughput of the various sample preparation procedures, the number of samples processed, and the time required are summarized in Table 5-1. By extrapolation, up to 1000 chip-ready cDNA probes can be prepared from bacterial stocks per week by one researcher using a single workstation.

To improve throughput, additional hardware can be added. For example, adding two more vacuum manifolds could accelerate either the plasmid or PCR isolation procedures. The expanded access area afforded by the Gripper and the software interface allow for other instrument integrations and controls such as a plate reader to further reduce user intervention. Potentially, inclusion of a thermocycler and plate reader would fully automate all steps following bacterial growth through plasmid

**Table 5-1.** Sample preparation procedures for DNA microarrays.

| Procedure | Samples Processed (N) | Procedure Time[a] (minutes) |
|---|---|---|
| Plasmid isolation | 192 | 114 |
| PCR reaction set-up | 384 | 20 |
| PCR clean up | 192 | 54 |
| Gel loading | 192 | 45 |
| DNA quantitation[b] | 192 | 72 |
| Redistribution[c] | 384 | 60 |

[a] After instrument setup.
[b] Using a four-tip MicroPROBE II.
[c] Four replicate 384-well plates prepared.

purification, DNA quantification, PCR setup, and PCR cleanup. This scheme has the advantage of minimal user intervention, but would be limited by available deck space, and would lower the duty cycle of the MultiPROBE II during thermocycling. Batch processing of similar samples allows for near constant use of the instrument, but requires greater attention from the user. Choosing between these different approaches will depend on the throughput desired and the resources available.

An additional constraint on the throughput of sample processing is the requirement for effective quality controls. Quality control measures must not be rate limiting, and are important in high-throughput situations. Electrophoresis of samples is an efficient and inexpensive way to monitor probe quality and quantity. To provide an automated solution to quality control, we have programmed the MultiPROBE II to load submerged agarose gels directly after plasmid isolation, PCR amplification, or PCR purification using commercial electrophoresis hardware.

We customized an Owl Separation Systems Model D3-14 gel electrophoresis unit and developed the gel pouring and positioning hardware such that the location of the wells in relation to the entire electrophoresis unit was consistent. This allowed for precise mapping of the poured agarose gel, and the gel wells, using WinPREP software. Figure 5-5 shows how the Owl electrophoresis unit is modified by adding Lucite retainers 140 mm apart to the inside of the buffer chamber, to hold the gel plate precisely in the middle of the chamber during the loading process (Figure 5-5a). The horizontal indexing of the gel tray is fixed using a spacer (1.5 mm wide) added to electrode side of gel tray (Figure 5-5b). The left and top sides of the buffer chamber are positioned onto the MultiPROBE II deck by pushing against standard plate adapter support tiles (Figure

(a)

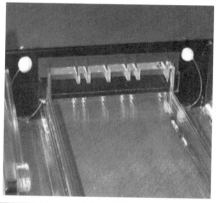

(b)

**Figure 5-5.** Hardware custom modifications for the Owl electrophoresis setup enabling the gels to be loaded automatically using MultiPROBE II workstations. (a) Beveled Lucite guides are glued inside the buffer chamber to precisely position the gel tray in the center. (b) The gel tray is indexed to the side of the buffer chamber using a 1.5 mm spacer. (c) The entire electrophoresis assembly is indexed onto the deck of the workstation by pushing two sides against standard plate support tiles. (d) The wells are indexed relative to the gel tray by positioning the combs to the top-left position during pouring of the agarose. (e) Image showing the MultiPROBE workstation loading a 100-well (two rows of 50 wells each) agarose gel, from within a PCR-purification protocol.

5-5c). The silicon-rubber feet of the gel unit provide adequate stability and further restraints are not required. To position the wells, 1, 2, 3, or 4 combs are marked and manually pressed into the top-left registration position prior to, and after pouring (Figure 5-5d). The gel must be poured at 55°C to prevent warping of the gel plate. These manipulations result in con-

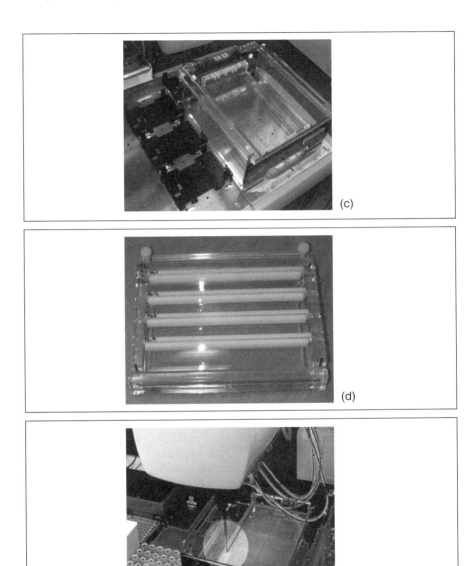

**Figure 5-5.** *Continued*

sistent positioning of the wells within the gel and consistent relationships between the gel and buffer tray, which is in a mapped location on the workstation deck. The entire gel unit is mapped on the deck using the WinPREP software labware definition library, and the well positions in the agarose gel are defined in a manner similar to other pieces of labware.

For gels with different numbers of wells (e.g., 100 vs. 200), each must have a separate labware definition.

To load samples, the electrophoresis unit can be closely integrated with sample preparation protocols. Figure 5-5e shows a four-tip Multi-PROBE II loading 96 samples and molecular weight markers onto an agarose gel immediately after preparing plasmid DNA. The samples are mixed with loading buffer and transferred to the agarose gel without changing tips. Note that this gel has been mapped as a 100-well gel. With care, gel loading is >99% successful. When a sample is not loaded properly, we found that it correlated with either a malformed pipette tip, or with improper loading of the tip onto the workstation. The MultiPROBE can also be used with reusable fixed "Versatips," which increased gel loading accuracy to >99.9%. (Further details can be found at http://homer.hsr.ornl.gov/CBPS/arraytechnology/gels.htm.)

Figures 5-3 and 5-4 show the results of our automated gel loading process. The mixing of 192 samples with gel loading buffer and individual sample loading requires approximately 45 minutes.

Electrophoresis units specifically designed for robotic workstations (E-Gel® 96 System; Invitrogen, Carlsbad, CA, USA) are available and can be integrated into the MultiPROBE through the WinPREP software. These units are designed to have the smallest possible "footprint" and take up the smallest possible amount of deck space. This is achieved by placing the wells in a staggered format, such that the DNA is electrophoresed between the wells of the next set of samples. Figure 5-6 shows the results of a 96-well PCR quality control analysis loaded by a MultiPROBE II workstation using the E-Gel 96 electrophoresis systems. The resulting image (Figure 5-6a) is confusing, and must be deconvoluted using E-Editor™ software supplied with E-gels for easy interpretation (Figure 5-6b). After deconvolution, the visualization of the probe DNA, including molecular weight markers, is similar to a conventional gel image.

Additional quality controls were designed to use the MultiPROBE capabilities. Fluorescence-based DNA quantitation was used to evaluate both PCR-produced and plasmid DNA in a 96-well format using a fluorescence-based plate reader. The PicoGreen® (Molecular Probes, now part of Invitrogen) reagent is highly specific for double-stranded DNA, and is able to measure the DNA even in the presence of residual mononucleotides (1, 18). Figure 5-7 shows representative results of PicoGreen analyses of probe cDNAs prepared by PCR. The average concentration in 100 µL of eluant is 289 ng/mL (standard deviation [SD] ± 87 ng/mL). These values can then be used to normalize the concentration of the probe prior to spotting onto the array. Setup and completion of the quantitation assay for 96 probe samples was accomplished in 40 minutes. Although complete automation is possible, plates were manually loaded into a HTS 7000 bioassay plate reader for measurements.

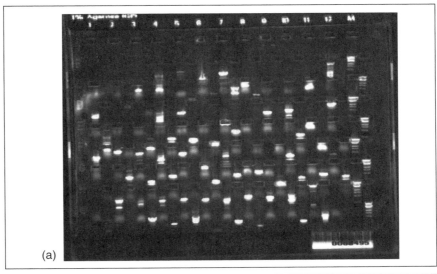

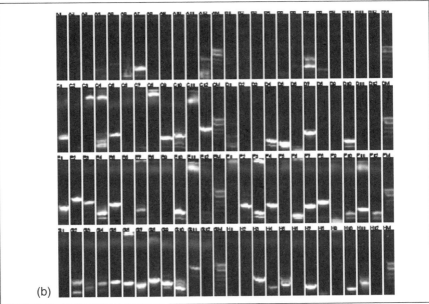

**Figure 5-6.** Demonstration of fully automated gel electrophoresis using the E-Gel® 96 System (Invitrogen). (a) Results of loading and running 96 PCR reactions using the MultiPROBE workstation. DNAs were visualized with Ethidium bromide staining and UV illumination. (b) The same image in panel (a) was deconvoluted using the E-series software supplied.

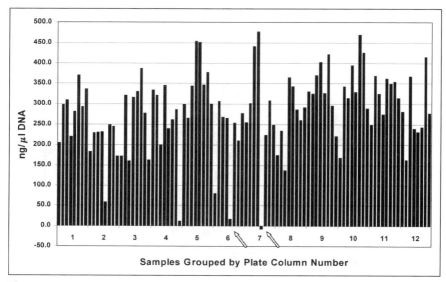

**Figure 5-7.** PicoGreen quantification of typical 96 PCR reactions prepared using a fully automated protocol on the MultiPROBE workstation. Bars represent the concentration of DNA in ng/µL for each probe prepared from the H3050 plate of the NIA 15K clone set of Tanaka et al. (20). Samples in the microplate are grouped by column number (1–12). The position of the negative (no DNA template) control samples are indicated with arrows. The data were exported to a spreadsheet for graphing, and could be loaded as a template file to direct WinPrep to normalize the concentrations of the samples.

For plasmid DNA, we similarly used the PicoGreen assay to quantitate plasmid DNA yields. The average plasmid DNA yield measured by PicoGreen assay from a 1.5 mL bacterial culture is approximately 22.4 µg (data not shown).

The HTS 7000 instrument also measures DNA by standard absorbance. To evaluate the quality of DNA samples, the plate reader was used to perform A260/A280 ratio measurements. The A260/A280 analyses of plasmid DNAs from Figure 5-3b is shown in Figure 5-8 (color Plate 1). The average value for these samples was 1.7.

The handling of large numbers of probes through multiple processes requires careful attention to sample tracking. Sample tracking must seamlessly follow changes in format between 96-well and 384-well microplates. Once the information tracking is robust, automated tracking of clone information during sample processing is much less likely to result in errors when compared to manual tracking. To collect sample tracking information from the MultiPROBE workstation, an internal positional database can be addressed using Microsoft Access. The internal database saves all information about plate names and clone positions as well. These

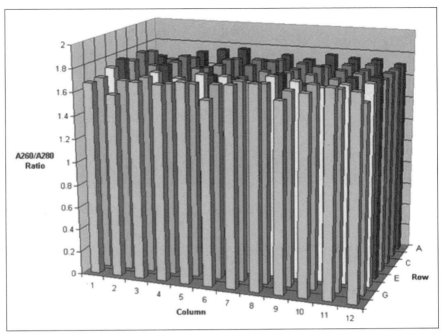

**Figure 5-8.** Chart showing the A260/A280 ratios of the 96 chip-ready PCR products from Figure 5-7. Values from the automated plate reader were exported to a spreadsheet for graphing. The spreadsheet could then be used to flag samples that do not meet quality controls. (See Plate 1.)

data can be exploited to automate tracking of samples, and to generate files for external applications used in cDNA chip production and analysis. We used database queries to extract the clone position information into tab-delineated files. These files were then supplemented with the original clone information to associate annotations with the new plate maps. Empty wells were tracked, and used for supplementing plates with controls or additional (new) specific gene sequences immediately prior to spotting. We also exported the information to a commercial software package, CloneTracker, which was included with the spotting instrument to automatically track samples from the original microplate to the final steps of microarray analysis. CloneTracker software requires a specific tab-delineated text file format to track samples spotted from plates onto microarray substrates; therefore, it was necessary to design database queries and filters that would extract plate and sample positional information as well as reformat the WinPREP data. This final step enables complete tracking between the parent cDNA clones and the individual microarray spots.

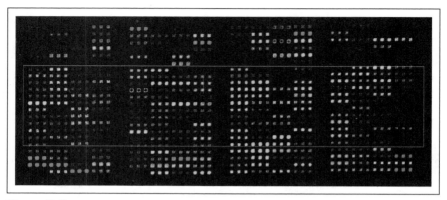

**Figure 5-9.** DNA differential expression microarray experiment using mouse kidney and mouse liver RNA targets. The probe cDNAs shown in Figure 5-4 are indicated by a gray box drawn in the middle of the array. In this experiment, hybridization to liver target RNA is shown in green, and hybridization to kidney target RNA is shown in red. (See Plate 2.)

Ultimately, the probes must be suitable for DNA microarray experimentation. Using the described protocols we prepared cDNA probes and performed differential gene expression analyses on tissues from mice (3; Hoyt et al., unpublished observations), and other organisms (8). A section of a cDNA array showing 352 unique cDNA probes spotted in triplicate is displayed in Figure 5-9 (color Plate 2). These images show the changes in gene expression observed when hybridized to mouse liver and kidney using 192 of cDNAs from Figure 5-4. This experiment was selected because of the inclusion of a known set of liver-specific genes, which are easily seen as bright green spots in Figure 5-9 and validate the results of the experiment. The bright fluorescence against a dark background shows excellent signal-to-noise ratio due to good surface chemistry saturation by the cDNA and effective hybridization of the bound probe.

## Discussion

Probe preparation for DNA microarrays typically involves multiple stages of sample preparation as outlined in Figure 5-1. In general, these steps include plasmid preparation, PCR amplification, PCR cleanup, and probe or product characterization. A single plasmid preparation should provide a sufficient amount of template for hundreds of PCR amplifications, and an individual PCR reaction is sufficient for preparing spots for hundreds of microarrays. When using a vacuum manifold system to prepare plasmid DNA, it is critical to properly dry and blot the columns prior to eluting the DNA. Residual wash solution contains salt that can contaminate the DNA and result in streaking during electrophoresis. Excessive exposure to alkaline conditions can cause denaturation, resulting in DNA that is permanently damaged and resistant to further enzymatic processing (3, 17). Eliminating the plasmid preparation steps can save time and money, but at the cost of reduced efficiency, and the loss of a valuable reagent (plasmid DNA and/or glycerol stocks) for sample archiving. A liquid handling system with an integrated gripper provides much greater flexibility when considering protocols for preparing samples. The Multi-PROBE II platform uses a remarkably adaptable gripper arm that provides rotational motion as well as for gripping of plates through either inside-out or outside-in motions. Most commercially available vacuum manifold systems could be adapted to the MultiPROBE deck and Gripper system (data not shown). Additionally, the extended access area enabled by the Gripper helps with integration of additional instrumentation. Finally, an integrated shaking platform provided important capabilities in mixing and resuspending samples. The flexibility allowed us to automate the majority of the processes outlined in Figure 5-1 using our existing equipment.

Some customized equipment was also fabricated from easily available materials to assist the buildup of a DNA processing pipeline. When loading samples onto agarose gels using the customized electrophoresis setup, faster loading times can be achieved by using gel combs that preserve 96-well spacing, which allows the MultiPROBE to perform simultaneous multichannel pipetting. Although automated sample loading is not faster than manual loading with a multichannel pipette, it eliminates the tedium and potential errors associated with manual loading, and allows the operator to attend to other machine functions. Commercially available 96-well electrophoresis units are faster and take up less deck space than the Owl electrophoresis unit described above. They are convenient options for very high-throughput and wealthy laboratories; however, the quality of the separation, the difficulty in deconvoluting images, and the price (including pre-poured gels in proprietary format) may make the customized Owl unit a better option for some facilities.

While absorbance or fluorescence can be used to quantitate plasmid DNA yields, the advantages of the fluorescence-based approach are that it has a wider measurement range, does not require special UV-transparent plates, and is specific for double stranded DNA. These advantages offset the additional cost associated with the PicoGreen reagent. Accurately quantitating DNA used for spotting enables the chip-ready probe DNA concentration to be standardized, either by dilution or drying, to provide consistent spot patterns. DNA concentrations outside the range of 25 to 300 µg/mL led to poor spots. Low DNA concentrations gave weak hybridization signals, and high DNA concentrations produced smearing ("comet tails") of the DNA spot during hybridization. DNA concentration determinations were not automated or integrated into the pipeline when using the plate reader because the workstation and the plate reader were controlled by software using different operating systems. Current instrumentation running the same operating systems would make these types of integrations easier. Our overall results for reading DNA concentrations coincide well with, and are complimentary to, visual results from agarose gels. However, agarose gels are advantageous because they identify probe preparations containing multiple DNA species. Multiple DNA bands cannot be resolved using fluorescence data. Also, after experience is gained, quantification of plasmid DNA is usually not necessary for setting up PCR. Generally, $1\,\mu L$ of the $100\,\mu L$ eluant from a plasmid preparation can be used as a template for PCR.

The lure of information provided by microarrays, capable of defining the functions of genes and systems, is powerful. For facilities to construct higher density cDNA microarrays, a difficult process of troubleshooting and optimization is inevitable to increase cDNA probe production. To prepare cDNA probes suitable for spotting, hundreds or thousands of unique plasmid clones must be converted into cDNA probes, and tracked through multiple stages and formats. Some programming and integration skills are needed to track samples using our automated system. As more software packages become available, the need for programming expertise is likely to diminish. Manually addressing high-throughput sample preparation without computer assistance is expensive, and can be dauntingly complex and time consuming. We have described our use of a single, but flexible, robotic workstation platform to address automation, quality controls, and sample tracking needs to produce thousands of clones at reasonable throughput. The MultiPROBE II HT EX liquid handling robot was capable of performing solid phase extractions, physically relocating labware, and proved versatile for various tasks required to produce DNA probes. Growth of clone libraries in bacteria, plasmid DNA preparation, PCR setup, PCR cleanup, and sample redistribution all benefited from the workstations capabilities. Customized accessory mapping allowed for conventional electrophoresis to be included in the system, and enabled facile and integrated quality control steps. Tracking of samples was possible by extracting information from the Microsoft Access database used by the WinPREP software, and we were able to automate the generation of text files required "downstream" by software for microarray analyses. Automated protocols using the workstation greatly reduced user intervention and associated errors while enhancing throughput.

## Acknowledgments

This manuscript has been authored by a contractor of the U.S.Government under contract DE-AC05-00OR22725. Accordingly, the U.S. Government retains a nonexclusive, royalty-free license to publish or reproduce the published form of this contribution, or allow others to do so, for U.S. Government purposes.

This research was sponsored by the Laboratory Directed Research and Development Program of Oak Ridge National Laboratory (ORNL), managed by UT-Battelle, LLC for the U.S. Department of Energy under Contract No. DE-AC05-00OR22725 and support from Packard Bio-Sciences, Inc.

## References

1. Ahn, S.J., Costa, J., and Emanuel, J.R. 1996. PicoGreen quantitation of DNA: effective evaluation of samples pre- or post-PCR. *Nucleic Acids Res* 24: 2623–2625.
2. Beg, O.U., and Holt, R.G. 1998. A cost-effective plasmid purification protocol suitable for fluorescent automated DNA sequencing. *Mol Biotechnol* 9: 79–83.
3. Birnboim, H.C., and Doly, J. 1979. A rapid alkaline extraction procedure for screening recombinant plasmid DNA. *Nucleic Acids Res* 7: 1513–1523.
4. Engelstein, M., Aldredge, T.J., Madan, D., et al. 1998. An efficient, automatable template preparation for high throughput sequencing. *Microb Comp Genomics* 3: 237–241.
5. Garner, H.R., Armstrong, B., and Kramarsky, D.A. 1992. High-throughput DNA preparation system. *Genet Anal Tech Appl* 9: 134–139.
6. Hegde, P., Qi, R.K., Abernathy, C., et al. 2000. A concise guide to cDNA microarray analysis. *BioTechniques* 29: 548–562.
7. Hoyt, P.R., Tack, L., Jones, B.H., et al. 2003. Automated high-throughput probe production for DNA microarray analysis. *BioTechniques* 34: 402–407.
8. Hoyt, P.R., Doktycz, M.J., Beattie, K.L., and Greeley, M.S. Jr. 2003. DNA microarrays detect 4-nonylphenol-induced alterations in gene expression during zebrafish early development. *Ecotoxicology* 12: 469–474.
9. Huang, G.M., Wang, K., Kuo, C., et al. 1994. A high-throughput plasmid DNA preparation method. *Anal Biochem* 223: 35–38.
10. Kirschner, L.S., and Stratakis, C.A. 1999. Large-scale preparation of sequence-ready bacterial artificial chromosome DNA using QIAGEN columns. *BioTechniques* 27: 72–74.
11. Konecki, D.S., and Phillips, J.J. 1998. TurboPrep II: an inexpensive, high-throughput plasmid template preparation protocol. *Biotechniques* 24: 286–293.
12. Neudecker, F., and Grimm, S. 2000. High-throughput method for isolating plasmid DNA with reduced lipopolysaccharide content. *BioTechniques* 28: 107–109.
13. Packard Liquid Handling. Application Note LHN-002. *Plasmid Purifications*. Packard BioScience Company.
14. Packard BioScience Company. Packard Specification Bulletin No M4197. 2001. *Gripper™ Integration Platform. Nucleic Acid Extraction Application*.

15. Pogue, R.R., Cook, M.E., Livingstone, L.R., and Hunt, S.W. 1993. Preparation of template for automated sequencing using QIAGEN resin. *BioTechniques* 15: 376–381.
16. Sambrook, J., Fritsch, E.F., and Maniatis, T. 1989. *Molecular Cloning. A Laboratory Manual,* 2nd ed. Cold Spring Harbor: Cold Spring Harbor Laboratory Press.
17. Sayers, J.R., Evans, D., and Thomson, J.B. 1996. Identification and eradication of a denatured DNA isolated during alkaline lysis-based plasmid purification procedures. *Anal Biochem* 241: 186–189.
18. Singer, V.L., Jones, L.J., Yue, S.T., and Haugland, R.P. 1997. Characterization of PicoGreen reagent and development of a fluorescence-based solution assay for double-stranded DNA quantitation. *Anal Biochem* 249: 228–238.
19. Skowronski, E.W., Armstrong, N., Andersen, G., et al. 2000. Magnetic, microplate-format plasmid isolation protocol for high-yield, sequencing-grade DNA. *BioTechniques* 29: 786–792.
20. Tanaka, T.S., Jaradat, A., Lim, M.K., et al. 2000. Genome-wide expression profiling of mid-gestation placenta and embryo using a 15,000 mouse developmental cDNA microarray. *Proc Natl Acad Sci U S A* 97: 9127–9132.

# 6

# Isolation of Nucleic Acids from Paraffin-Embedded Archival Tissues and Other Difficult Sources

Evgeny N. Imyanitov, Evgeny N. Suspitsin,
Konstantin G. Buslov,
Ekatherina Sh. Kuligina,
Evgeniya V. Belogubova, Alexandr V. Togo,
and Kaido P. Hanson

*N.N. Petrov Institute of Oncology, St. Petersburg, Russia*

Many applications of DNA and RNA analysis require the ability to work with suboptimal tissue sources. In medical research, the availability of archival paraffin-embedded tissue collections provides invaluable potential resources for well-designed, large-scale, retrospective studies. Pathological specimens can also be used for routine diagnostic purposes, mainly for PCR detection of infectious agents or analysis of cancer-related genetic events (21, 23, 28). Furthermore, DNA testing has become an indispensable part of the forensic expertise of biological remains (12). In addition, some recent historical and evolutionary studies have been based on the genetic analysis of archaeological findings (15).

The potential complicating factors for these applications include low quantity of preserved nucleic acids, severe degradation of the genetic material, significant damage of DNA and RNA chemical structure, admixture of biological contaminants, as well as the presence of various PCR inhibitors (13, 20, 21, 28). Because of these difficulties, many specialists remain reluctant to take on the task of genetically analyzing these so-called "difficult" tissue sources. However, a two-decades-long history of the activities in this field has undoubtedly proven that, with the development of modern protocols, the analysis of residual genetic material has become a straightforward and reliable procedure. Perhaps the most spectacular success story is the recent identification of the bones belonging to the Russian Tsar Nicholas II and family, executed in 1918, whose remains had been subjected to an intentional heavy chemical and physical destruction, and then buried in soil for more than sixty years (9). Molecular archaeologists and paleontologists have managed not only to decipher DNA sequences of extinct species, such as Neanderthal hominids and mammoth (11, 18), but also to gain some important scientific information from the analysis of secondary DNA sources, such as food remains, bacteria populating ancient humans and other living beings, and animals domesticated by prehistoric tribes (15). DNA testing led to definite identification of the pathogen that caused the devastating Medieval era Black Death epidemic in the 1300s in Europe. Five agents (anthrax, typhus, tuberculosis, hemorrhagic fever, and plague) were suspected in the deaths of between 17 and 28 million people during that time. Genetic analysis has identified *Yersinia pestis*, the plague pathogen, in the bodies of victims of this epidemic (15). Czarnetzki et al. (6) examined DNAs from as many as 7063 archaeological human remains and confirmed the rare frequency of Down syndrome in ancient Europe.

Currently, all the approaches involving genetic analysis of nucleic acids obtained from difficult tissue sources are based on the PCR technique. Archival formalin-fixed paraffin-embedded tissues are used as a source of DNA and RNA for both medical research and diagnostic purposes. Other relevant applications, such as forensic expertise, evolutionary studies, and archaeological investigations, require only DNA analysis. To date, only short DNA or RNA fragments are accessible for routine, reproducible PCR amplification (1, 20, 23, 28). The most likely cause of artifacts is the prior contamination of analyzed tissue by "irrelevant" biological material. Although some procedures allow a reduction in the consequences of the contact between target tissue and biological admixtures, or at least to estimate the probability that the obtained result is false, none of the measures offers an absolute guarantee (15, 19, 24).

This chapter describes general principles of the genetic analysis of difficult tissue sources. In addition, we present some basic protocols as well as share our experiences in this area.

## DNA Isolation

Procedures for DNA isolation can be classified into three different groups:

1. Methods involving direct addition of DNA-containing tissue lysates to PCR reaction

2. "Classical" DNA extraction protocols based on the use of organic solvents (phenol and chloroform)
3. DNA purification procedures employing DNA-binding substances (15, 23).

The use of tissue lysates has both advantages and disadvantages. Direct lysis usually takes significantly less time than relatively complicated multi-step DNA isolation procedures. The risk of losing DNA is usually negligible; however, omitting the DNA purification step may result in the reduced availability of DNA to the interaction with Taq polymerase. Tissue lysates often contain impurities (i.e., heme compounds, collagen, denatured proteins, small degraded DNA fragments), which inhibit PCR reaction (2, 17, 25). Although even simple boiling of the tissue in water may be sufficient for DNA release, most researchers prefer intensive proteinase K digestion of the specimens in PCR-compatible detergent solution (Triton X-100, Tween 20, NP-40) prior to heating of samples (23) (see also Protocol 1). It is believed that proteinase K pretreatment not only improves the DNA solubility by disruption of structural tissue components, but also destroys many PCR inhibiting proteins (21). Heating is an essential step because it inactivates proteinase K and thus protects Taq polymerase; in addition, it was shown that heating may increase the yield of DNA product (27).

Phenol-chloroform extraction is currently the standard protocol for DNA isolation from difficult sources (see Protocol 2). Contrary to the above-mentioned single-step approach, this method is compatible with the exhaustive lysis of the target tissue using strong detergent, for example, SDS containing buffer. Furthermore, pure DNA provides better PCR results for some tissue sources (5), probably due to partial elimination of enzymatic inhibitors. This method is also preferable if the double-stranded structure of DNA has to be preserved. Phenol-chloroform extraction is widely used, especially for the tissues with low cellularity and therefore low DNA concentration (i.e., bones). Precipitation of residual amounts of nucleic acids with ethanol may be inefficient and, therefore, an addition of glycogen carrier is highly recommended (23). As an alternative to organic extraction, many researchers use commercial kits that are based on the tissue lysis in guanidinium thiocyanate buffer and subsequent DNA purification involving DNA-binding substances (e.g., silica beads, columns, and membranes). It is suggested that the purity of DNA obtained with the latter approach is better compared to the phenol-chloroform protocol (15, 30). However, DNA-binding particles bind only fragments with a length exceeding 100 bp, which may not be advantageous when handling difficult tissue sources (15).

### Summary of Our Experience

We routinely use archival formalin-fixed paraffin-embedded tissues as a source of DNA in many research and diagnostic applications. Our experience overlaps with that of others, especially with the guidelines presented by Poljak et al. (23), and is summarized in the following sections.

### Shorten the Fragment Size

In comparison with tedious and time-consuming attempts to optimize the DNA extraction protocol, it is significantly more rewarding simply to shorten the expected fragment size while planning the PCR design. Some published recommendations propose 200 bp or even 300 to 400 bp as the upper limit for the PCR product length when the routine analysis of archival tissue is planned (10). Given significant inter- and intra-laboratory variations in the fixation procedures, storage time, tissue cellularity, among other factors, we suggest that amplification of fragments below 150 bp (preferably around 110–120 bp) is always advantageous in terms of reliability and reproducibility. This is similar to the data described in other studies (23, 28).

### Use Direct Tissue Lysates

We certainly favor the use of direct tissue lysates instead of DNA purification. In our hands, DNA isolation did not result in the improvement of PCR yield or success rate; similar observations have been reported in other studies (5, 7, 23).

### Dilute the DNA

Contrary to our experience, some believe that the difficulties of DNA analysis from archival tissues are attributed to the low concentration of DNA; therefore, some researchers attempt to add more DNA lysate to PCR reaction. However, we find that the dilution of DNA (1:10, or even 1:50–1:100) is more helpful, apparently because of the lower amount of PCR inhibitors. This is also confirmed by data presented in other studies (23, 28).

### "Hot-Start" Technique at the Denaturing Step

A "hot-start" format of PCR is absolutely essential for the success of amplification. The "hot-start" technique implies that one of the critical components of PCR reaction (most often, Taq polymerase) is added to the reaction at the denaturation step. This simple modification significantly improves the specificity of the first round of amplification, because it prevents the elongation of primers that mis-anneal to the template at the room temperature during the assembling of the PCR cocktail (22). Earlier "hot-start" procedures required either manual addition of Taq polymerase to the PCR tubes staying in the thermocycler at high temperature, or use of heat-prone wax barrier separating PCR mix and the enzyme prior to the denaturation step. Relatively recent introduction of heat-activated Taq polymerases (e.g., AmpliTaq Gold from Applied Biosystems, Inc., Foster City, CA, USA; or Platinum Taq from Invitrogen, Carlsbad, CA, USA) has made the "hot-start" protocol much more convenient and elegant (15).

## Different PCR Reaction Mixture Composition

Composition of PCR reaction mixture for the analysis of archival DNA may differ from that for intact chromosomal DNA. It is not unusual to observe that amplification of partially degraded DNA requires elevated concentration of magnesium chloride from a typical 1.5–2.0 mM to 2.5–4.0 mM.

## Adjust the PCR Cycling Conditions

Many difficulties of PCR optimization process are related to the fact that during the initial PCR cycles, critical reagents (i.e., primers, thermostable polymerase, etc.) are in excess, thus favoring a nonspecific DNA synthesis, while the final rounds of the reaction occur in a shortage of the mentioned components, which compromises the PCR yield. We use a very simple adjustment of the PCR cycling conditions that sometimes increases the success rate of amplification from 15% to 30% to a more satisfactory 80% to 90%. We often start PCR with very short annealing and synthesis steps (1 second each), and program thermocycler with the steadily increasing times for the duration of both these phases (+1 second with each new cycle). With this approach, typical PCR conditions would be as follows: 95°C for 30 seconds; annealing step, 57°C for 1 second + 1 second per each new cycle; synthesis step, 72°C for 1 second + 1 second per each new cycle. If the number of cycles is 40, the duration of both annealing and synthesis in the last cycle will be 41 seconds, and this significantly exceeds times in the initial stages of PCR.

## Archival DNA Needs More Rounds of Amplification

PCR for degraded archival DNA requires 5 to 15 more cycles than the analysis of intact genetic material (28), that is, while typical PCR amplification from high molecular weight DNA requires 30 to 35 cycles, archival DNA is usually subjected to 40 to 45 rounds of amplification.

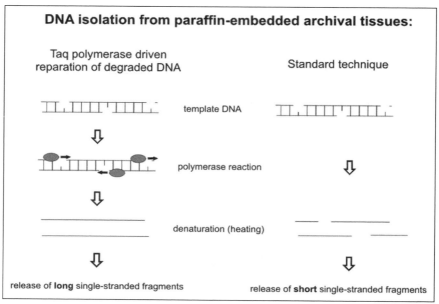

**Figure 6-1. Principle of the DNA reconstruction protocol.** Because the poor quality of archival DNA is at least in part attributed to the single-strand breaks, this template can be substantially restored by filling the nicks in the polymerase reaction. This procedure may substantially increase the length of DNA fragments, which increases the PCR success rate and allows the amplification of larger products.

## Partial Restoration of Degraded DNA

We have developed an approach allowing the partial restoration of degraded DNA (14) (Protocol 3, below). The poor quality of archival for-malin-fixed DNA is at least partially attributed to the single-strand breaks, and filling-in the nicks can substantially restore the template integrity by the polymerase reaction prior to DNA isolation (Figure 6-1). This simple modification substantially increases the success rate of PCR and allows amplification of larger fragments (Figure 6-2). The concept of Taq-polymerase driven DNA restoration was subsequently confirmed in an analysis of postmortem tissues (3). Bonin et al. (4) quantitatively com-pared the success rate of PCR for DNA isolated from Bouin's fixed tissues with and without the restoration step. The fragment with the size of 100 bp was amplified in 18 of 20 (90%) cases that were not subjected to the DNA restoration (−), whereas the use of the above-mentioned DNA-reconstructing protocol (+) for the same tissue samples led to the 100% success rate (20/20 cases). For longer DNA fragments, the data are as follows: 15/20 (75%−) and 20/20 (100%+) for the fragment of 204 bp; 0/20

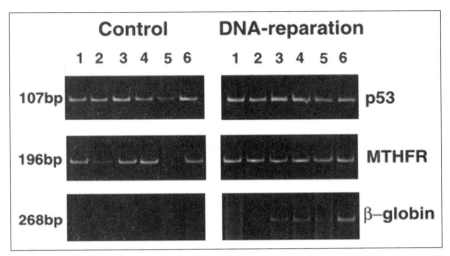

**Figure 6-2.** Effects of Taq polymerase-driven restoration of paraffin-embedded formalin-fixed tissues to derive DNA. Six randomly selected archival breast cancer tissue sections (lanes 1–6) were subjected to DNA isolation in duplicate. The processing of the samples had been done as described in the Protocol 3, without (left) or with (right) addition of Taq polymerase. The 10 μL PCR reactions included 1 μL DNA-containing tissue lysate, 0.5 units heat-activated Taq DNA polymerse, 1× PCR buffer, 1.5 mM MgCl₂, 200 μM dNTP, 1 μM each primer, and were carried out for 40 cycles (95°C for 35 seconds, 57°C for 45 seconds, 72°C for 45 seconds) after an initial activation of the polymerase at 95°C for 10 minutes. The products were separated in the polyacrylamide gel. The primers were 5'-AGG TCT GGT TTG CAA CTG GG-3' and 5'-GAG GTC AAA TAA GCA GCA GG-3' for p53 gene, 5'-TGA AGG AGA AGG TGT CTG CGG GA–3' and 5'-TGG CCA GCG CGG TGA GAG TG-3' for MTHFR gene, and 5'-CAA CTT CAT CCA CGT TCA CC-3' and 5'-GAA GAG CCA AGG ACA GGT AC-3' for β-globin gene. Used with permission from *BioTechniques* (14).

(0%–) and 7/20 (35%+) for the 291 bp product; and 0/20 (0%–) and 2/20 (10%+) for the DNA sequence with the length of 333 bp.

## RNA Isolation

RNA is extremely susceptible to degradation by RNAses and, therefore, even the isolation of RNA from fresh tissue samples is considered to be a tricky and troublesome procedure (21). Not surprisingly, most researchers simply do not believe that it is possible to extract RNA from archival material. However, this is contrary to our experience and some other published data (16, 29). As the formalin is a potent RNAse inhibitor (20), standard morphological tissue fixation protocol allows short RNA fragments to remain intact for many years. Therefore, the RNA analysis of paraffin-embedded material is no less complicated than DNA testing, and can be routinely used in medical research and diagnostics.

There are two alternative RNA isolation procedures. The first method is similar to the RNA isolation from fresh tissues and involves lysis of the specimen in the guanidinium thiocyanate buffer (21). Several independent reports and our experience indicate that the second method—proteinase K enabled lysis in the presence of SDS—provides better results (16, 29) (see Protocol 4). Ethanol precipitation of RNA is significantly more reliable if glycogen is added as a carrier (29). RNA extraction is followed by cDNA synthesis, which can utilize either random primers or specific oligonucleotides corresponding to the target sequence.

### Summary of Our Experience

The success rate is surprisingly high (more than 90% for some archival collections), although the differences in tissue processing between various morphological laboratories appear to be more crucial than for DNA analysis.

As in the case of DNA, it is critical to reduce the PCR fragment length to 110 to 120 bp.

Unlike of DNA isolation, careful rehydration of formalin-fixed tissue sections turned out to be essential in our hands. Specht et al. (29) presented similar data; however, their study results contrast with the data published by Korbler et al. (16).

If simultaneous DNA analysis is required, there is usually no need to separately extract DNA and, unless RNAse-free DNAses are used, RNA extracts contain noticeable amounts of DNA. Of course, this fact has to be considered while planning the PCR design. The sites of primer annealing should be separated by introns; otherwise, it will be impossible to discriminate between RNA- and DNA-derived PCR products.

## Protocols

### Protocol 1. Preparation of DNA-Containing Lysates from Formalin-Fixed Paraffin-Embedded Tissue Sections

Protocol 1 is based on the principles described in earlier studies (7, 23, 26).

1. Place 5 to 10 μm thick tissue section (50–100 mg) into the 1.5 mL Eppendorf tube.
2. Tissue deparaffinization: Add 500 μL of xylene, vortex, incubate for 15 to 30 minutes at room temperature, and then remove xylene carefully. Repeat this step.

3. Add 500 μL of 96% ethanol, vortex, incubate for 15 to 30 minutes at room temperature. Remove ethanol by pipette or spin down the tissue debris if the section has lost its integrity.
4. Add 100 μL of the lysis buffer (10 mM Tris-HCl, pH 8.3, 1 mM EDTA, 0.5% NP-40, 0.5% Tween-20).
5. Boil at 100°C for 5 minutes in a heat-block in order to partially solubilize and disintegrate the tissue components.
6. Add proteinase K up to 500 μg/mL, incubate at 60°C overnight. Note that some research groups use even higher concentration of proteinase K, or repeatedly add fresh enzyme during prolonged incubation (8, 23).
7. Add the chelating resin Chelex-100 up to 5%, boil the sample at 100°C for 10 minutes in a heat-block in order to inactivate proteinase K.
8. The obtained lysate is suitable for PCR. Dilute the sample 1:10 in water, and use 1 μL for each PCR reaction.

## Protocol 2. Phenol-Chloroform Extraction of DNA from Formalin-Fixed Paraffin-Embedded Tissue Sections

The following procedure is according to protocol described in reference 5, with slight modifications.

1. Deparaffinize tissue sections as described in Protocol 1.
2. Add 500 μL of the lysis buffer (10 mM Tris-HCl, pH 8.0, 1 mM EDTA, 1% SDS, 500 μg/mL proteinase K), incubate overnight at 60°C.
3. Add an equal volume of phenol, chloroform, and isoamyl alcohol mixture (25:24:1). Vigorously vortex the suspension for 15 to 30 minutes, separate aqueous and organic phases by centrifugation, and carefully transfer the upper phase to a new Eppendorf tube.
4. Repeat the extraction by the equal volume of the mixture of chloroform and isoamyl alcohol (24:1). After centrifugation, carefully transfer the upper phase to the new Eppendorf tube.
5. Precipitate the nucleic acid from the aqueous phase by the addition of sodium acetate to 0.25 M, 1 μL of glycogen carrier solution (20 mg/mL), and 2 volumes of ethanol. After incubation at −20°C for at least one hour, precipitate the DNA by centrifugation at 15,000× *g* for 10 minutes at 0° to 5°C.
6. Wash the pellet with 70% ethanol to remove salts and dissolve the DNA in 20 to 100 μL of sterile water.

## Protocol 3. Taq Polymerase-Driven Restoration of Partially Degraded DNA from Formalin-Fixed Paraffin-Embedded Tissue Sections

This procedure is described in reference 14.

1. Deparaffinize tissue sections as described in Protocol 1.
2. Transfer the deparaffinized section into an Eppendorf tube containing 500 μL of 1× TE-buffer (10 mM Tris-HCl, 1 mM EDTA, pH 8.0)
3. Add proteinase K to 20 μg/mL, incubate for 20 minutes at room temperature to make DNA molecules available for subsequent polymerase reaction. Note: Excessive proteinase K digestion may result in tissue disintegration and subsequent loss of DNA.
4. Carefully remove traces of proteinase K by three washes in 1× TE buffer. Note that some tissues, for example lung or liver specimens, are too fragile, so the repetitive rinsing in washing solution may result in the loss of material. In such cases, overnight treatment in low concentrations of proteinase K (1–2 μg/mL) and omitting the washing step are suggested.
5. Add 100 μL of PCR-like mixture (10 mM Tris-HCl [pH 8.3], 1.5 mM MgCl$_2$, 2% Triton X-100, 200 μM each dNTP) and incubate at 1 hour at 55°C to achieve partial solubilization of the tissue.
6. Add Taq polymerase up to 0.1 U/μL, incubate the sample for 20 minutes at 72°C.
7. Repeat steps 5 to 8 from Protocol 1.

## Protocol 4. RNA Extraction from Formalin-Fixed Paraffin-Embedded Tissues

This procedure is done according to reference 29.

1. Deparaffinize tissue sections as described in Protocol 1.
2. Rehydrate the section by sequential incubation in 96%, 80%, and 70% ethanol for 15 minutes each. Note that in our experience, this step is critical, although some laboratories omit rehydration without significant effect on their results (16).
3. Add 400 μL of the lysis buffer (10 mM Tris-HCl, pH 8.0, 0.1 mM EDTA, 2% SDS, 500 μg/mL proteinase K). Incubate at 60°C for 12 to 16 hours until the tissue has completely disintegrated.
4. Add equal volume of water-saturated acidic phenol and 0.3 volume of chloroform. Vigorously vortex for 10 minutes.
5. Centrifuge at 15,000× $g$ for 15 minutes at 0° to 5°C.
6. Place the tubes on ice. Carefully transfer the supernatant into a new microcentrifuge tube. Add 0.1 volume of 3 M sodium acetate

(pH 4.0) and 0.3 volume of chloroform. Vigorously vortex for 10 minutes.

7. Centrifuge at 15,000× $g$ for 15 minutes at 0° to 5°C.

8. Place the tubes on ice. Carefully transfer the supernatant into a new microcentrifuge tube. Add 1 µL of glycogen carrier solution (20 µg/mL) and 1 volume of isopropanol. Incubate at −20°C for at least 3 hours.

9. Concentrate the precipitate by centrifugation at 15,000× $g$ for 15 minutes at 0° to 5°C.

10. Remove isopropanol. Wash the pellet in 70% ethanol for 5 to 10 minutes.

11. Remove the 70% ethanol, dry the pellet at 37°C. Dissolve the RNA in 10 µL of RNAse-free water for 10 minutes at 65°C. Store samples at −20°C.

As mentioned above, the direct lysis of tissue lysates (Protocol 1) appears to be preferable in most instances. If high purity of DNA is indeed necessary, or target tissue has low DNA concentration, then the DNA extraction procedure (Protocol 2) is recommended. DNA reconstruction method (Protocol 3) may be recommended in those instances where the improvement of success rate of PCR is highly desirable, or in case of a desperate need to obtain a PCR fragment from a truly important or unique tissue sample. Currently, RNA isolation procedures do not include too many alternative approaches and, thus, the only Protocol 4 is recommended for all circumstances.

## Acknowledgments

The work is supported by the International Association for the Promotion of Co-operation with Scientists of the New Independent States (NIS) of the former Soviet Union (INTAS grant 03-51-4234) and Russian Foundation for Basic Research (RFBR grant 04-04-49060).

## References

1. Abrahamsen, H.N., Steiniche, T., Nexo, E., et al. 2003. Towards quantitative mRNA analysis in paraffin-embedded tissues using real-time reverse transcriptase-polymerase chain reaction: a methodological study on lymph nodes from melanoma patients. *J Mol Diagn* 5: 34–41.
2. Akane, A. 1996. Hydrogen peroxide decomposes the heme compound in forensic specimens and improves the efficiency of PCR. *BioTechniques* 21: 392–394.
3. Bonin, S., Petrera, F., Niccolini, B., and Stanta, G. 2003. PCR analysis in archival postmortem tissues. *Mol Pathol* 56: 184–186.

4. Bonin, S., Petrera, F., Rosai, J., and Stanta, G. 2005. DNA and RNA obtained from Bouin's fixed tissues. *J Clin Pathol* 58: 313–316.

5. Cao, W., Hashibe, M., Rao, J.Y., et al. 2003. Comparison of methods for DNA extraction from paraffin-embedded tissues and buccal cells. *Cancer Detect Prev* 27: 397–404.

6. Czarnetzki, A., Blin, N., and Pusch, C.M. 2003. Down's syndrome in ancient Europe. *Lancet* 362: 1000.

7. De Lamballerie, X., Chapel, F., Vignoli, C., and Zandotti, C. 1994. Improved current methods for amplification of DNA from routinely processed liver tissue by PCR. *J Clin Pathol* 47: 466–467.

8. Diaz-Cano, S.J., and Brady, S.P. 1997. DNA extraction from formalin-fixed, paraffin-embedded tissues: protein digestion as a limiting step for retrieval of high-quality DNA. *Diagn Mol Pathol* 6: 342–346.

9. Gill, P., Ivanov, P.L., Kimpton, C., et al. 1994. Identification of the remains of the Romanov family by DNA analysis. *Nat Genet* 6: 130–135.

10. Greer, C.E., Peterson, S.L., Kiviat, N.B., and Manos, M.M. 1991. PCR amplification from paraffin-embedded tissues. Effects of fixative and fixation time. *Am J Clin Pathol* 95: 117–124.

11. Hagelberg, E., Thomas, M.G., Cook, C.E. Jr., et al. 1994. DNA from ancient mammoth bones. *Nature* 370: 333–334.

12. Hochmeister, M.N. DNA technology in forensic applications. 1995. *Mol Aspects Med* 16: 315–437.

13. Hoss, M., Jaruga, P., Zastawny, T.H., et al. 1996. DNA damage and DNA sequence retrieval from ancient tissues. *Nucleic Acids Res* 24: 1304–1307.

14. Imyanitov, E.N., Grigoriev, M.Yu., Gorodinskaya, V.M., et al. 2001. Partial restoration of degraded DNA from archival paraffin-embedded tissues. *BioTechniques* 31: 1000–1002.

15. Kaestle, F.A., and Horsburgh, K.A. 2002. Ancient DNA in anthropology: methods, applications, and ethics. *Am J Physiol Anthropol Suppl* 35: 92–130.

16. Korbler, T., Grskovic, M., Dominis, M., and Antica, M. 2003. A simple method for RNA isolation from formalin-fixed and paraffin-embedded lymphatic tissues. *Exp Mol Pathol* 74: 336–340.

17. Kosel, S., Grasbon-Frodl, E.M., Arima, K., et al. 2001. Inter-laboratory comparison of DNA preservation in archival paraffin-embedded human brain tissue from participating centres on four continents. *Neurogenetics* 3: 163–170.

18. Krings, M., Stonem, A., Schmitz, R.W., et al. 1997. Neanderthal DNA sequences and the origin of modern humans. *Cell* 90: 19–30.

19. Marota, I., Basile, C., Ubaldi, M., and Rollo, F. 2002. DNA decay rate in papyri and human remains from Egyptian archaeological sites. *Am J Phys Anthropol* 117: 310–318.

20. Masuda, N., Ohnishi, T., Kawamoto, S., et al. 1999. Analysis of chemical modification of RNA from formalin-fixed samples and optimization of molecular biology applications for such samples. *Nucleic Acids Res* 27: 4436–4443.

21. Mies, C. 1994. Molecular biological analysis of paraffin-embedded tissues. *Hum Pathol* 25: 555–560.

22. Nuovo, G.J., Gallery, F., MacConnell, P., et al. 1991. An improved technique for the in situ detection of DNA after polymerase chain reaction amplification. *Am J Pathol* 139: 1239–1244.
23. Poljak, M., Seme, K., and Gale, N. 2000. Rapid extraction of DNA from archival clinical specimens: our experiences. *Pflügers Arch* 439(Suppl 3): R42–R44.
24. Pusch, C.M., Bachmann, L., Broghammer, M., and Scholz, M. 2000. Internal Alu-polymerase chain reaction: a sensitive contamination monitoring protocol for DNA extracted from prehistoric animal bones. *Anal Biochem* 284: 408–411.
25. Scholz, M., Giddings, I., and Pusch, C.M. 1998. A polymerase chain reaction inhibitor of ancient hard and soft tissue DNA extracts is determined as human collagen type I. *Anal Biochem* 259: 283–286.
26. Sepp, R., Szabo, I., Uda, H., and Sakamoto, H. 1994. Rapid techniques for DNA extraction from routinely processed archival tissue for use in PCR. *J Clin Pathol* 47: 318–323.
27. Shi, S.R., Cote, R.J., Wu, L., et al. 2002. DNA extraction from archival formalin-fixed, paraffin-embedded tissue sections based on the antigen retrieval principle: heating under the influence of pH. *J Histochem Cytochem* 50: 1005–1011.
28. Shibata, D. 1994. Extraction of DNA from paraffin-embedded tissue for analysis by polymerase chain reaction: new tricks from an old friend. *Hum Pathol* 25: 561–563.
29. Specht, K., Richter, T., Muller, U., et al. 2001. Quantitative gene expression analysis in microdissected archival formalin-fixed and paraffin-embedded tumor tissue. *Am J Pathol* 158: 419–429.
30. Yang, D.Y., Eng, B., Waye, J.S., et al. 1998. Technical note: improved DNA extraction from ancient bones using silica-based spin columns. *Am J Phys Anthropol* 105: 539–543.

# 7

# Comparison of Sequencing Data for DNAs Prepared Using TempliPhi Technology and Using Traditional Preparation Protocols

Richard Sheldon, Lloydia Reynolds,[1] Markryan Dwyer,[1] and Jan Kieleczawa
*Biological Technologies Department, Wyeth Research, Cambridge, MA*

The TempliPhi amplification system centers on a well-characterized, high fidelity bacteriophage polymerase ($\Phi 29$) capable of quickly amplifying DNA from linear, plasmid, and genomic sources (1, 9). There are several applications of this attractive technology in capillary-electrophoresis sequencing protocols that can potentially increase efficiency in each phase of the process from DNA isolation to data processing (2). Thus, by reducing the time from source DNA to edited sequence, combined with lower cost, fewer human-hours, and little waste, TempliPhi is positioned to make a significant impact on a sequencing laboratory's efficiency, especially in a high-throughput environment (1, 2, 9). In addition, in a core DNA sequencing laboratory, the TempliPhi can be used to "rescue" low-quality or scarce templates (L. Haines and J. Kieleczawa, unpublished data).

The TempliPhi strategy utilizes a <u>r</u>olling <u>c</u>ircle <u>a</u>mplification (RCA) mechanism and phosphorothioate-modified random hexamers that are not degraded by $\Phi 29$'s proofreading and strand-displacement capabilities (1, 9). The random priming allows for this polymerase to synthesize complimentary strands at multiple replication forks originating on the same template; amplification rates are further increased when free primers and polymerase bind to newly displaced strands. Exponential amplification

---

[1]LR and MD were participants of a Wyeth-sponsored summer program.

allows for minimal amount of starting material (1–100 pg) that includes prepared DNA, plasmids, bacterial cells from colonies, glycerol stocks, or liquid cultures as well as genomic and clinical specimens of finite quantity (1, 5, 7–9, 12). This highlights another aspect of the TempliPhi platform. Potentially, it is a single procedure for providing sequencing-grade DNA from a variety of sources, which is a boon to any support center or core DNA sequencing facility that strives for uniformity of process. Comparing these potential benefits to current output standards, it is obviously important to incorporate the TempliPhi technology into any sequencing laboratory. For another view of possible application and implementation of the TempliPhi-based DNA amplifications, the reader is encouraged to review Chapter 9 in this volume.

This chapter compares the quality of DNA prepared using TempliPhi-driven amplification and the DNA prepared with two classical isolation methods. In most experiments, three different starting DNA sources were used for these comparisons: glycerol stocks, colonies and purified plasmid DNA. The two primary criteria used for this evaluation are quality read length and the pass rate. In addition, we further optimized the TempliPhi condition when the starting materials are colonies on an agar plate.

## Materials and Methods

### DNA Isolation from Saturated Cultures

For comparison purposes, two commercially available DNA preparation kits were used in these investigations: a manually performed, alkaline lysis-based, 96-well filter plate method (from Edge BioSystems, Gaithersburg, MD, USA), and the Qiagen BioRobot 8000-based, 96-well plate QIAprep method (from Qiagen, Inc., Valencia, CA, USA). To provide a similar starting point for the TempliPhi and filter plate preparation methods, common bacterial cultures were initiated from a single colony or glycerol stocks in Erlenmeyer flasks with 350 mL of Terrific Broth (TBr), 2× YT or Luria Broth (LB). All media were prepared according to standard molecular biology protocols (10) containing 100 µg/mL ampicillin or zeocyn, and incubated overnight, with agitation, at 300 rpm, 37°C, in an air shaker (model 25D; New Brunswick Scientific, New Brunswick, NJ, USA). Later, 1.0 mL of culture was transferred to each well of a deep plate, pelleted, and processed as recommended by manufacturers. DNA was eluted in 100 µL of 10 mM Tris/HCl, pH 8.5 for the Qiagen method and in 100 µL of 10 mM Tris/HCl, 0.01 mM EDTA, pH 8.0 buffer for Edge Biosystems method, sequenced and stored at −20°C for further experiments, if needed.

### TempliPhi DNA Amplification

Unless otherwise stated the DNA amplification method followed the protocol provided with TempliPhi kit (Amersham Biosciences, Piscataway, NJ, USA, now part of GE Healthcare).

The starting sources of DNA were (a) glycerol stocks were produced by touching with a toothpick and transferred to a sample buffer, (b) each colony was placed into 5 μL of water and thoroughly vortexed, and (c) purified plasmid DNA.

Briefly, into a chilled, 96-microwell plate, approximately 0.1 μL of one of the above template and 2.5 μL of sample buffer were added and gently mixed. Plates were then sealed with rubber mates, denatured (for various times at 95° or 98°C) and cooled to 4°C on a PTC-225 cycler (MJ Research, Waltham, MA, USA), removed and spun down. Then, 2.5 μL of a reaction buffer/enzyme mix was added to samples and incubated at 30°C for 4 to 12 hours and stored at −20°C until sequencing step was performed (typically the next day).

The following plasmid DNAs were used in the tests: pCDNA3.1/GS plasmid in GeneHog™ *Escherichia coli*, pcDNA3.1 Hygro (both purchased from Invitrogen, Carlsbad, CA, USA) and pGem3zf (purchased from Promega, Madison, WI, USA).

### Characterization of DNA

Using a multichannel pipette, 1 μL of eluted or amplified DNA was combined with 8 μL TEsl (10 mM Tris-HCl, 0.01 mM EDTA, pH 8.0) and 1 μL 10× DNA loading buffer (9), gently mixed and spun. In addition, pGem3zf standard (0.2 mg/mL) was similarly prepared along with a low molecular mass DNA ladder (Invitrogen, Inc., Carlsbad, CA, USA). The samples were run on 1%, 96-well E-gel agarose system (Invitrogen) following the manufacturer's conditions and visualized using Kodak's EDAS 290 digital imaging system (Eastman Kodak Company, Rochester, NY, USA).

### DNA Sequencing

The DNA sequencing reactions were assembled on ice by combining approximately 200 ng of templates in 6 μL of 10 mM Tris-HCl, 0.01 mM EDTA, pH 8.0, with 1 μL of 5 μM primer in either 96-well plates or 8/12 PCR tube strips (200 μL). Samples were tightly sealed and heat-denatured (6) for five minutes at 98°C, unless stated otherwise, on a PTC-225 cycler (purchased from MJ Research). Following quick spin, samples were placed on ice and 3 μL of diluted (0.8 mL of ABI PRISM BigDye™ Terminator Cycle Sequencing Ready Reaction Kit mix V3.0 and 0.4 mL of 5×

BigDye dilution buffer) was added to each well. The samples were again re-sealed, gently vortexed, and spun down before being returned to the thermocycler for cycle sequencing using the following cycling protocol: [96°C for 10 seconds, 50°C for 5 seconds, and 60°C for 4 minutes] × 25.

Sequencing reactions were purified of excess dye and salts by passing through a Millipore filter plate (catalog # MAHVN45; Millipore) packed with G-50 Sephadex fine beads (Amersham Biosciences). The flow-through was heat denatured for two minutes at 90° to 95°C and loaded onto the ABI3700 DNA analyzer for processing using the manufacturer's guidelines. The ABI3700 DNA analyzer, dye-terminator mix, and 5× dilution buffer were purchased from Applied Biosystems, Inc. (Foster City, CA, USA).

The quality of sequencing data was evaluated using phred $Q \geq 20$ read length values (3, 4) and the pass rate. The sample was "passed" if the continuous $Q \geq 20$ read length was at least 100 bases. The $Q \geq 20$ values and pass rate was the average of 8 to 32 samples processed under the same conditions; the number of tested samples depended on the type of experiment and is indicated in legends for each figure.

## Results and Discussion

To compare TempliPhi-derived and "traditionally" prepared DNAs for overall sequence quality and pass rate, glycerol stock containing a pCDNA3.1/GS plasmid was used to initiate overnight cultures in three culture media: 2× YT, LB, and TBr. The saturated cultures were distributed into 32 wells of a 96-well plate and used for DNA preparation. The number of such identical plates was prepared for easy comparison of various DNA isolation methods. Additionally, identical plates were used to prepare glycerol stocks by combining 0.5 mL of 50% glycerol and 0.5 mL of an appropriate growth culture and stored at –80°C. DNA preparations were sequenced immediately or stored at –20°C until next day.

### Media Effects on the DNA Quantity

Visualization of the 1% agarose gels showed marked differences in the DNA yield between the different growth media and preparation methods. As expected, the enriched 2× YT and TBr media yielded more DNA than the LB medium for both the EdgeBiosytems and QIAprep methods. Of the two, the automated QIAprep method produced higher concentrations of DNA and was more consistent across the wells. However, the TempliPhi amplified DNA for 12 hours was much more uniform regardless of the media type and the nature of a starting point (Figure 7-1). Similar

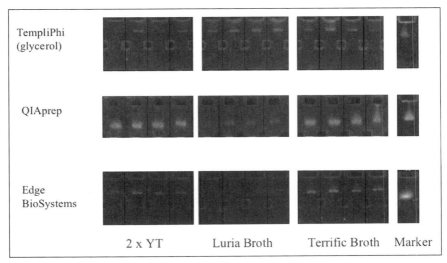

TempliPhi
(glycerol)

QIAprep

Edge
BioSystems

2 x YT          Luria Broth          Terrific Broth     Marker

**Figure 7-1.** **Agarose gel analysis of DNAs prepared using three different methods.** DNAs were prepared as described in the Materials and Methods section and 1 μL aliquots were run on an E-gel electrophoresis system. The *top panel* shows "TempliPhied" DNA from glycerol stocks prepared from three different culture media. *Middle and bottom panels* show DNAs isolated from overnight cultures using QIAprep and EdgeBiosystems protocols, respectively. The panels on the right hand side are pGem3zf control DNAs (200 ng/lane).

trends have been observed elsewhere in regards to uniformity of amplification of DNA from glycerol stocks (2).

This result is an illustration of the differences between the two classes of protocols. As pointed out earlier, the limiting reagent to the amount of DNA a TempliPhi reaction produces over a fixed time is primarily dependent on the amount of available nucleotides and polymerase added to the reaction mixture. On the other hand, for the plasmid DNA preparations, the number of plasmids in the bacterial culture caps the ultimate amount of DNA that can be extracted. More importantly, the success rate using TempliPhi protocol has been shown to be less dependant upon plasmid copy number and relative population of bacterial cells, and more dependent on the presence of template in liquid cellular culture (2).

### Media Effects on a Phred $Q \geq 20$ Score

As with the media effect on DNA yield, media type also affects the overall sequencing success rates and quality scores (Figure 7-2). The QIAprep method yielded higher quality scores than the other two methods tested (Figure 7-2a) while the reaction pass rates remained comparable (Figure 7-2b).

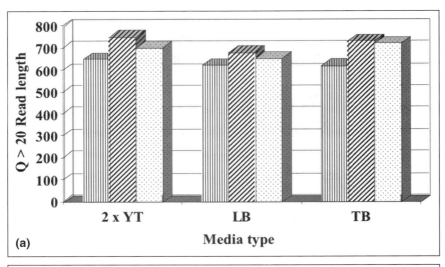

(a)

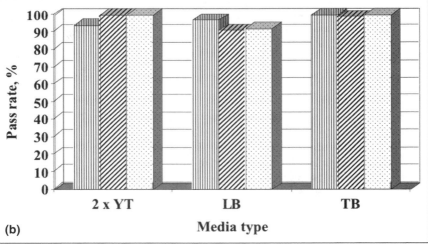

(b)

**Figure 7-2.** Assessing read length (a) and pass rate (b) for DNA samples prepared from various sources. Each data point is the average 8 to 24 individual reads. Symbols are (bars with perpendicular lines ‖) represent data for Templiphi preparations; (bars with // lines) are data for DNAs prepared with Qiagen method; and (bars with dots) are data for DNAs prepared using EdgeBiosystems kit. In most high-throughput sequencing centers, the typical pass rate is around 85% to 90%. Higher pass rate presented in panel b is the result of the experimental design (see Materials and Methods section for more detail).

Even after accounting for DNA concentration differences, templates prepared from the LB medium yielded reduced quality sequencing results as measured by both pass rate and $Q \geq 20$ score compared to the data for DNAs obtained from enriched media. This broth-to-broth variation, however, was less pronounced for TempliPhi-amplified DNAs with glycerol stocks as starting points. The data presented here suggest that Qiagen method in combination with rich media yields more uniform DNA concentrations, leading to highly consistent sequencing results. Similarly, the "TempliPhied" DNAs from various glycerol stocks result in most consistent sequencing data. These two observations can be the basis for setting up more efficient workflow in a sequencing laboratory that is also engaged in preparation of DNA templates. The choice of a DNA preparation protocol can depend on the nature of the starting point, as ultimately the uniformity of DNA concentrations and their quality leads to better sequencing results because less attention will have to be devoted to out-of-range samples. In the end, the more outliers in a process, the more likely that the overall plate-wide sequence quality will be diminished (2). Therefore, by reducing variability among samples throughout the entire sequencing process, these methods show the potential for increasing a laboratory's overall productivity. Further refining of these procedures is important for greater optimization of the sequencing effort. Of particular interest is the fact that high quality sequence is generated without the need of a lengthy DNA preparation process, which leads to faster delivery of sequencing data.

### Effects of Denaturation Temperature and Time on Sequence Quality of TempliPhi-Derived DNA Templates from Colonies

In initial experiments (denaturation time vs. starting source of TempliPhi-derived DNAs) the least consistent data was generated when colonies from the agar plates were the starting source for TempliPhi-derived DNA templates. In the attempt to optimize "the colony source starting point," we carried out experiments with various denaturation times at two different temperatures. The rationale for such an attempt can be derived from data presented in Kieleczawa's study (6). Initially, the purified pGem3zf plasmid was denatured in sample buffer at 95° and 98°C (Figure 7-3). As evident from Figure 7-3, the transition from the supercoiled form to single-stranded (ss) form is faster at 98°C; the majority of supercoiled form is converted to ss form just after five minutes at 98°C, while taking 7.5 to 10 minutes at 95°C. We used the same approach for initial denaturation of colonies picked from agar plates: Figure 7-4a shows that the optimal denaturation time is 4 to 5 minutes at either 95° or 98°C (as opposed to just 3 min of denaturation at 95°C recommended in TempliPhi protocol). For comparison, the denaturation profile is also shown when

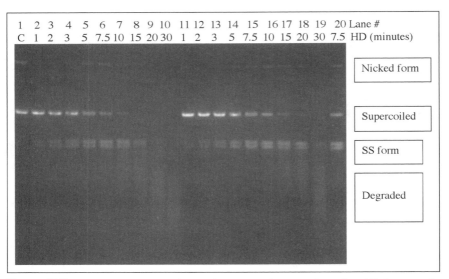

**Figure 7-3. Denaturation profiles of pGem3zf at 95°C and 98°C in sample buffer from a TempliPhi kit.** Two hundred nanograms (0.5 µL) of DNA were added to TempliPhi sample buffer (9.5 µL) and heat-denatured at 95°C and 98°C. The denaturation times are indicated above the picture. Lanes 2–10 show denaturation at 98°C and lanes 11–19 show denaturation at 95°C. Following the denaturation, samples were placed on ice till the end of series. One µL of 10× agarose loading buffer (10) was added to each tube and samples were electrophoresed on 1% agarose gel (1× TAE buffer, 0.5 µg/mL of ethydium bromide, 100 mA, 2 hours). C indicates control DNA (non-denatured pGem3zf) and lane #20 shows the typical denaturation of pGm3zf in 10 mM Tris/Cl, 0.01 mM EDTA buffer at 98°C. HD is heat denaturation.

the starting source of DNA is the liquid culture (Figure 7-4b); in this case, our experiments confirm the recommendations accompanied the TempliPhi kit (11) and other published data (2, 8, 9). We believe that the longer denaturation times needed to obtain better sequencing quality data with colonies as a starting source of DNA are linked to more effective production of TempliPhi-accessible form of DNA.

## Summary

This chapter reviewed the quality of DNA templates generated using various preparation methods, primarily from the DNA sequence quality point of view. As expected, the quality of sequencing data (using read lengths and pass rate criteria) generated with templates purified with classical methods remains superior, but the invested cost and time can be a

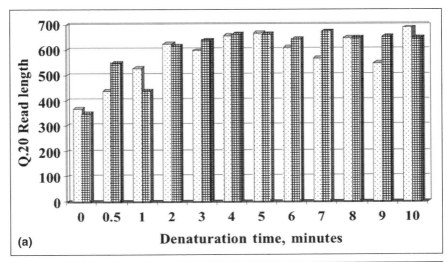

(a)

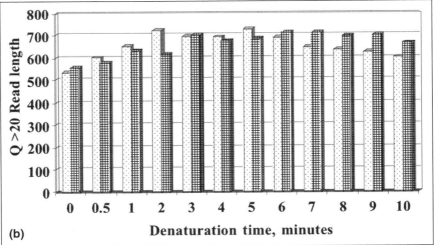

(b)

**Figure 7-4.** The Q ≥ 20 read lengths for sequencing samples prepared using TempliPhi kit from colonies (a) and from liquid culture (b). Following the re-suspension of colonies or liquid culture in sample buffer, the mixtures were heat-denatured for various times at either 95°C (dotted bar) or 98°C (crosshatched bar), "TempliPhied" and sequenced, and analyzed as described in the Materials and Methods section. Each data point is the average 32 samples.

limiting factor in many applications. Often, the longest possible read length and the highest pass rate are not the primary study objectives; for example, library screening may require just 100 to 400 bases of good quality data, and acquired in the fastest time from DNA source to sequence data, it is sufficient to make on go/no-go decisions. The Templiphi technology may be one of those applications. It is straightforward, cost-effective, fast, and amenable for automation. It can also be applied when the amount of initially available DNA template is limited or is of poor quality. However, caution must be exercised when other, more sophisticated template requirements are needed, for example, when an exact reproduction of the starting DNA template is necessary. We believe that this and related technologies will find many applications in modern molecular laboratories, which are continuously striving to introduce new protocols for faster and more efficient processing of biological samples.

## References

1. Dean, F.B., Nelson, J.R., Giesler, T.L., and Lasken, R.S. 2001. Rapid amplification of plasmid and phage DNA using Phi29 DNA polymerase and multiply-primed rolling circle amplification. *Genome Res* 11: 1095–1099.
2. Detter, J.C., Jett, J.M., Lucas, S.M., et al. 2002. Isothermal strand-displacement amplification applications for high-throughput genomics. *Genomics* 80: 691–698.
3. Ewing, B., Hillier, L., Wendl, M., and Green, P. 1998. Base calling of automated sequencer traces using Phred. I. Accuracy assessment. Genome Res 8: 175–185.
4. Ewing, B., Hillier, L., Wendl, M.C., and Green, P. 1998. Base-calling of automated sequencer traces using Phred. II. Error probabilities. *Genome Res* 8:186–194.
5. Hosono, S., Faruqi, F., Dean, F.B., Du, Y., et al. 2003. Unbiased whole-genome amplification directly from clinical samples. *Genome Res* 13: 954–964.
6. Kieleczawa, J. 2005. Controlled heat-denaturation of DNA plasmids. In: Kieleczawa, J., ed. *DNA Sequencing: Optimizing the Process and Analysis.* Sudbury: Jones and Bartlett Publishers; 1–10.
7. Lasken, R.S., and Egholm, M. 2003. Whole genome amplification: abundant supplies of DNA from precious samples or clinical specimens. *Trends Biotechnol* 21: 531–535.
8. Mamone, A. 2003. Representational amplification of genomic DNA. *Life Sci News* 14: 4–6.
9. Nelson, J.R., Cai, Y.C., Giesler, T.L., et al. 2002. TempliPhi, Φ29 DNA polymerase based rolling circle amplification of templates for DNA sequencing. *Biotechniques* 32:S44–S47.
10. Sambrook, J., and Russell, D.W. 2001. *Molecular Cloning,* 3rd ed. Cold Spring Harbor, NY: Cold Spring Harbor Laboratory Press.
11. TempliPhi 100 Amplification Kit. 2002. *Instructions.* Sunnyvale, CA: Amersham Biosciences Corp.
12. Vincent, M., Xu, Y., and Kong, H. 2004. Helicase-dependent isothermal DNA amplification. *EMBO Rep* 5: 795–800.

# 8

# Whole Genome Amplification Using the Multiple Displacement Amplification Reaction

**Roger S. Lasken**
*Center for Genomic Sciences, Allegheny-Singer Research Institute, West Penn Allegheny Health System, Pittsburgh, PA, USA*

The goal of whole genome amplification is to generate large quantities of DNA from a starting DNA template. Unlike the polymerase chain reaction (PCR), which targets a short segment of the template for amplification using defined primers, whole genome amplification ideally replicates the entire template as evenly as possible. Random and degenerate primers are used to initiate DNA synthesis from any DNA template present in the reaction. Like PCR, whole genome amplification has found a multitude of laboratory applications as researchers have explored ways to take advantage of amplified DNA. Obviously, one use is to obtain a large supply of DNA from minute amounts of starting material. Hundreds of micrograms of amplified DNA can be obtained from a few microliters of blood from a finger stick or a few cells in a buccal swab (18), enabling large epidemiologic and genetic association studies to be carried out with these less intrusive methods for blood and cell sample collecting. Whole genome amplification has also allowed restoration of depleted DNA archives even when only trace amounts remain (24). While no method is yet capable of the fidelity with which cells replicate their chromosomes—a feat requiring highly complex and redundant biochemical pathways (20)—newer methods are now delivering remarkably faithful amplification in vitro. The Multiple Displacement Amplification Method (MDA) (9–11, 18, 22, 24) preserves an estimated 99.98% of the genome (29) with high sequence fidelity based on mutation analysis (28). For depleted DNA archives, often irreplaceable and collected at great expense, MDA is an important method to extend the lifetime of research projects. Even when genomic DNA can be readily obtained from fresh blood samples, new microarray technologies that consume large quantities of DNA can require MDA to generate sufficient supplies.

Another use for MDA is simply as a DNA sample preparation method even when DNA is not in short supply. Because MDA reaction assembly is easily automated with liquid handling instruments, and filtration or centrifugation steps are usually not required to clean up the amplified DNA, it is a convenient method to prepare DNA for subsequent analysis. Traditional methods of obtaining DNA from cells, such as phenol extraction and bead- and cartridge-based purification products, are time consuming and difficult to automate. However, amplified DNA is enriched relative to cellular and other contaminants and can be added directly to downstream sequencing and genotyping reactions. For example, plasmid DNA can be amplified from *E. coli* cultures by MDA (10) and robotically delivered directly to high throughput sequencing reactions using a commercial product called TempliPhi® (Amersham Biosciences, now a part of GE Healthcare, Piscatawy, NJ, USA). Genome sequencing facilities are now carrying out tens of thousands of these reactions in a single day with large cost and labor savings (11).

Perhaps the ultimate challenge is to amplify the genome of a single cell. While still not successful in every reaction, MDA has advanced even this capability to the point where many new research strategies are now possible. Amplified DNA from a single blastomere (16) has been used for genotyping, and has the potential for in vitro fertilization clinics to use as template in tests for genetic diseases. MDA from single bacterial cells (31) has been demonstrated and can be used for genomic sequencing from unculturable species. In both of those studies, some DNA sequence was lost during the amplification from a single cell. This presumably results from damage to some of the DNA during cell lysis or loss of DNA during handling. Also, increased amplification bias occurred, where some sequences are more highly represented than others, possibly resulting from stochastic effects of amplifying from a single genome copy. Small variations early in the amplification may be magnified throughout the reaction. More than a billion-fold amplification occurred from about five femtograms of *E. coli* DNA (one genome copy) to about a 25 µg yield of amplified DNA (31). This extremely large amount of amplification might be expected to introduce more variation in the frequency of different sequences. Quantitative PCR demonstrated a several thousand-fold amplification bias between 10 loci tested compared to only a sixfold bias when starting from many template copies (see below). Nevertheless, several laboratories are using MDA to obtain initial genomic sequence from single cells of unculturable microbes obtained from soil, marine, and other environments. An estimated 99% of all microbial species have not yet been identified. Combined with improved approaches to isolate individual microbial cells, MDA promises to enable genomic analysis of these species without the need to develop culturing methods.

## A Brief History of Whole Genome Amplification Methods

Replication of double stranded DNA is no simple task. Reconstitution of the cellular system for DNA replication in vitro can require more than 20 purified proteins (20). Origin recognition proteins initiate separation of the two DNA strands, helicases and DNA binding proteins contribute to propagation of replication forks, and protein complexes made of numerous subunits prime the synthesis with RNA primers and tightly clamp the DNA polymerases on to the DNA template.

PCR was the first simple laboratory method that could exponentially replicate double-stranded DNA (27, 32) for use in biotechnology. Heating to 95°C melts the double helix into two single strands that are then available for annealing with the primers. Cooling allows annealing of the oligonucleotide primers and extension by a DNA polymerase. However, this returns the DNA to a double stranded form, requiring cycling to another 95°C step, cooling to anneal more primers, and so on through multiple cycles.

Pioneering whole genome amplification methods carried out PCR with random primers in a method called primer extension preamplification (PEP) (41), or with degenerate primers in degenerate oligonucleotide-primed PCR (DOP) (7, 38). These powerful methods amplified DNA even from single cells (reviewed in 39), including sperm (2, 8, 34, 41) and blastomeres (21, 35, 40), for preimplantation diagnosis. The amplified DNA does not provide complete coverage of the entire genome (1, 17, 38), but is valuable for many genotyping applications. DOP is widely used to synthesize labeled probes for use in fluorescence in situ hybridization (FISH) and comparative genome hybridization (CGH) (40). A variation of PEP called tagged PCR (15, 37) uses primers with random 3′ ends, as in PEP, but with a constant 5′ end that provides a primer annealing site for subsequent PCR amplification. Linker adapter PCR is an alternative method to DOP and PEP in which genomic DNA (gDNA) is digested and short linkers are ligated to the ends providing a single annealing site for PCR priming (26, 33). Linker adapter PCR eliminates some of the amplification bias associated with thermal cycling where sequence has a large effect on both the denaturation of template and the Tm of primers. While linker adapters provide more uniform priming, only a limited subset of sequence is amplified and the method requires considerably more steps. A proprietary method of linker adapter PCR called OmniPlex® (Rubicon Genomics, Ann Arbor, MI, USA) has been tested for use in genotyping (3). A newer version of PEP called improved PEP (iPEP) (12) uses two

DNA polymerases: Taq DNA polymerase and a proofreading enzyme such as Vent (New England Biolabs, Beverly, MA, USA) or Pfu DNA polymerase (Stratagene, La Jolla, CA, USA). Analogous to "long PCR" (4), the proofreading polymerase uses its 3′–5′ exonuclease activity to remove misincorporated nucleotides that slow the progression of Taq DNA polymerase. iPEP increased genotyping accuracy (42) compared to PEP, which is reported to result in some loss of heterozygosity for single cells by unequal amplification of one allele over the other (14, 30, 39). As with other PCR-based whole genome amplification methods (19), iPEP only generates short DNA products. iPEP amplified DNA could only be used for PCR genotyping assays that targeted segments of <400 bp. Even though it is based on "long PCR," iPEP products apparently become shorter and shorter with random priming to regions internal to amplicons (6).

## Multiple Displacement Amplification

MDA is the first whole genome amplification method that does not depend on PCR. Exponential amplification of the DNA template is achieved in an isothermal reaction carried out at 30°C. MDA uses a very unusual DNA polymerase from the bacteriophage Phi29 that has exceptionally tight binding to the DNA template and strong "strand displacing" activity. The reaction begins with extension of random primers (Figure 8-1a). As the polymerase extends primers, it also displaces downstream DNA strands from the template (Figure 8-1b). The Phi29 DNA polymerase is particularly proficient at this concurrent DNA synthesis and strand displacement of downstream products. The displaced single stranded DNA presents more sites for annealing of the random primers and the reaction quickly becomes exponential (Figure 8-1c) by a mechanism called hyperbranching (25). The hyperbranched structure has many "replication forks" at which the DNA polymerase invades double stranded template by displacing the complementary strand. In the cell, DNA helicases and single stranded DNA binding proteins assist in advancing a replication fork (20). It is the remarkably tight DNA binding and "processivity" of the Phi29 DNA polymerase that makes strand displacement sufficiently efficient for MDA to occur without the assistance of these other proteins. *Processivity* is defined as the number of nucleotides that the DNA polymerase can add on to the primer before the polymerase dissociates from the DNA. For example, the large fragment of *E. coli* DNA polymerase I (Klenow fragment) can only add about 10 nucleotides to the primer in one binding event with the template. The Phi29 DNA polymerase averages an amazing 70,000 nucleotides before coming off the template (5), making it the most processive DNA polymerase, in the

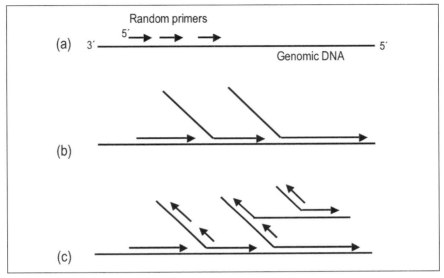

**Figure 8-1.** **Multiple displacement amplification reaction.** (a) DNA synthesis is primed by random hexamers. (b) Strand displacement synthesis occurs as the Phi29 DNA polymerase extends primers while concurrently displacing downstream products. (c) Exponential amplification occurs by a "hyperbranching" mechanism.

absence of accessory proteins, ever described. Its processivity and strand displacement synthesis are thought to be the keys to maintaining efficient hyperbranching and generating long DNA products. The amplified DNA is an average 12 kb in length (10) and ranges from about 2 kb to over 100 kb, which is an advantage over the PCR-based whole genome amplification methods that generate DNA of only about 100 to 400 bp.

The MDA reaction also depends on maintaining very high random primer concentrations (usually 50 µM) that can blanket the DNA template. Early attempts at MDA generated only low levels of amplification. It was discovered that the Phi29 DNA polymerase itself was degrading the primers as the reaction proceeded (10). The polymerase has a 3′–5′ exonuclease activity that confers high fidelity to the DNA products. This so-called proofreading activity removes misincorporated nucleotides (e.g., a G inserted opposite a template T) and results in a low mutation rate of only one error in $10^6$ to $10^7$ nucleotides (13). However, polymerases often have proofreading activities that avidly attack single stranded DNA, degrading them from the 3′ end. The solution to the problem was to protect the primers with thiophosphate-modified nucleotides on the 3′ end (10). With the random primers protected from degradation, exponential amplification was achieved.

Another advantage of the MDA method is that almost the entire DNA template is well represented and with low amplification bias between different regions of the genome. In a study of 47 single copy loci in the human genome, one on the p and q arm of each chromosome, all were well represented in the amplified DNA. Furthermore, they were still present at between about one half to three copies per genome equivalent of amplified DNA (Figure 8-2). While some amplification bias had occurred, over this sixfold range from 0.5 to 3 copies per genome, this was a dramatic improvement over the PCR-based whole genome amplification methods, which have drastic variation in the representation of different sequences (Figure 8-3) (9). MDA's low amplification bias, coverage of more than 99.9% of the genome, and the great length of the amplified DNA has dramatically advanced our ability to replicate DNA for biotechnology applications (23).

Another form of amplification bias is critical if the DNA is to be used as template for genotyping assays. Both chromosomes of a diploid organism, such as humans, must be evenly represented in order to carry out accurate genotyping. A heterozygous individual could be inaccurately scored as homozygous simply because only one of the chromosomes was well represented. Several studies indicate that genotyping of MDA amplified DNA is about 99.9% accurate (3, 18, 29). It might be hoped that MDA would amplify the maternal and paternal alleles evenly for single nucleotide polymorphisms (SNPs) and point mutations, because the surrounding sequence is usually identical. SNPs and point mutations should experience identical local sequence effects on template melting and priming efficiency. It is possible that when extensive haplotype sequence differences do occur near a point mutation or SNP of interest, this might result in biased amplification and account for some of the subsequent genotyping errors observed. There are also reports that amplification efficiency is reduced near the ends of templates such as near telemers (22).

## Laboratory Protocols for Whole Genome Amplification with MDA

### Sources of DNA Templates

MDA can be used to amplify small quantities of extracted DNA or to amplify DNA directly from biological specimens such as cultured cells or blood. When starting from purified DNA templates, the quality of the DNA is important. Samples that are highly degraded to less than 2 to 4 kb in length (and lacking at least a small amount of longer DNA strands) may yield only partial recovery of the genome. Nevertheless, when no other alternative is available, MDA can be useful even for partially

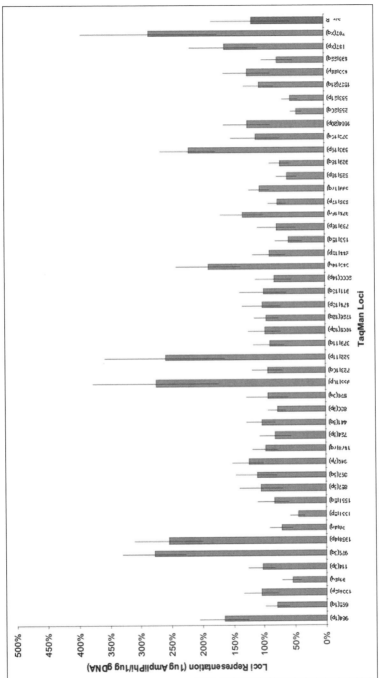

**Figure 8-2.** **Amplification bias analysis for MDA by quantitative TaqMan PCR assays for 47 human loci.** Loci representation is expressed as the percent of locus copies in the amplified DNA relative to the starting DNA template. Each bar represents the average of amplifications of 44 different individual's DNA. A value of 100% indicates that the sequence is present in the same copy number per genome in the amplified DNA that occurs in the starting genomic DNA template. The average of all of the 47 loci was 117% (far right bar). *Source:* Reprinted from [18] with permission of *Genome Research,* Cold Spring Harbor Laboratory Press.

115

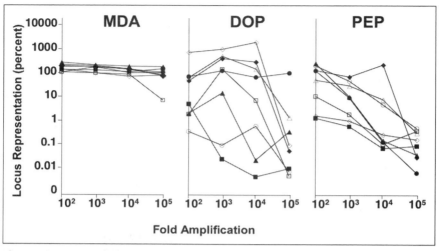

**Figure 8-3.** **Effect of amplification on gene representation bias.** The relative representation of eight loci was determined using TaqMan quantitative PCR for DNA amplified by the MDA, DOP, and PEP methods as indicated. The X-axis represents the fold amplifications compared to the starting DNA template; the Y-axis is the locus representation. The results for eight loci are indicated as follows: (◇) CXCR5; (△) connexin40; (□) MKP1; (○) CCR6; (◆) acidic ribosomal protein; (▲) CCR1; (■) cJUN; (●) CCR7. Reprinted from reference 9 with permission of the National Academy of Sciences, USA.

damaged and degraded samples. In the case of DNA that has been stored for a long period of time, it is advisable to start with more input DNA, for example 100 ng, as apposed to the normally recommended 10 to 30 ng. The amplified DNA should be tested for detection or genotyping of several genes to determine the quality of recovered sequence before use in more extended studies.

MDA is also very valuable for whole genome amplification directly from clinical and other biological samples. In many cases, the highest quality amplified DNA is obtained by direct addition of cells to the reaction. Extraction and purification of DNA from blood and buccal swabs prior to use in MDA not only requires more steps, but also introduces more opportunities for DNA loss, damage, and contamination. In the case of biological specimens, MDA serves two functions. First, large amounts of DNA can be obtained from minute specimens. Just a few microliters of blood can be used to generate hundreds of micrograms and even milligrams of amplified DNA, depending on the reaction volume. This can be critical for studies such as global SNP screening and microarray strategies that require large amounts of DNA. Second, the MDA reaction serves as a means to obtain highly enriched DNA without the need for other purification steps. The whole genome amplification enriches the DNA relative to cellular debris, inhibitors such as nucleases and hemoglobin, growth media, or other contaminants found in crude clinical or field samples. As a general method for DNA sample preparation, MDA yields DNA template that can often be used directly in genotyping or other assays with no need for clean up steps.

## MDA Reaction Protocols

Several different MDA protocols have been published (9–11, 22), and reaction components are available from several different commercial sources. The three general requirements for MDA are: (1) an appropriate source of DNA template; (2) a means to denature DNA template either with alkaline (9, 18) or high temperature (95°C) (10) treatment; and (3) the Phi29 DNA polymerase and reaction buffers, protected primers (10), and dNTP substrate to support DNA synthesis. One typical protocol, based on alkaline denaturation of template, is given below (23) as an example.

### *Required Reagents and Materials*

The Denaturation Solution consists of 400 mM KOH, 10 mM EDTA. Include 100 mM dithiothreitol (DTT) for cell lysis for blood, cultured cells, or other biological specimens. It is recommended that DTT be omitted for buccal swabs as better results were sometimes observed. The Denaturation Solution can be stored for one week at room temperature.

   The Neutralization Buffer consists of 800 mM Trizma hydrochloride (Tris-HCl; e.g., Sigma catalog number T-5941; Sigma Chemical Co., St. Louis, MO, USA) approximately pH 4 (unadjusted).

   Other required reagents:

* Phi29 DNA polymerase
* Nuclease-free water
* 4× MDA reaction mix consisting of: 150 mM Tris-HCl, pH 7.5, 200 mM KCl, 40 mM $MgCl_2$, 80 mM ammonium sulfate, 4 mM each of dATP, dGTP, dCTP, dTTP, 0.2 mM random hexamer (protected with thiophosphate modification on the two 3′-terminal nucleotides (10))
* Microcentrifuge tubes or microtiter plates
* Thermal cycler, water bath, or heating block

   Assemble reactions in 0.2 mL tubes or microtiter plate wells. For whole genome amplification using purified human DNA, it is recommended that at least 10 ng DNA template be added (about 3000 copies of the human genome). When required, less template can also be amplified; however, some loss of sequence from the amplified DNA is possible. If the DNA is suspected of being partially degraded, addition of more DNA (30–100 ng) may improve results.

Add reagents in the following order at room temperature:

- 3 μL DNA (10–100 ng), or cellular sample
- 3.5 μL Denaturation Solution

Mix and incubate for three minutes at room temperature.

- 3.5 μL Neutralization Buffer, mix, use in MDA reaction within one hour.
- 27 μL H$_2$O
- 12.5 μL 4× MDA reaction mix
- 0.5 μL Phi29 DNA polymerase

Mix well.

The final reaction volume is 50 μL. (Note that the protein concentrations and unit definitions of the Phi29 DNA polymerase can vary between suppliers. Always use the manufacturer's recommended amounts. For multiple reactions, the Phi29 DNA polymerase can be added, on ice, to the 4× MDA reaction mix, if it is added just prior to use. Mix well as the polymerase is usually in 50% glycerol.)

- Incubate for 6 to 16 hours at 30°C (minimum of 8 hours to assure maximum yields)
- Stop the reaction by heating to 65°C, 3 minutes.

This step is essential in order to inactivate the DNA polymerase and its associated 3′-5′ exonuclease activity which can degrade the amplified DNA if not inactivated.

### Notes

Even in the presence of excess dNTP substrate, MDA reactions tend to be self-limiting with DNA yields of about 0.8 μg/μL ± 15%. Therefore, the yield of amplified DNA is limited by the reaction volume with a 100 μL reaction generating about 80 μg, and a 10 mL reaction about 8 mg of DNA. The yield of DNA product plateaus in about six hours. For convenience, reactions can be set up in the afternoon and left at 30°C overnight for 16 hours. Use of a thermal cycler minimizes evaporation, and conveniently allows the programming of the 30°C incubation, the 65°C termination step after 16 hours, and storage at 4°C until the sample is retrieved from the thermal cycler. Longer incubation times for terminating the reaction at 65°C may be required for larger reaction volumes to bring them to sufficient temperature to inactivate the polymerase. For example, for a 10 mL preparative reaction carried out in a Falcon tube, use 10 minutes in a 65°C water bath. Store amplified DNA at −20°C. For long-term storage, it should be aliquoted into multiple tubes to reduce degradation from freeze/thaw cycles (36).

## Commercial Whole Genome Amplification Kits

The Phi29 DNA polymerase purity and concentration are critical to achieving peak performance in MDA. For convenience, commercial MDA-based whole genome amplification kits are available from several sources and are validated by their manufacturers for optimal reaction conditions. The MDA reaction protocols below are based on reagents obtained from Qiagen (REPLI-g® whole genome amplification kit) and the volumes and conditions used follow the manufacturer's instructions for this kit.

### *Whole Genome Amplification Using Purified Genomic DNA: 50 µL MDA Reaction Volume*

This protocol is optimized for whole genome amplification from a >10 ng genomic DNA template. The template DNA should be suspended in TE (10 mM Tris/Cl, pH 7.5, 1 mM EDTA, pH 8.0) at a concentration >4 ng/µL. Smaller amounts of starting material can be used if the DNA is of sufficient quality. For best results, the template DNA should be >2 kb in length with some fragments >10 kb.

All buffers and reagents should be vortexed before use to ensure thorough mixing.

### *Things To Do Before Starting*

- Prepare sufficient amounts of the Denaturation Solution and Neutralization Buffer for the number of MDA reactions being carried out. For example, for 500 µL of Denaturation Solution, add 40 µL of 5 M KOH and 10 µL of 0.5 M EDTA (pH 8) into 450 µL deionized water. The Denaturation Solution can be stored for one week at room temperature. The container must be tightly sealed to avoid neutralization with $CO_2$.
- Set a water bath, heating block, or thermocycler to 30°C.
- The DNA polymerase should be thawed on ice (see step 5). All other components can be thawed at room temperature.

- Denaturation Solution should not be stored longer than one week.
- All buffers and reagents should be vortexed before use to ensure thorough mixing.

## Procedures

1. Place 2.5 µL template DNA (>4 ng/µL) into a microcentrifuge tube.

The template DNA should be suspended in TE at a concentration >4 ng/µL.

2. Add 2.5 µL Denaturation Solution to the DNA. Mix by vortexing and centrifuge briefly.
3. Incubate the samples at room temperature for three minutes.
4. Add 5 µL Neutralization Buffer to the samples. Mix by vortexing and centrifuge briefly.
5. Thaw the Phi29 DNA polymerase on ice. Thaw all other components at room temperature, vortex, and then centrifuge briefly.
6. The concentrated reaction buffer may form a precipitate after thawing. The precipitate will dissolve by vortexing for 10 seconds.
7. Prepare a master mix on ice (Table 8-1). Mix and centrifuge briefly.

Note: It is important to add the master mix components in the following order. After the addition of water and reaction buffer, briefly vortex and spin down the mixture before adding the DNA polymerase. Mix well following the addition of the DNA polymerase. The master mix should be kept on ice and used immediately upon addition of the DNA polymerase.

8. Add 40 µL of the master mix to 10 µL of denatured DNA (step 4).
9. Incubate at 30°C for 6 to 16 hours.

Optimal results are achieved using an incubation time of approximately eight hours or up to 16 hours in an overnight reaction.

10. Inactivate REPLI-g DNA polymerase by heating the sample for three minutes at 65°C.
11. Store amplified DNA at 4°C for short-term storage or −20°C for long-term storage.

Amplified DNA should be treated as genomic DNA with minimal freeze-thaw cycles. Storage of nucleic acids at low concentration over a long period of time may result in acid hydrolysis. We therefore recommend storage of nucleic acids at a concentration of at least 100 ng/µL.

**Table 8-1.** Master mix for commercial REPLI-g® whole genome amplification kit.

| Component | Volume per 50 µL Reaction |
|---|---|
| Nuclease-free water | 27.0 µL |
| Repli-g buffer, 4× | 12.5 µL |
| REPLI-g DNA polymerase | 0.5 µL |
| Total volume | 40.0 µL |

Volumes shown are for one 50-µL reaction and must be scaled up for the number of reactions to be performed or for larger volume amplifications.

### Whole Genome Amplification from Blood or Cells: 50 µL MDA Reaction Volume

#### Important Points to Consider Before Starting

- The protocol is optimized for 0.5 µL whole blood or cell material (e.g., sorted cells, tissue cultured cells, etc.). The cell concentration should be >600 cells/µL.
- The DNA polymerase should be thawed on ice (see step 8). All other components can be thawed at room temperature.
- Denaturation Solution should not be stored longer than one week.
- All buffers and reagents should be vortexed before use to ensure thorough mixing.

#### Things to Do Before Starting

Prepare 500 µL (or an amount appropriate for the number of MDA reactions) of Denaturation Solution by adding 40 µL 5 M KOH and 10 µL 0.5 M EDTA (pH 8) into 450 µL deionized water.

Note: The Denaturation Solution can be stored for one week at room temperature. The vial must be tightly sealed to avoid neutralization with $CO_2$.

Set a water bath or heating block to 30°C.

#### Procedure

1. Prepare sufficient Denaturation Solution for the total number of whole genome amplification reactions.
2. Place 0.5 µL cell material (>600 cells/µL) or 0.5 µL blood into a microcentrifuge tube. A DNA control reaction can be set up using 10 ng (1 µL) for the control gDNA template.

3. Add 2.5 µL PBS to the sample.
4. Add 3.5 µL Denaturation Solution. Mix by vortexing and centrifuge briefly.
5. Incubate for 10 minutes on ice.
6. Add 3.5 µL Neutralization Buffer. Mix by vortexing and centrifuge briefly.
7. Thaw the DNA polymerase on ice. Thaw all other components at room temperature, vortex, and then centrifuge briefly. The reaction buffer may form a precipitate after thawing. The precipitate will dissolve by vortexing for 10 seconds.
8. Prepare a master mix, adding components in the order listed in Table 8-1. After the addition of water and REPLI-g buffer, briefly vortex and spin down the mixture before addition of DNA polymerase. The master mix should be kept on ice and used immediately upon addition of DNA polymerase.
9. Add 40 µL of the master mix to 10 µL of denatured DNA (step 6).
10. Incubate at 30°C for 6 to 16 hours.

Optimal yields are assured using an incubation time of at least eight hours.

11. Inactivate the DNA polymerase by heating at 65°C for three minutes.
12. Store amplified DNA at 4°C for short-term storage or −20°C for long-term storage.

Amplified DNA should be treated as genomic DNA with minimal freeze/thaw cycles. Storage of nucleic acids at low concentration over a long period of time may result in acid hydrolysis. We therefore recommend storage of nucleic acids at a concentration of at least 100 ng/µL.

## Troubleshooting Guide

1. There is reduced or no high molecular weight product in agarose gel, but the DNA yield in positive control is approximately 40 µg/50 µL reaction volume (a successful reaction yield), and, thus, the reaction failed. There is a possible inhibitor in the gDNA template. Clean up or dilute the gDNA and re-amplify.
2. Downstream application results are not optimum. Sensitive downstream applications may require DNA cleanup after the REPLI-g reaction.

### Genomic DNA Protocol

1. Reduced or no locus representation in real-time PCR analysis but DNA yield is approximately 40 µg.
   a. gDNA template is degraded. Use intact template. Use larger amount of gDNA.
   b. Ineffective denaturation of template gDNA. Use fresh 5 M KOH stock solution to prepare the denaturation solution.
2. Allele drop-out observed in genotyping assay but DNA yield is approximately 40 µg.
   a. gDNA template is degraded. Use intact gDNA template. Use larger amount of gDNA.
   b. Ineffective denaturation. Use fresh 5 M KOH stock solution to prepare denaturation solution.

### Blood and Cell Protocol

1. There is a reduced or no locus representation in real-time PCR analysis, but the DNA yield is approximately 40 µL.
   a. There is a higher than normal concentration of heparin used as blood anticoagulant. Dilute the heparin-treated blood up to five-fold using 1× PBS.
   b. There is an ineffective lysis of cells. Use fresh 5 M KOH stock solution to prepare denaturation solution.
2. Allele drop-out is observed in genotyping assay, but the DNA yield is approximately 40 µg.
   a. Higher than normal concentration of heparin used as blood anticoagulant. Dilute the heparin-treated blood up to fivefold using 1× PBS.
   b. Ineffective lysis of cells. Use fresh 5 M KOH stock solution to prepare denaturation solution.

## Determination of Concentration and Yield of DNA

A 50 µL REPLI-g reaction typically yields approximately 40 µg of DNA regardless of the amount of template DNA, allowing direct use of the amplified DNA in most downstream genotyping experiments. However, if a more accurate quantification of DNA is required, it is important to utilize a DNA quantification method that is specific for double-stranded DNA, because REPLI-g Kit amplification products contain unused reaction primers. PicoGreen® reagent (Invitrogen, Carlsbad, CA) displays enhanced binding to double-stranded DNA and may be used, in conjunction with a fluorometer, to quantify the double-stranded DNA product. For best results, the sample should be diluted with 2 volumes of water and thoroughly mixed prior to addition of PicoGreen.

### Whole Genome Amplification from Other Biological Sample Types

Paraffin embedded tissues do not support whole genome amplification by MDA, probably due to the cross-linking and short length of the DNA. Some protocols for other sample types may be found in the literature (23) or are available from MDA reagent manufacturers.

### Acknowledgment

Qiagen's permission to reprint Repli-g protocols and their troubleshooting guide are gratefully acknowledged.

## References

1. Arnheim, N. 1992. Preimplantation genetic diagnosis—a rolling stone gathers no moss! *Hum Reprod* 7: 1481.
2. Arnheim, N., Calabrese, P., and Nordborg, N. 2003. Hot and cold spots of recombination in the human genome: the reason we should find them and how this can be achieved. *Am J Hum Genet* 73: 5–16.
3. Barker, D.L., Hansen, M.S., Faruqi, A.F., et al. 2004. Two methods of whole-genome amplification enable accurate genotyping across a 2320-SNP linkage panel. *Genome Res* 14: 901–907.
4. Barnes, W.M. 1994. PCR amplification of up to 35-kb DNA with high fidelity and high yield from lambda bacteriophage templates. *Proc Natl Acad Sci U S A* 91: 2216–2220.
5. Blanco, L., Bernad, A., Lazaro, J.M., et al. 1989. Highly efficient DNA synthesis by the phage phi 29 DNA polymerase. Symmetrical mode of DNA replication. *J Biol Chem* 264: 8935–8940.
6. Buchanan, A.V., Risch, G.M., Robichaux, M., et al. 2000. Long DOP-PCR of rare archival anthropological samples. *Hum Biol* 72: 911–925.
7. Cheung, V.G., and Nelson, S.F. 1996. Whole genome amplification using a degenerate oligonucleotide primer allows hundreds of genotypes to be performed on less than one nanogram of genomic DNA. *Proc Natl Acad Sci U S A* 93: 14676–14679.
8. Cullen, M., Perfetto, S.P., Klitz, W., et al. 2002. High-resolution patterns of meiotic recombination across the human major histocompatibility complex. *Am J Hum Genet* 71: 759–776.
9. Dean, F.B., Hosono, S., Fang, L., et al. 2002. Comprehensive human genome amplification using multiple displacement amplification. *Proc Natl Acad Sci U S A* 99: 5261–5266.
10. Dean, F.B., Nelson, J.R., Giesler, T.L., and Lasken, R.S. 2001. Rapid amplification of plasmid and phage DNA using Phi 29 DNA polymerase

and multiply-primed rolling circle amplification. *Genome Res* 11: 1095–1099.

11. Detter, J.C., Jett, J.M., Lucas, S.M., et al. 2002. Isothermal strand-displacement amplification applications for high-throughput genomics. *Genomics* 80: 691–698.

12. Dietmaier, W., Hartmann, A., Wallinger, S., et al. 1999. Multiple mutation analyses in single tumor cells with improved whole genome amplification. *Am J Pathol* 154: 83–95.

13. Esteban, J.A., Salas, M., and Blanco, L. 1993. Fidelity of phi 29 DNA polymerase. Comparison between protein-primed initiation and DNA polymerization. *J Biol Chem* 268: 2719–2726.

14. Foucault, F., Praz, F., Jaulin, C., et al. 1996. Experimental limits of PCR analysis of (CA)n repeat alterations. *Trends Genet* 12: 450–452.

15. Grothues, D., Cantor, C.R., and Smith, C.L. 1993. PCR amplification of megabase DNA with tagged random primers (T-PCR). *Nucleic Acids Res* 21: 1321–1322.

16. Handyside, A.H., Robinson, M.D., Simpson, R.J., et al. 2004. Isothermal whole genome amplification from single and small numbers of cells: a new era for preimplatation genetic diagnosis of inherited disease. *Mol Hum Reprod* 10: 767–772

17. Hawkins, T.L., Detter, J.C., and Richardson, P.M. 2002. Whole genome amplification—applications and advances. *Curr Opin Biotechnol* 13: 65–67.

18. Hosono, S., Faruqi, A.F., Dean, F.B., et al. 2003. Unbiased whole-genome amplification directly from clinical samples. *Genome Res* 13: 954–964.

19. Kittler, R., Stoneking, M., and Kayser, M. 2002. A whole genome amplification method to generate long fragments from low quantities of genomic DNA. *Anal Biochem* 300: 237–244.

20. Kornberg, A., and Baker, T.A. 1991. *DNA Replication.* New York: W.H. Freeman.

21. Kristjansson, K., Chong, S.S., Van den Veyver, I.B., et al. 1994. Preimplantation single cell analyses of dystrophin gene deletions using whole genome amplification. *Nat Genet* 6: 19–23.

22. Lage, J.M., Leamon, J.H., Pejovic, T., et al. 2003. Whole genome analysis of genetic alterations in small DNA samples using hyperbranched strand displacement amplification and array-CGH. *Genome Res* 13: 294–307.

23. Lasken, R.S., Hosono, S., and Egholm, M. 2004. Multiple displacement amplification (MDA) of whole human genomes from various samples. In: Demidov, V.V., Broude, N.E., eds. *DNA Amplification: Current Technologies and Applications.* Norfolk, UK: Horizon Bioscience; 267–290.

24. Lasken, R.S., and Egholm, M. 2003. Whole genome amplification: abundant supplies of DNA from precious samples or clinical specimens. *Trends Biotechnol* 21: 531–535.

25. Lizardi, P.M., Huang, X., Zhu, Z., et al. 1998. Mutation detection and single-molecule counting using isothermal rolling-circle amplification. *Nat Genet* 19: 225–232.

26. Ludecke, H.J., Senger, G., Claussen, U., et al. 1989. Cloning defined regions of the human genome by microdissection of banded chromosomes and enzymatic amplification. *Nature* 338: 348–350.

27. Mullis, K.B., and Faloona, F. 1987. Specific synthesis of DNA in vitro via a polymerase-catalyzed chain reaction. *Method Enzymol* 155: 335–350.

28. Nelson, J.R., Cai, Y.C., Giesler, T.L., et al. 2002. TempliPhi, phi29 DNA polymerase based rolling circle amplification of templates for DNA sequencing. *Biotechniques* Suppl: 44–47.

29. Paez, J.G., Lin, M., Beroukhim, R., et al. 2004. Genome coverage and sequence fidelity of 29 polymerase-based multiple-strand displacement whole-genome amplification. *Nucleic Acids Res* 32: e71.

30. Paunio, T., Reima, I., and Syvanen, A.C. 1996. Preimplantation diagnosis by whole-genome amplification, PCR amplification, and solid-phase minisequencing of blastomere DNA. *Clin Chem* 42: 1382–1390.

31. Raghunathan, A., Ferguson, H.R., Bornarth, C., et al. 2005. Genomic DNA amplification from a single bacterium. *Appl Environ Microbiol* 71: 3342–3347.

32. Saiki, R., Scharf, S., Faloona, F., et al. 1985. Specific synthesis of DNA in vitro via a polymerase-catalyzed chain reaction. *Science* 230: 1350–1354.

33. Saunders, R.D., Glover, D.M., Ashburner, M., et al. 1989. PCR amplification of DNA microdissected from a single polytene chromosome band: a comparison with conventional microcloning. *Nucleic Acids Res* 17: 9027–9037.

34. Schmitt, K., Lazzeroni, L.C., Foote, S., et al. 1994. Multipoint linkage map of the human pseudoautosomal region, based on single-sperm typing: do double crossovers occur during male meiosis? *Am J Hum Genet* 55: 423–430.

35. Snabes, M.C., Chong, S.S., Subramanian, S.B., et al. 1994. Preimplantation single-cell analysis of multiple genetic loci by whole-genome amplification. *Proc Natl Acad Sci U S A* 91: 6181–6185.

36. Steinberg, K.K., Sanderlin, K.C., Ou, C.Y., et al. 1997. DNA banking in epidemiologic studies. *Epidemiol Rev* 19: 156–162.

37. Sun, F.N., Arnheim, N., and Waterman, M.S. 1995. Whole genome amplification of single cells: mathematical analysis of PEP and tagged PCR. *Nucleic Acids Res* 23: 3034–3040.

38. Telenius, H., Carter, N.P., Bebb, C.E., et al. 1992. Degenerate oligonucleotide-primed PCR: general amplification of target DNA by a single degenerate primer. *Genomics* 13: 718–725.

39. Wells, D., and Sherlock, J.K. 1998. Strategies for preimplantation genetic diagnosis of single gene disorders by DNA amplification. *Prenat Diagn* 18: 1389–1401.

40. Wells, D., Sherlock, J.K., Handyside, A.H., et al. 1999. Detailed chromosomal and molecular genetic analysis of single cells by whole genome amplification and comparative genomic hybridisation. *Nucleic Acids Res* 27: 1214–1218.

41. Zhang, L., Cui, X., Schmitt, K., et al. 1992. Whole genome amplification from a single cell: implications for genetic analysis. *Proc Natl Acad Sci U S A* 89: 5847–5851.

42. Zheng, S., Ma, X., Buffler, P.A., et al. 2001. Whole genome amplification increases the efficiency and validity of buccal cell genotyping in pediatric populations. *Cancer Epidemiol Biomarkers Prev* 10: 697–700.

# Solid-Phase Nucleic Acid Purification Using Magnetic Beads

**9**

## Kevin McKernan and Erik Gustafson
*Agencourt Bioscience Corporation, Beverly, MA, USA*

Nucleic acid isolation is a fundamental requirement for molecular biology research and molecular diagnostics. Prior to amplifying, sequencing, transfecting, or cloning DNA, the DNA must be extracted from the sample of interest and ideally purified from contaminating background nucleic acids, protein, and other cell components. The purity requirements vary tremendously depending on the downstream applications. For example, in some cases involving PCR, the purity requirement may be minimal. On the other hand, plasmids purified from *E. coli* for gene therapy are closely monitored for the level of endotoxin and must be below 300 endotoxin units per milligram of DNA. In the case of RNA purification, RNA integrity and RNAse inhibitors must be considered. For DNA sequencing, it has been shown that slightly denatured or even degraded DNA will sequence more effectively than supercoiled DNA (29). As a result, rarely is one purification product optimized for every need.

## Solid-Phase Nucleic Acid Purification

Traditional methods for nucleic acid purification were derived from organic chemistry and utilized different solvents and phase extractions (45). These early techniques were cumbersome and utilized hazardous organic materials such as phenol and chloroform. They are still utilized today, but have largely been replaced by commercial kits that use column-based purification techniques. Such column-based methods are simple to use and provide high quality, reproducible results. Purification with ion-exchange chromatography columns requires binding, washing, and elution in buffers of differing ionic strengths. Purification over silica-based resins requires binding in the presence of high molar chaotropic salts, followed by wash and elution in low ionic strength buffer. Labor savings and ease of use offset the slightly higher cost per preparation (prep) of these methods. They are ubiquitous in labs routinely processing small numbers of samples per day. For specific applications that require higher throughput methods, such as the Human Genome Project (27, 53), vacuum filtration plates were the next revolutionary leap forward and brought plasmid purifications into 96-well format. Nevertheless, it became challenging to automate massively parallel filtration robotics or centrifugation robotics. System requirements for genome center scale purification required one preparation per second throughput or approximately 40 plates per hour. As a result, either plate-based centrifugation approaches or magnetic bead-based purification platforms became the mainstay in the genome facilities. Magnetic beads offered many advantages in terms of speed, automation, simplicity of plate traffic, and homogeneous formats (very few serial reaction steps required). A popular magnetic bead technology utilized during the human genome project was called Solid Phase Reversible Immobilization (SPRI). Developed by Hawkins et al. (14, 21, 22) at the Whitehead Institute, Massachusetts Institute of Technology (MIT), SPRI became the main purification platform at three of the four genome centers in the United States (Broad Institute in Cambridge, MA; Washington University, St. Louis, MO; and Joint Genome Institute, Walnut Creek, CA). The predominant factors in SPRI's success in the genome facilities are attributed to not only cost per preparation, but also the cost of the equipment required to perform the process and the attainable throughputs when properly automated.

Scaling up the DNA cloning, isolation, purification, and sequencing process to tackle the ambitious goal of sequencing every base in the human genome began with the methods widely available at the time. To prepare plasmid DNA for sequencing, traditional alkaline lysis preps utilize a process that must first make the solution alkaline and subsequently neutralize the pH. These two steps cannot be combined into one step due to their different pH requirements. In addition, the neutralization step creates a very challenging congealed flocculent solution that must be removed either by filtration or centrifugation. This adds an additional plate to the process whose bar code needs to be tracked, and its inventory and procurement managed. Samples must be cleanly transferred to this new plate free of contamination, and sample identity maintained. This additional step has process ramifications, as robotic protocols need to be developed and sophisticated error recovery scripts implemented to avoid sample identity corruption in the event of any robotic crashes or interruptions. SPRI allows a homogeneous solution where the lysis step and the solid support DNA binding can occur simultaneously without any flocculent generation. If desired, the growth, purification, and sequencing can occur in the same plate resulting in a single plate pipeline. Because most genome centers prefer to retain an archival plate and perform forward and reverse sequencing, a single plate pipeline is less ideal, but for certain single direction application like SAGE (serial analysis of gene expression) sequencing it may be beneficial.

Substantial progress has been made on the traditional alkaline lysis protocols. The Sanger Centre (see Chapter 10) and Celera (25) have made significant advancements, compressing this system into 384-well plates and utilizing a centrifugation step to achieve the size selection for the plasmid DNA by pelleting the genomic DNA in the neutralized flocculent. Despite this single manual step, the process is highly parallel: 10 to 20 384-well plates can be centrifuged in 10 minutes. However, this comes at the cost of more source and destination plates or "plate traffic" through the robotic systems. For genome centers, this isn't a costly transaction because the plastics are fairly inexpensive, but as sequencing becomes more of a diagnostic tool, homogeneous assays (assays that occur in one tube and present little chance of sample identity ambiguity) are clear winners. Bar code tracking is also impacted with multiple source and destination plates, and this has cost consequences on the design of any laboratory information management system (LIMS) and robotic platform systems.

## Advantages of Magnetic Bead DNA Purification in a Large Scale Sequencing Pipeline

### Throughput and Automation

To address throughput, the cost of the robotics or capital expenditure ("capex") required for a given preparation throughput is a prime concern, as an army of robots can make the least effective protocol appear to be very high throughput at the cost of floor space and robotic support. Unlike reagents that are spent as they are used, robotics are amortized every minute of every day and usually over three years. As a result, a non-utilized or broken robot that is making no product is still amortized during its unproductive days. If the instruments cannot be surged (run at 2× the speed) to replace the lost production capacity, the downtime increases the cost per prep substantially. This is a very important detail in a sequencing lab where close to half of the cost of sequencing is equipment amortization. Likewise, any robot that is entirely dependent on another step upstream of it is at risk of being idle, and in fact is functionally vulnerable to single points of failure. Pipelines that do not mitigate single points of failure in a given process will have a higher equipment amortization cost than anticipated, because a single robotic failure has the potential to arrest the entire pipeline. As a result, there is a tremendous amount of savings that can be gained by selecting processes that have simple and redundant robotic requirements. During the course of the human genome project at the Whitehead Institute, over $2 million worth of equipment was required to attain one preparation per second throughput with SPRI technology. Today, the chemistry has matured and the purification of fosmids and 5 kb insert plasmid can be performed at the rate of 4000 clones per hour on a single 384-head instrument like a Beckman FX (Beckman Instruments, Fullerton, CA, USA) for approximately $110,000 in equipment cost.

To further illustrate the cost savings a SPRI platform affords, consider the SPRI-based SprintPrep plasmid purification procedure. This purification procedure utilizes *E. coli* cells that are grown in 384-thermal-cycling plates for about 10 hours. These cultures are directly purified without cell pelleting or media removal at a throughput of eleven 384-well plates per hour per Beckman FX. The only manual intervention required is moving the plates from the shaker to the prepping robot. Assuming that the cost for the Beckman FX is $110,000 and the cost for a shaker/incubator is $5000/unit, the amortization over a three-year period with the instrument in effective operation for 6 hours per day and 200 days per year is less than one penny per preparation with the newer chemistry.

## SPRI versus Multiple Displacement Amplification (MDA) for Plasmid Preparation

The only other plasmid purification preparation system that approaches this speed and robotic efficiency is MDA with products such as TempliPhi (see Chapters 7 and 8). TempliPhi doesn't require cell culturing, but takes just as long to amplify the DNA in vitro as the in vivo *E. coli* polymerase machinery. There are several distinctions in the products generated by these two methods.

Firstly, plasmid DNA replicated in *E. coli* is functional. It can be isolated and quantified. It can be immortalized in a cell by transformation, transfection, and cloning, and can subsequently be used as a template to drive expression of cloned genes in vivo or in vitro. DNA generated from a TempliPhi reaction is comprised of long, linear copies of the template DNA, both double stranded and single stranded, initiated at random places along the template. A single unit of plasmid DNA cannot be isolated from these products for use in any of the functions mentioned above without further processing. TempliPhi-derived products cannot be reliably quantified spectrophotometrically as the amplified DNA is a mixture of target and amplified background or hexamers. The DNA is also heterogeneous not only in size but in content, as the hexamers are known to form multimers or concatenated primer dimers during amplification (31).

Secondly, plasmid isolations are size selective. The DNA purified is enriched for the target nucleic acid of interest. TempliPhi-derived products are complete, nonspecific nucleic acid amplifications, or whole sample amplifications. Unlike PCR amplification, it does not reduce the complexity of the isolated nucleic acids and, thus, is challenging to utilize for low copy clones like fosmids and BACs where the target DNA is a minority species in the cell next to the genomic DNA.

Finally, most genome centers desire an archive plate of the clones (especially for fosmids and BACs). In the case of TempliPhi, a glycerol plate must be grown for 10 to 16 hours and then the plate must be TempliPhi-amplified for an additional 10 to 16 hours. With the SPRI process, the glycerol can be obtained by simply growing 80 μL of cells and removing 20 μL for SPRI while leaving the remaining 60 μL for glycerol addition and storage. Because the SPRI purification process takes about 15 minutes per plate compared to 10 to 16 hours for a TempliPhi-driven reaction, the DNA can be processed and sent to a detector a full day in advance. This has a profound impact when building a just-in-time production facility. Elimination of buffers in the process and reduction of cycle time are paramount to the successful management of a production facility.

## Universal Applicability of Magnetic Bead-Based Pipelines

An interesting benefit discussed later in the text is seen in DNA sequencing laboratories that build magnetic bead pipelines. Magnetic bead protocols are available for many nucleic acid isolation procedures, such as purification of PCR products, DNA from whole blood, viral DNA and RNA from serum, or RNA for microarray labeling. Since consumable items like plasticware and filter plates don't change very much, the same robotic equipment can be used for plasmid DNA purifications or the other applications mentioned above. A magnetic bead platform presents a tremendous amount of flexibility over using three or four different techniques such as TempliPhi-based protocols for plasmid isolation, ExoSAP for PCR purification, and filtration or spin column stations for RNA purification. SPRI allows a single core nucleic acid engine to be built, and the infrastructure and troubleshooting overhead required in any given molecular biology laboratory can be simplified by centralizing the technology platform for all nucleic acid isolation.

Magnetic beads are also positioned to hold the future in the diagnostic use of nucleic acid isolation because reactions can be performed in a completely closed system without the need to ever transfer a sample to a new tube or expose the sample to the serial addition of multiple reagents.

## Description of SPRI Technologies

There are several magnetic bead-based purification systems on the market today. They broadly fall into two categories that differ with respect to the functional group present on the bead: silica-based magnetic beads and carboxy-based magnetic beads. The former are generally comprised of silica-based resins impregnated with magnetite (iron oxide-$Fe_3O_4$). The resulting amorphous particles respond to a magnetic field and utilize the same binding chemistry as that required for non-magnetic silica resins. That is, nucleic acids bind to the silica beads in the presence of high molar chaotropic salts like NaI, guanidine hydrochloride, urea, and guanidine isothiocyanate, followed by elution in water. The carboxy-based magnetic beads utilize SPRI chemistry, which is differentiated from silica based purification in several measures. SPRI purification is accomplished by binding nucleic acids to carboxyl-coated solid supports like beads using various inert crowding reagents, including alcohols, polyethylene glycol (PEG), polyvinylpyrrolidone-40, and others. The molecular weights of these crowding reagents can be optimized to produce low viscosity solutions with substantial precipitating power. Size-specific nucleic acid isolation can be performed by either adjusting the concentration of the crowding reagent, the molecular weight of the crowding reagent, or adjustment to salt, pH, polarity, or hydrophobicity of the solution. Large DNA molecules will easily be crowded out of solution at low concentrations of salt and crowding reagent, whereas the smaller size species require higher concentrations of crowding reagents. Silica-based purification does not afford such a range of size selectivity.

Carboxy groups also play a key role in effective elution. Unlike the Si-OH groups of silica beads, carboxy groups have a pKa of 4.7, so they are negatively charged at neutral pH. This is a key feature to the "Reversible Immobilization" because DNA is negatively charged, and in the absence of any crowding reagents or salt, DNA repels itself from these beads at neutral pH. This is critical for complete elution so that the adsorbed nucleic acids can be released from solid supports. Many solid supports will bind DNA very effectively, but reversing the binding can be inefficient under conditions that do not damage DNA. Utilizing acidic or basic conditions to elute DNA from a solid support should be avoided since these treatments can produce abasic sites and depurinate DNA or nick DNA (45).

Bead geometry is an important feature with SPRI particles. Automation is facilitated when beads remain in solution for long periods of time. When attempting to form such a colloidal suspension, size and uniformity of beads are important parameters. Large beads (>5 µm) are difficult to keep suspended in solution and tend to settle or precipitate out of solution. This is problematic if a consistent delivery of magnetic material

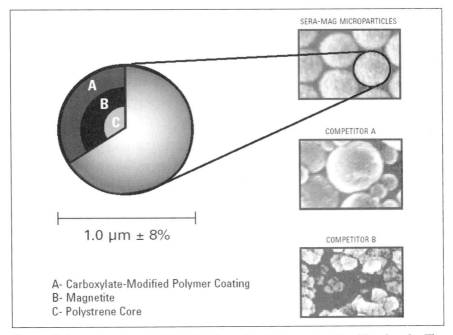

SERA-MAG MICROPARTICLES

COMPETITOR A

COMPETITOR B

1.0 μm ± 8%

A- Carboxylate-Modified Polymer Coating
B- Magnetite
C- Polystrene Core

**Figure 9-1.** Comparison of SPRI carboxy beads to magnetic silica beads. The top bead picture is a SPRI carboxy bead. The lower pictures show various magnetic silica beads available on the market. As can be seen, the carboxy beads are very uniform.

or solid support is needed, and as a result this requires additional equipment to keep the beads in suspension. Beads that have a wide size distribution can be problematic because on a per gram basis, the smaller beads have geometrically more surface area than the larger ones, and yet the larger ones are selected for the magnetic separation. A solution of beads containing 50% of small beads (by weight) is likely to produce reduced yields as these smaller beads will have lower affinity for the magnetic field yet retain $\pi r^2$ more surface area and DNA. Uniform bead sizes are critical for reproducible results. Very large beads (>20 μm) are generally avoided since they are not only difficult to keep in solution, but are difficult to manufacture and retain a complete coating of the iron oxide core. Iron oxide leaching can be very problematic for downstream enzymatic assays, because it can chelate $MgCl_2$ and poison a PCR reaction. Rather than having iron oxide impregnated throughout the particle, as in magnetic silica beads, SPRI beads have a magnetic iron oxide core completely covered by a polymer coating, which prevents leaching of iron into the medium. Figure 9-1 contains electron microscope images of various beads.

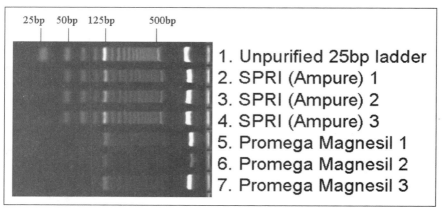

**Figure 9-2.** SPRI-based products can size select a 50 mer from a 25 mer as seen in lanes 1–4. Silica bead–based kits appear to capture less material from equal percentages of product regardless of size and fail to capture any small products. Procedures were performed according to manufacturer's instructions to purify a 25-bp ladder.

Carboxy beads appear to have better size discrimination when attempting to selectively separate a nucleic acid of a specific molecular weight. This can be seen when comparing sequencing reaction cleanup kits or PCR cleanup kits, where size selectivity is critical and very narrow size selections must be made between nucleotides and 30 mers (Figure 9-2).

In addition, it is common for the carboxy-based methods to capture two to three times more product over their silica counterparts. When comparing state of the art silica spin gel filtration columns like Qiagen's DyeEx spin column, the CleanSeq carboxy beads provide over tenfold more signal on an ABI 3730xl sequencer (Figure 9-3). This is likely the combined result of binding two- to threefold more DNA and also performing a more thorough desalting of the sample. Adequate desalting of a sample has a profound impact on the electrokinetic injection of a capillary sequencer. Similar results have also been seen with RNA capture. SPRI carboxy-based beads have proved to capture more RNA and DNA material than their competitive silica-DEAE spin columns (Figure 9-7c and Figure 9-14). It is difficult to ascertain whether this is a result of the silica functional group or the design of the solid support, because there are no silica magnetic beads that have the same functional group density and geometry as commercially available carboxylated beads. The binding capacity of the SPRI beads is demonstrated to be large. This appears in part to be due to a bimodal binding mechanism. Steadily increasing the amount of DNA in a fixed volume reveals that SPRI recovers 100% of the

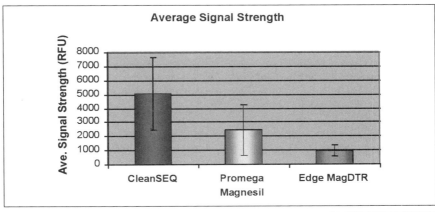

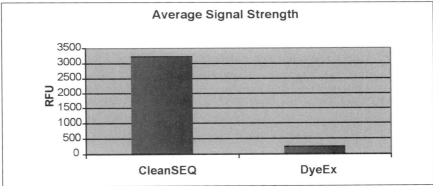

**Figure 9-3.** SPRI methods have a higher binding capacity than competitor methods such as silica magnetic beads and filtration-based solid support purification kits. CleanSEQ is a SPRI-based dye-terminator removal kit. DyeEx, Magnesil, and MagDTR are Qiagen, Promega, and EdgeBiosystems dye terminator removal kits, respectively. BigDye sequencing reactions were performed as described in the materials and methods section.

product up to a certain level of input, then recovers 70% of the product thereafter, up to the highest amount tested (Figure 9-4). This phenomenon is very reproducible and represents a capacity of at least 4 µg of DNA binding per 1 µg of beads.

Both silica and carboxy functional groups can be "charge switched" (3, 22, 37, 48). *Charge switching* refers to the altering of pH to achieve different charge densities on the solid phase. Carboxy beads have a pKa of 4.7, which means they have 50% of the negative carboxy groups potonated at a pH of 4.7. Varying the pH during binding or elution may afford some benefits, but is usually challenging to utilize as a sole means to achieve size selection of different nucleic acids. Hawkins demonstrates SPRI

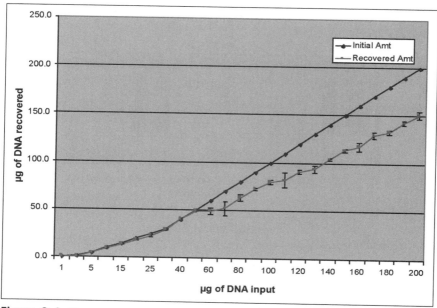

**Figure 9-4. Example of the high binding capacity of SPRI beads.** Increasing amounts of 2-kb-average-size sheared genomic DNA in a 20-µL volume were bound and recovered using the SPRI-based AMPure kit. There is 100% recovery up to 50µg of DNA, and then a steady recovery of 70% of the DNA up to 200µg, the highest amount tested.

utilizing an elution buffer of Tris-HCl (pH 8.0) and a PEG/NaCl solution with a pH of 5 (14, 21, 22). The lower the pH of the binding reagent, the more neutral the COOH functional group becomes. Alternatively, the higher the pH, the more negatively charged the group becomes. This allows favorable adsorption and release of DNA to and from the magnetic beads. In fact, preferred embodiments of the charge switch technology described rely on the presence of carboxy functional groups to help elute DNA at a higher pH (3). However, to size-select nucleic acids, a hydrophobicity and polarity change of the solution is necessary to drive different-sized nucleic acids to the beads, which is achieved with alcohols, ethylene glycols, or PEG and salts; use of pH alone makes it difficult to size-select to the beads.

The separation of nucleic acids based upon charge alone will not provide effective size-selective purification. This can be illustrated with nucleotide triphosphates, which are highly charged components of PCR reactions that are present in high molar quantity. They are effectively retained during charge switch purification, perhaps at the expense of the actual PCR product. This is easily seen when using charge switch tech-

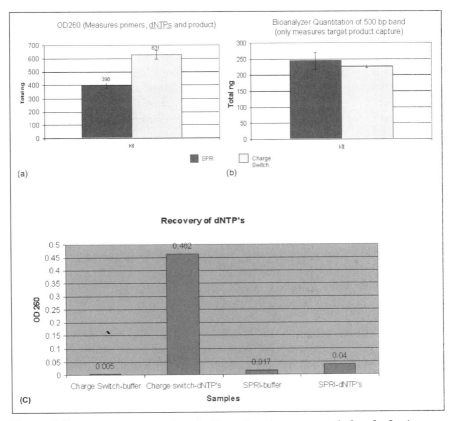

**Figure 9-5.** **Effect of pH on the binding of various magnetic beads.** Invitrogen charge switch beads, which use only pH shifts, fail to provide any size discrimination. SPRI beads, which utilize slight pH shifts, and PEG precipitation gradients allow high recovery and size discrimination. (a) Quantitation of PCR products indicates higher recovery by charge switch using absorbance. (b) SPRI shows a higher recovery using electrophoresis on a chip. (c) Purification of PCR reactions containing just buffer or dNTPs demonstrates that charge switch technology retains a significant amount of dNTPs. Kits were tested according to the manufacturer's instructions.

niques to purify PCR products and comparing recoveries obtained based upon UV spectroscopy or newer techniques, like the 2100 Agilent Bioanalyzer (Agilent Technologies, Waldbroom, Germany). Absorbance measures not only recovered DNA, but also unincorporated nucleotides and primers, while the Bioanalyzer measures only DNA of a specific size. By comparing the data from these two methods, it can be seen that the absorbance component (OD) reading is contributed by molecules other than the PCR product, and in fact is due to retained nucleotides (Figure 9-5).

### Critical Factors to Magnetic Bead-Based Assays

A key benefit of solution-suspended solid phase particles is that the solid phase is free to move throughout the solution and greatly accelerate the binding time. Fixed solid supports (like filter plates) have a challenging time reaching all volumes of the reaction and thus DNA is lost simply due to its failure to contact the support. As a result, bead assays deliver very reproducible results over wide dynamic range of sample concentrations.

Fixed supports also lose sample volume due to fluid retention in the filter. However, the described benefits of magnetic particles come with some deficits. Because the solid support is not fixed, it can agglutinate or clump, complicating its manipulation; this is the most common problem encountered when converting, for example, from filter-based assays to magnetic bead-based protocols. The predominant cause of agglutination is protein adsorption to the beads. This is particularly problematic when purifying DNA from a highly concentrated cell mass like *E. coli* or animal tissue preparations. Agglutination occurs when the proteins vastly outnumber the beads, and proteins begin cross-binding to multiple beads and coalescing beads in solution (data not shown). Agglutination is easily overcome by increasing either the percentage of solid phase or the surfactant in solution (Tween 20 or SDS detergents are preferred). For very protein rich tissue samples, one can resort to protein denaturing agents such as urea, GITC, or Proteinase K digestion to prevent bead agglutination. Once the effects of proteins on the beads are negated and those of SPRI-based protocols are evaluated (or understood), it can be tuned to extract nucleic acids from almost any tissue or cell type.

## Selected Applications of SPRI Technology

### Plasmid Purification

As described in a previous section, a SPRI prep has been designed that purifies plasmid DNA directly from *E. coli* growth cultures without first pelleting the cells. For this method of plasmid purification from *E. coli*, particular attention needs to be paid to the cell culture conditions. Simply adjusting the bead concentration to vastly outnumber the amount of protein in a given cell culture could present a dangerous strategy. Undergrown cultures will provide too little DNA, while overgrown cultures can exhibit properties that damage DNA or create highly variable input to a sequencing pipeline. We have focused the variability one sees in cell culture by optimizing several key parameters.

# Color Addendum

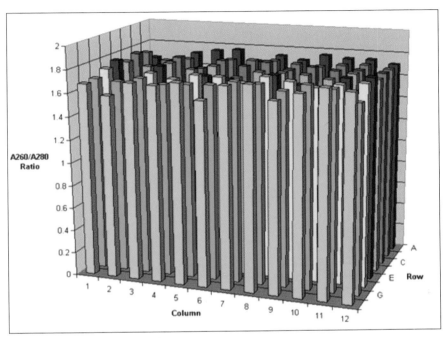

**Plate 1.** (Figure 5-8; page 79)

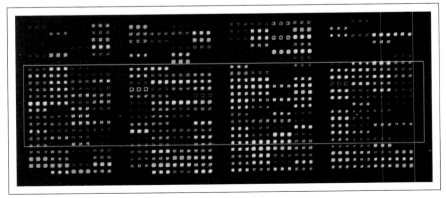

**Plate 2.** (Figure 5-9; page 80)

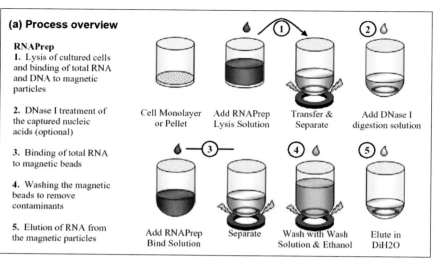

**Plate 3.** (Figure 9-7a; page 142)

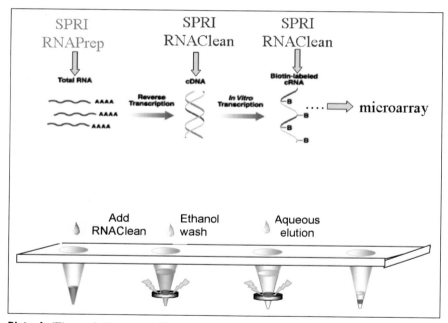

**Plate 4.** (Figure 9-11; page 148)

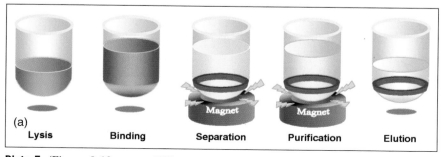

**Plate 5.** (Figure 9-13a; page 152)

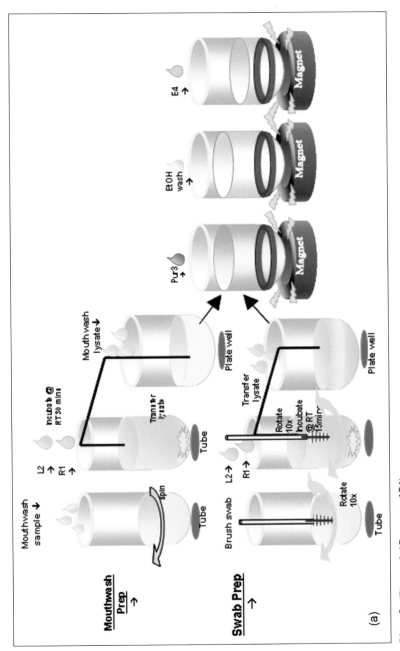

**Plate 6.** (Figure 9-15a; page 154)

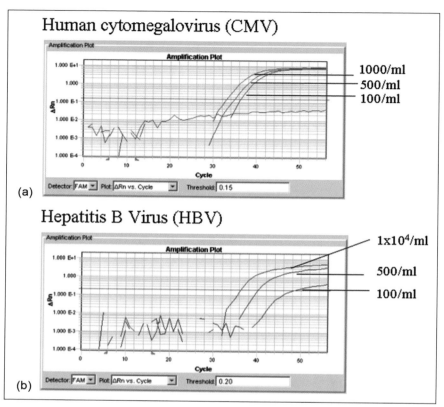

**Plate 7.** (Figure 9-17a and b; page 157)

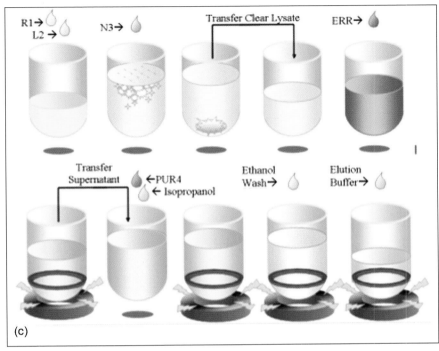

**Plate 8.** (Figure 9-20c; page 162)

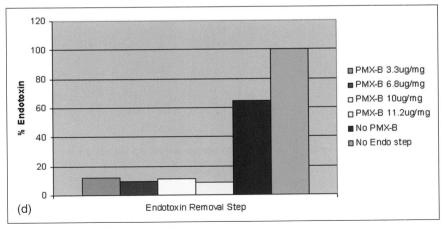

**Plate 9.** (Figure 9-20d; page 162)

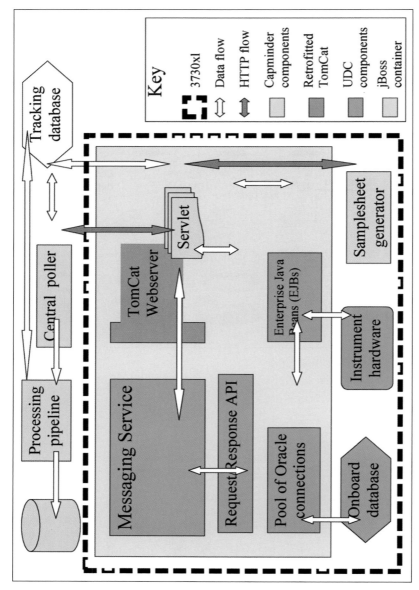

**Plate 10.** (Figure 10-15; page 197)

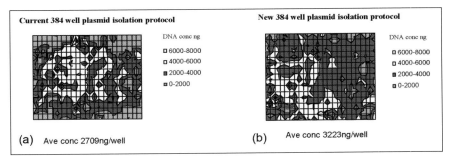

**Plate 11.** (Figure 10-10a and b; page 186)

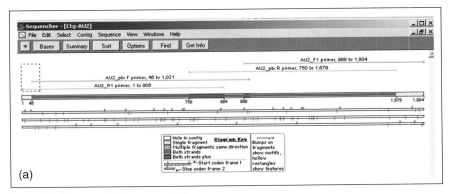

**Plate 12.** (Figure 17-7a; page 326)

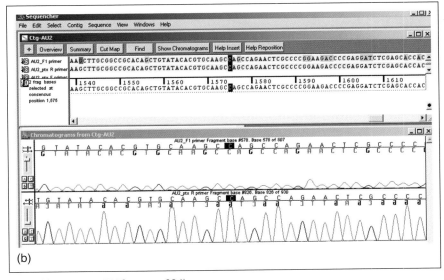

**Plate 13.** (Figure 17-7b; page 326)

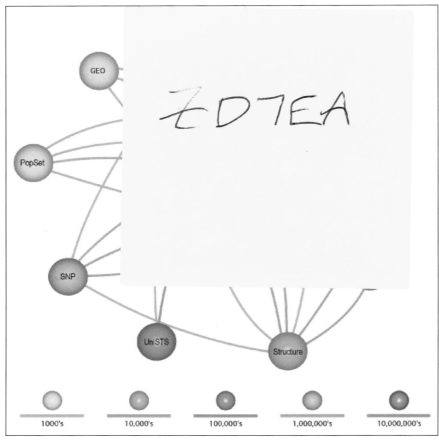

**Plate 14.** (Figure 17-13; page 341)

It is well known that the size of a cloned insert in a plasmid can affect the cell's growth rate (17). This can be visualized by the variable colony size on any given lawn of a library with no size selection. To mitigate this effect, we have implemented vectors that reduce transcription into the vector while tightly controlling our library insert size. It is not always possible to attain a narrow insert size distribution when constructing clone libraries, and this is particularly true when only small quantities of DNA are available. As a result, the common culture broths are particularly problematic given the variable growth rates of each clone. Both LB and 2xYT (45) are not buffered mediums, which results in quickly growing cells that rapidly turn the culture more acidic. As the pH of the culture shifts, cells begin to lyse and release nucleases. Terrific Broth (TB) is a buffered medium (45) that allows the culture to continue growing longer and develop higher densities of cell growth. It is common to get more DNA from a TB growth, however, the sequencing data are of lower quality when compared to those DNA prepared from cell cultures grown in LB or 2xYT (Figure 9-6). At Agencourt, we observed that the phosphates in the TB medium are the source of the problem. Most solid supports designed to capture DNA will also capture phosphates. Phosphates are known to chelate $MgCl_2$ and hinder a sequencing or PCR reaction (42). Replacing the buffer in the media to a non-phosphate based buffer presented a rich media that produces more DNA that sequences equally well on a per nanogram basis (11). This is critical for adequate induction of fosmid DNA.

Another feature to address is the anaerobic growth. Most 384-well cell cultures are in a mosaic state between anaerobic and aerobic growth. Growth media containing pyruvate can enable *E. coli*'s facultative anaerobic state and enhance DNA recovery (10).

The final adjustment is to increase the temperature of growth. It is well known that pUC replication is temperature-sensitive and the optimal copy number is achieved at 42°C, because this is the $T_m$ for the RNAI primer. This final adjustment provides higher DNA yields and more consistent well-to-well recoveries. Increased evaporation can be mitigated with proper humidity control of the growth.

With these three adjustments, we are able to generate very reproducible growth densities and healthy cultures that when purified will sequence with remarkable success. This is important, as dead cells have just as much protein as healthy cells and only the healthy cells have undegraded DNA worth purifying. With healthy cells, one can systematically tune the appropriate number of beads required to reduce agglutination and produce superior purification conditions.

## RNA Purification

RNA isolation requires techniques that protect the RNA before, during and after isolation. RNA is subject to degradation from ubiquitous

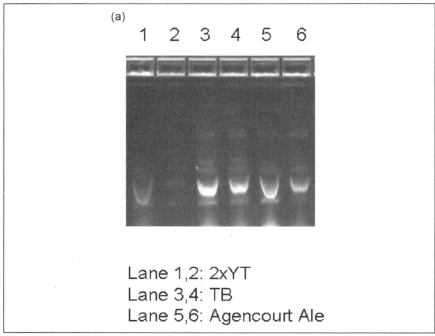

(a)

1 2 3 4 5 6

Lane 1,2: 2xYT
Lane 3,4: TB
Lane 5,6: Agencourt Ale

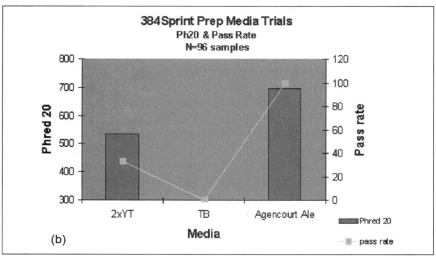

(b)

**Figure 9-6.** (a) Agarose gel demonstrating DNA yield from identical clones in three different medias utilizing Agencourt's SprintPrep reagents. (b) Sequencing results are best with phosphate-free-buffered media.

RNases before and after cell lysis. It is also subject to rapid hydrolysis, especially at higher temperature and in the presence of divalent cations (40). Correct sample handling at all stages is critical to obtaining high quality RNA. Universal precautions such as dedicated equipment, sterile environment, RNase free solutions and plasticware are all standard and familiar to those routinely working with RNA. RNA lysis buffers generally contain a strong denaturant such as guanidine isothiocyanate (GITC) which rapidly inactivates RNases by solubilizing cellular protein and other cell components. Original methods for routine isolation of cellular total RNA involve solubilizing cell or tissues in 4 M GITC buffer, then isolating RNA through differential centrifugation in CsCl (7). A method developed in 1987 by Chomczynski rapidly supplanted techniques that employed ultracentrifugation (8, 9). A reagent consisting of a mixture of acid phenol, guanidine isothiocyanate, and chloroform (marketed as Tri-Reagent or TriZol by Molecular Research Center Inc., Cincinnati, OH, USA) is used to solubilize tissues or cells. RNA separates in the aqueous phase, while DNA and proteins separate into the organic phenol phase at acid pH. RNA is precipitated with ethanol and resuspended in RNase free water. This method works well, but involves the use of caustic, hazardous chemicals and requires hands-on, time-consuming precipitation steps. It is not amenable to purification in a multi-well format. In addition, trace organic solvents can adversely affect the yield and quality of the RNA (51).

Solid phase techniques, on the other hand, are used to remove the organic extraction phase of the RNA isolation procedure and allow multiple parallel processing. Column purification involving silica resin is the most widely used method of solid-phase RNA extraction and is compatible with the traditional RNA isolation buffers. However, it still requires centrifugation or vacuum filtration to perform, which makes it more difficult to automate. SPRI technology, heretofore used mainly for DNA purification, is also very well suited for capturing intact, high quality RNA; it allows for much easier sample processing than column based methods as there is no centrifugation or vacuum filtration involved. It is very amenable to processing isolation of RNA in multi-well plates from starting materials such as cultured cells, tissues, and blood (Figures 9-7 [color Plate 3] and 9-8). Recovery is linear over a 5-log range (Figure 9-9). Contaminating DNA, usually present in any RNA isolation technique, can be optionally removed by a DNase I digestion directly on the column or on the beads in most solid phase techniques.

There are many ways to assess the RNA quality; an optical density reading with an OD260/280 ratio of 2 indicates pure RNA. Formaldehyde gels or even standard agarose gels are used to evaluate ribosomal 28S and 18S intensities as well as for signs of degradation indicated by smearing. One method that is rapidly becoming popular is analysis by

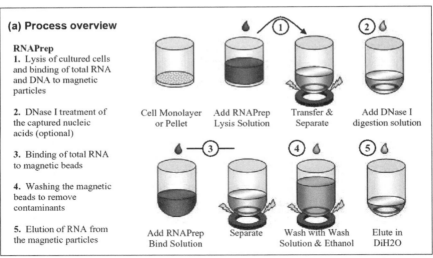

**(a) Process overview**

**RNAPrep**

1. Lysis of cultured cells and binding of total RNA and DNA to magnetic particles

2. DNase I treatment of the captured nucleic acids (optional)

3. Binding of total RNA to magnetic beads

4. Washing the magnetic beads to remove contaminants

5. Elution of RNA from the magnetic particles

Cell Monolayer or Pellet

Add RNAPrep Lysis Solution

Transfer & Separate

Add DNase I digestion solution

Add RNAPrep Bind Solution

Separate

Wash with Wash Solution & Ethanol

Elute in DiH2O

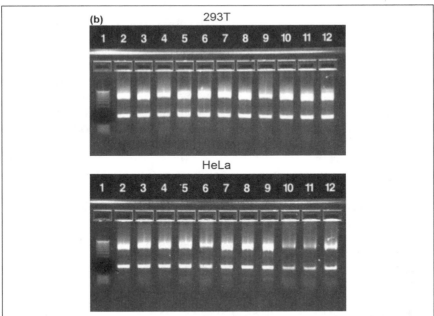

**(b)** 293T

1 2 3 4 5 6 7 8 9 10 11 12

HeLa

1 2 3 4 5 6 7 8 9 10 11 12

**Figure 9-7.** (a) Schematic diagram showing process for total RNA isolation from cultured cells using SPRI (see Plate 3). (b) 293T or HeLa cells were cultured and isolated directly from a 96-well tissue culture dish using the Beckman FX automated platform and SPRI RNAPrep. Twelve wells picked at random were run on an agarose gel.

142

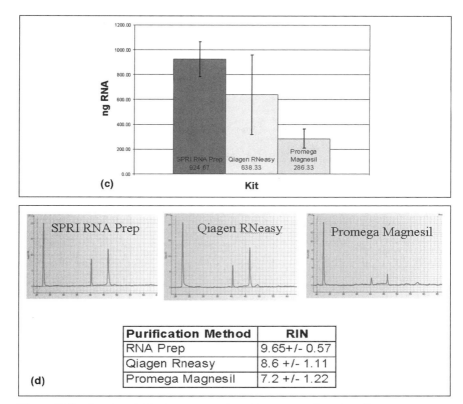

**Figure 9-7.** *Continued* (c) RNA was isolated from $1 \times 10^5$ 293T cells using various kits according to manufacturer's instructions. The SPRI technology captured 1.5 to 3 times more RNA than other competing technologies. RNeasy and Magnesil are filter and magnetic-bead-based competitor methods, respectively, to the SPRI RNAPrep. (d) RNA quality as determined by Agilent Bioanalyzer analysis including calculated RNA Integrity Number (RIN) scores.

electrophoresis on a chip using the Agilent Bioanalyzer 2100. Electropherograms are used to calculate RNA quality by examining the 28S and 18S ratio using area under peaks. In addition, the Agilent system has a software option including a complex algorithm that analyzes many regions of the electropherogram and calculates an RNA integrity score (RIN number) (36). This returns a quality score between 1 and 10, with 10 being the highest quality RNA. When comparing SPRI to RNA purification methods such as TriZol, silica columns, silica magnetic beads, or charge switch methods, SPRI compares favorably by any method of RNA quality or quantity used (Figures 9-7 and 9-10).

A growing RNA isolation application is for microarray analysis, both for upstream RNA sample prep, and for labeled probe generated from the

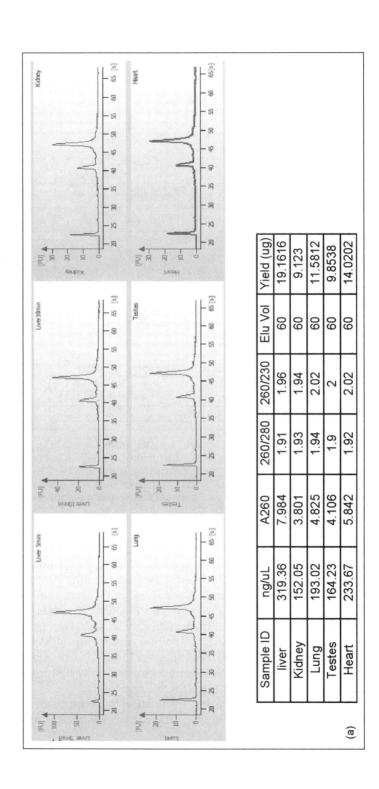

| Sample ID | ng/uL | A260 | 260/280 | 260/230 | Elu Vol | Yield (ug) |
|-----------|-------|-------|---------|---------|---------|------------|
| liver | 319.36 | 7.984 | 1.91 | 1.96 | 60 | 19.1616 |
| Kidney | 152.05 | 3.801 | 1.93 | 1.94 | 60 | 9.123 |
| Lung | 193.02 | 4.825 | 1.94 | 2.02 | 60 | 11.5812 |
| Testes | 164.23 | 4.106 | 1.9 | 2 | 60 | 9.8538 |
| Heart | 233.67 | 5.842 | 1.92 | 2.02 | 60 | 14.0202 |

(a)

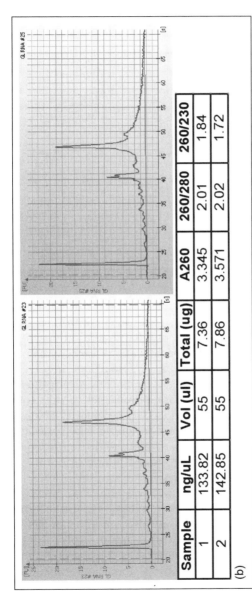

| Sample | ng/uL | Vol (ul) | Total (ug) | A260 | 260/280 | 260/230 |
|--------|-------|----------|------------|-------|---------|---------|
| 1 | 133.82 | 55 | 7.36 | 3.345 | 2.01 | 1.84 |
| 2 | 142.85 | 55 | 7.86 | 3.571 | 2.02 | 1.72 |

(b)

**Figure 9-8.** (a) RNA from several different rat tissue types was isolated in a multiwell plate using SPRI methods. (b) Total RNA from 10mL of Paxgene stabilized blood (2.5 mL of whole blood) was isolated using SPRI methods.

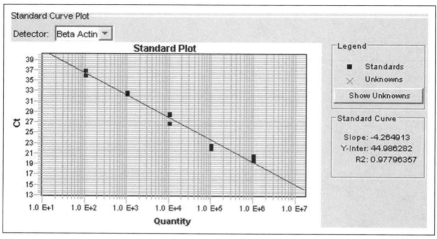

**Figure 9-9.** SPRI RNA isolation linear over a 5 log range. 293T cells were diluted and total RNA isolated. Recovery of RNA using SPRI is linear from $1 \times 10^6$ cells down to 100 cells determined by quantitative RT-PCR assay of the beta actin gene.

RNA sample. There are many microarray options, but home brew methods where cDNAs or oligos are spotted onto glass slides still remain popular because content can be controlled and they are generally cheaper. There is a shift toward using commercial arrays for quality control, reproducibility, and content. A discussion of microarray manufactured options is beyond the scope of this text, but reviews on this subject are available (46). Measuring RNA expression profiles with microarrays involves making a probe from the mRNA. Labeled probes can be cDNA, amplified cRNA, or amplified cDNA. Direct labeling methods, which produce a cDNA probe by reverse transcription of mRNA in the presence of labeled nucleotides, require a lot of mRNA. Such methods are used when a lot of sample is available and in dual labeling experiments with home brew spotted arrays. Because many applications produce limited amounts of starting RNA, linear amplification methods, where the amplified RNA is representative of the original sample, are becoming popular (54). In most procedures, mRNA is not even isolated, rather labeling or amplification occurs directly from total RNA. The most widely used method of RNA amplification is the Eberwine T7-based RNA amplification protocol (16, 43). Here, the total RNA is used as a source material to copy mRNA into double stranded cDNA using standard procedures. Incorporated in the cDNA is a T7 polymerase promoter site, used to drive in vitro transcription of labeled cRNA, which is fragmented and used to probe the microarray. High recovery is essential for this application as higher yield at either the cDNA or cRNA step will translate into more probe. This is important

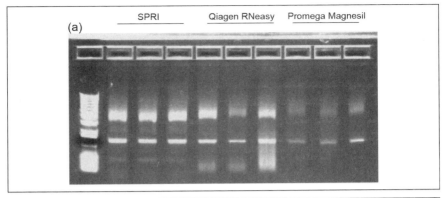

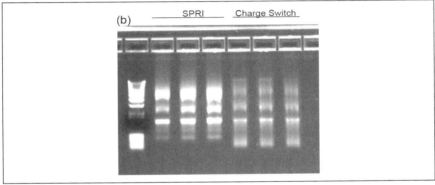

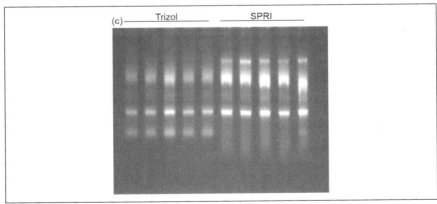

**Figure 9-10.** Gel electrophoresis of total RNA isolated with SPRI from $1 \times 10^5$ 293T cells compared to (a) Qiagen's silica columns and Promega's magnetic silica beads and (b) Invitrogen charge switch magnetic beads. (c) $1 \times 10^4$ 293T cells were isolated with Trizol and RNAPrep. No DNAse I step was included in the RNAPrep in this experiment.

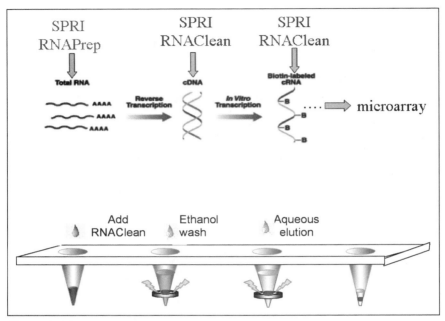

**Figure 9-11.** SPRI RNAClean reagent can be used to clean up both the cDNA and cRNA reactions of the Eberwine antisense RNA amplification protocol. (See Plate 4.)

for processing of smaller amounts of input samples generated from techniques like laser capture microdissection. Again, silica-based columns are commonly used for this application, but as 96-well microarrays become common, magnetic bead-based solutions will have distinct advantages. One advantage of using SPRI reagents is that the cDNA and cRNA purification steps use the same reagent and are identical (Figure 9-11; color Plate 4). Further, SPRI magnetic bead-based techniques have consistently shown twofold greater cRNA yield over silica column based purifications (Figure 9-12). Multiwell microarrays are predicted to become indispensable to obtain gene expression profiles in high-throughput target validation using compound libraries and gene knockdown experiments using siRNA. SPRI is well suited for automation of this process in 96-well format, including the upstream total RNA preparation and the probe purification, and can greatly facilitate high throughput screening of expression profiling for drug target validation.

## Genomic DNA Isolation

Current manual methods of DNA isolation are reliable and robust for small numbers of samples. However, these conventional procedures are

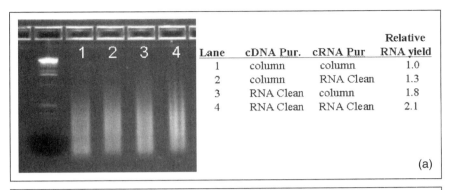

| Lane | cDNA Pur. | cRNA Pur | Relative RNA yield |
|---|---|---|---|
| 1 | column | column | 1.0 |
| 2 | column | RNA Clean | 1.3 |
| 3 | RNA Clean | column | 1.8 |
| 4 | RNA Clean | RNA Clean | 2.1 |

(a)

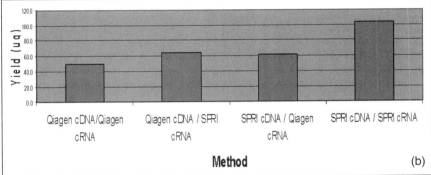

(b)

**Figure 9-12.** (a) cRNA product was produced from 1 μg of rat thymus total RNA using the MessageAmp aRNA kit (Ambion). SPRI clean up was substituted for the standard silica column clean up as indicated. cRNA yields are doubled when using SPRI in both clean-up steps. (b) cRNA was produced from 5 μg rat liver RNA in the Eberwine T7 RNA amplification procedure using the Invitrogen cDNA synthesis kit and Ambion IVT kit, substituting Qiagen cDNA clean up and RNeasy columns with SPRI RNA Clean. cRNA yields are increased twofold when SPRI is used at both clean-up steps. Purification was performed according to manufacturer's instructions.

labor-intensive and time-consuming and susceptible to contamination from different sources. The need for genomic DNA (gDNA) sample preparation is steadily increasing as efforts to analyze DNA to discover biomarkers for genetic diseases increases. In addition, molecular diagnostic applications, genotyping for disease association, forensics applications, as well as efforts to isolate and characterize genomic DNA from microorganisms in body fluids, will increasingly rely on higher throughput methods for isolation of genomic DNA. Isolation of nucleic acid from cells and tissue is more difficult than isolation from in vitro reactions such as PCR or in vitro transcription (IVT), because of the complexity of the lysate that contains high concentrations of proteins, high molecular weight

DNAs, RNAs, carbohydrates, lipids, and salts. Therefore, the purification of genomic DNA from cells requires a different approach than the better known methods of isolation of plasmid DNA. The latter relies on a denaturing environment to trap high molecular weight gDNA in an insoluble complex with proteins and carbohydrates, leaving the covalently-closed circular plasmids in solution. Obviously, this is not suitable for genomic size DNA. Traditional gDNA isolation techniques use detergents and or strong denaturants to lyse membranes and denature proteins (45), and typically, a lysate-clearing step is necessary to remove proteins and other contaminants. Using high salt conditions such as NaCl or ammonium sulfate is effective at precipitation and removing proteins. More typical, though, are organic extraction techniques using phenol/chloroform, which are also quite effective at removing total protein. However, both approaches require tedious extraction and spin separation steps as well as ethanol precipitation and handling hazardous chemicals. In particular, these traditional techniques are not very amenable to automation.

Solid phase techniques can be applied to improve processing of larger sample numbers for genomic DNA isolation applications. These techniques are in need of improvements if larger numbers of samples are to be processed. However, such a complex starting material presents real problems for solid phase purification with both filter and magnetic bead-based systems. Proteins in particular tend to make lysates viscous, and can clog filters and crosslink beads, making them difficult to handle for efficient isolation of nucleic acids. Hybrid techniques that rely on organic extraction followed by solid-phase purification techniques work well, but do not remove the laborious upstream sample processing steps. To remove proteins, the most widely practiced method is to include protein digestion in the lysis step. Protein degrading enzymes such as proteinase K are almost universally employed to hydrolyze proteins, making a cleared lysate that is easier to handle and increase yields (6). Since proteinase K will also hydrolyze enzymes utilized in downstream manipulations of DNA (restriction analysis, PCR, sequencing, etc.), proteinase K needs to be heat-inactivated. More proteinase K will be required with whole blood purification as a large percentage of the cells in whole blood (red blood cells) only provide mitochondrial DNA. These red blood cells are also a source of cell debris and proteins. Serum, on the other hand, contains lower levels of proteins, few lysed cells, bacterial and viral DNA, and overall less cellular debris to be removed from the purification. As a result, techniques that can purify whole blood are usually adequate for serum, with the caveat that serum purification demands high sensitivity as the amount of DNA in serum is very low. Many viral content assays demand DNA or RNA purification protocols containing below 100 virus copies per milliliter of serum. In these applications, it is important to avoid any purification conditions that might damage the limited DNA or RNA that is available.

Blood is the most widely used source to obtain gDNA. Most blood DNA isolation techniques can be described as a variation of one of two general methods. Both start with isolation of nuclear material. In the first, whole blood is collected and separated into plasma, buffy coat, and red blood cells by centrifugation. The DNA containing white blood cells are then physically isolated from the buffy coat and used as the starting material for DNA isolation by cell lysis in a detergent buffer. In a second general protocol, a nonionic detergent is added to whole blood to lyse the cell membrane, followed by pelleting of intact nuclei. In either method, the nuclei are lysed in a buffer containing NaCl and EDTA, and the nuclear protein fraction is digested with proteinase K buffer containing SDS. Proteins are then removed by either salting out or phenol chloroform extraction. The resulting supernatants are ethanol precipitated to obtain the final DNA. As neither of these methods is amendable easily to automation, a third method—isolation directly from whole blood—is preferable. In magnetic bead based solid phase extractions such as SPRI, small volumes of fresh or frozen blood (50–200 μL) can be used directly in a lysis, bind, wash and elute protocol with good yields (Figure 9-13). It is important for isolation protocols to remove blood components such as hemoglobin, lactoferrin, and immunoglobulin G as well as anticoagulants like heparins and EDTA, which can inhibit downstream PCR (2). High quality DNA can be obtained using SPRI from fresh or frozen blood collected in an anticoagulate. The DNA is sufficiently clean to utilize in downstream PCR amplification, and can be stored at −20°C for extended periods without losing activity in PCR (Figure 9-13). In addition, the SPRI blood DNA isolation procedure returns a higher yield of DNA compared to competing DNA isolation procedures (Figure 9-14; color Plate 5).

Another source of gDNA is from the isolation of DNA from buccal cells (19). Because genomic DNA is identical whether it comes from blood cells or cheek cells, buccal cell isolation is a viable alternative to isolation from blood. Buccal cell DNA is used for many diagnostic applications such as epidemiologic studies and paternity testing. In fact, in its annual report summary for testing, the American Association of Blood Banks, Parentage Testing Standards Program Unit reported that in 2003 almost 92% of all samples processed were from buccal swabs, with the remaining 8% being from blood. There are several advantages to buccal cell DNA isolation over blood. First, no needles are involved, so it is less invasive and painless. Typically, cotton swabs or brushes are used to remove cell material from the inside of the cheek. This is helpful when isolating from children and for getting repeat samples from individuals. DNA can even be self-harvested by the individual. A second advantage is the reduced

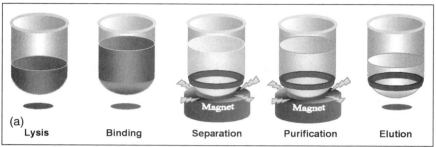

(a)

Lysis   Binding   Separation   Purification   Elution

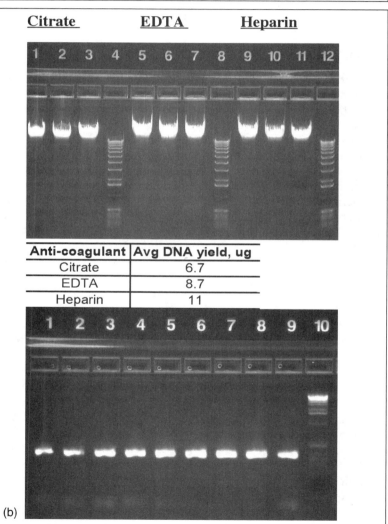

Citrate     EDTA     Heparin

| Anti-coagulant | Avg DNA yield, ug |
|----------------|-------------------|
| Citrate | 6.7 |
| EDTA | 8.7 |
| Heparin | 11 |

(b)

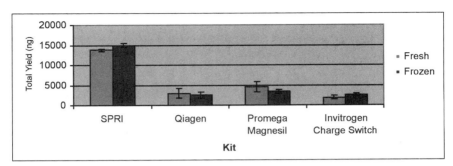

**Figure 9-14.** Comparison of gDNA yields from 200 μL of fresh human blood preserved in EDTA using SPRI and competitor kits. All isolations were performed according to manufacturer's protocol.

risk for blood borne hazards during collection due to the absence of needles. A third advantage is that the sample is stable at room temperature for extended periods of time unlike blood samples that need to be processed immediately or frozen. A fourth advantage is that buccal cell collection can be less costly than obtaining DNA from blood in large epidemiologic studies (1). A disadvantage to buccal cell collection is that the method yields less DNA than blood. However, a buccal collection method rapidly gaining popularity involves collecting DNA by rinsing the mouth with mouthwash (23). Following a quick spin, the pellet is harvested in lysis buffer. The yields from this method are much higher than swab and provide sufficient DNA for many downstream PCR applications. Again, solid phase techniques such as magnetic bead isolations provide easy automation for processing higher throughput numbers. A SPRI method has been developed to collect DNA from both mouthwash and swabs (Figure 9-15; color Plate 6).

The practice of preserving tissue and tumor biopsy samples by formalin fixation and paraffin embedding has been around long before most molecular biological techniques were in widespread use. These samples, originally saved to perform traditional cytological staining, may be useful for retrospective studies to correlate disease phenotype and outcome with the genotype. Because of the widespread use of paraffin preservation even today, methods are being developed to efficiently extract nucleic acids for genotyping. Isolation of DNA from these samples involves harsh organic

---

◄─────────────────────────────────────────

**Figure 9-13.** (a) Schematic diagram of SPRI blood DNA isolation protocol. (See Plate 5.) (b) SPRI was used to isolate genomic DNA from 200 μL of fresh horse blood collected in three different anticoagulants. Gel of isolated DNA; yields are indicated (top). 1 μL of DNA was used in a 10-μL PCR with horse transferrin primers (bottom). No inhibition of PCR was observed.

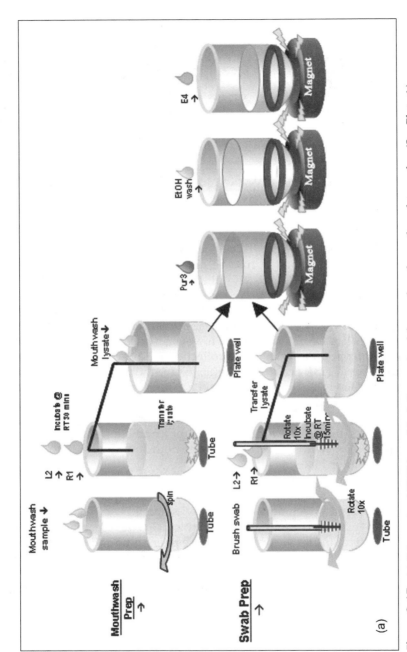

**Figure 9-15.** (a) Schematic of SPRI buccal isolation protocol for swab and mouthwash samples. (See Plate 6.)

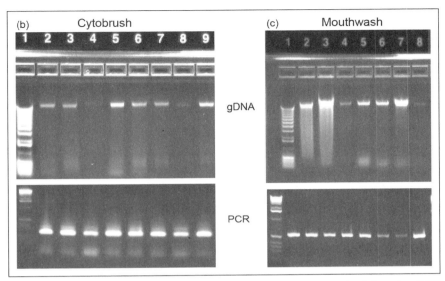

**Figure 9-15.** (b) Genomic DNA isolated with the cytobrush method showing PCR with human cytoskeletal β-actin primers. (c) Genomic DNA isolated with the mouthwash method showing PCR with human ADP ribosylation factor-1 primers.

extractions with xylene to remove the paraffin, followed by a proteinase K digestion, and subsequent phenol : chloroform extraction and ethanol precipitation. These toxic chemicals present many waste disposal issues including storage and cost (see Chapter 6). DNA yields are generally low after extraction from paraffin and the DNA is often damaged by the formalin fixation process, which can cause DNA adduct formation resulting in inter- and intra-strand crosslinking to DNA and crosslinking of DNA to proteins (26). This may limit the success of a downstream PCR application or at the very least preclude amplification of larger amplicons. Solid phase techniques can significantly shorten DNA isolation from paraffin and eliminate the need for xylene extraction. In the SPRI process, paraffin samples can be added directly to a lysis buffer containing proteinase K. The 55°C incubation will melt the paraffin, eliminating the need for organic extraction. Following protein digestion, the sample is mixed with a solid-phase binding buffer containing magnetic beads, and then washed and eluted (Figure 9-16). With the size selection techniques available with SPRI, it may also be possible to enrich for the larger molecular weight fragments and improve the success rate of PCR.

The largest market segment for molecular diagnostic tests is in the area of infectious diseases. Viral monitoring and testing is critical for monitoring the blood supply and for routine clinical testing for causative agents as well as viral load monitoring and genotyping of chronic infections such as human immunodeficiency virus (HIV) and hepatitis C virus

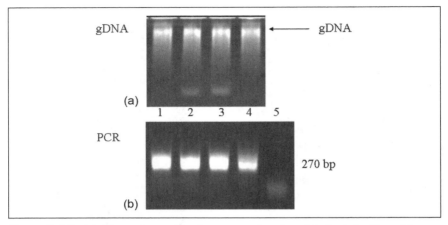

**Figure 9-16.** (a) 10 mg of paraffin tissue were incubated in lysis buffer with proteinase K, then added to SPRI binding buffer, recovering about 200 ng of genomic DNA, lanes 1–4. (b) The genomic DNA was amplified with human B-actin primers to produce a 270-bp product, lanes 1–4. Lane 5 is the negative control.

(HCV). Solid phase techniques can offer the sensitivity needed for clinical and diagnostic applications while providing ease of automated format in a closed system. Viral nucleic acid can be isolated using SPRI techniques with sensitivity to detect down to at least a hundred copies per milliliter of viruses from plasma using real time or quantitative PCR and 50 copies per milliter using a PCR endpoint assay (Figure 9-17; color Plate 7). Several medium-throughput fully enclosed automated systems are coming onto the market for use in molecular diagnostics for clinical testing (Vidiera system from Beckman-Coulter, Fullerton, CA, USA; Magtration 12GC system from PSS Bio Instruemets, Pleasanton, CA, USA). Such systems allow minimal sample handling and intervention for processing, which leads to safer working conditions and less human error. Magnetic beads provide an effective automation vehicle, while maintaining excellent quality and sensitivity. They also make the automation less wieldy to produce, as no vacuum filtration or centrifugation is required in the automated steps.

## Extending the Applications for SPRI: Bifunctional SPRI Beads

SPRI technology can be enhanced to extend the useful range of purification options available. Through carboxydiimide coupling, other functional groups can be covalently coupled to the beads through the surface carboxy groups. Due to the density of carboxyl groups on the SPRI bead surface, this does not involve every group. In fact, sufficient carboxy

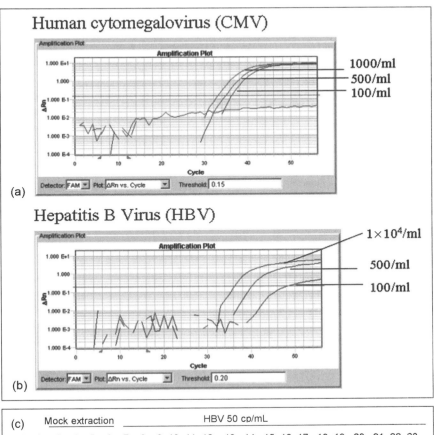

**Human cytomegalovirus (CMV)**

(a)

1000/ml
500/ml
100/ml

**Hepatitis B Virus (HBV)**

(b)

$1 \times 10^4$/ml

500/ml

100/ml

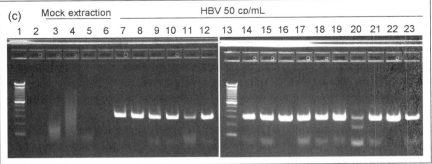

(c) Mock extraction      HBV 50 cp/mL

1 2 3 4 5 6 7 8 9 10 11 12 13 14 15 16 17 18 19 20 21 22 23

**Figure 9-17.** Serial dilutions of virus were made into 200 µL human plasma, and then extracted with a SPRI viral nucleic acid isolation kit. Specific primers were used in a TaqMan qPCR or qRT-PCR assay and compared to standard curves to determine copy number for (a) human cytomegalovirus (see Plate 7) and (b) hepatitis B virus. (c) Plasma alone (mock, lanes 3–6) or HBV virus diluted in human plasma to 50 copies per mL were extracted with SPRI. One half of the eluted product (10 µL) was used in a 50-cycle, gene-specific PCR and analyzed by gel electrophoresis. Sixteen out of 16 virus samples were positive (lanes 7–12 and 14–23). Lane 2 is a PCR negative control.

**157**

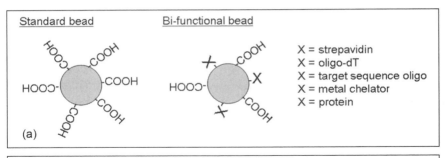

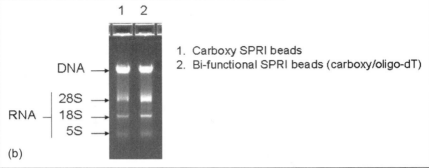

**Figure 9-18.** (a) Schematic of SPRI carboxy beads and SPRI bi-functional beads. (b) Total nucleic acid was isolated from 293T-cell lysates using SPRI carboxy and SPRI bi-functional beads to illustrate that the bi-functional beads are fully active in SPRI.

functional groups remain following carboxydiimide coupling to afford effective SPRI nucleic acid purification, without any apparent loss in yield (Figure 9-18). Such dual functionality can be utilized in many ways to provide a means for one step applications in which binding of nucleic acid can occur via SPRI, and a second functional group can be utilized to remove or retain another target molecule.

This heterobifunctional technology is currently being expanded for its application to mRNA purification from total RNA, beta globin mRNA depletion from total RNA, transcript depletion or enrichment, duplex sequencing reactions, parallel protein and DNA purification, and viral concentration and purification.

An example of a bifunctional bead application is one-step purification of mRNA from a cell or cell lysate. In this application, a cell lysate is made and total nucleic acid is bound to SPRI beads, which also contain covalently attached oligo-dT as a second functional group. Following the SPRI isolation, the beads are eluted in a low ionic strength buffer, which promotes elution of total nucleic acid and subsequent binding of

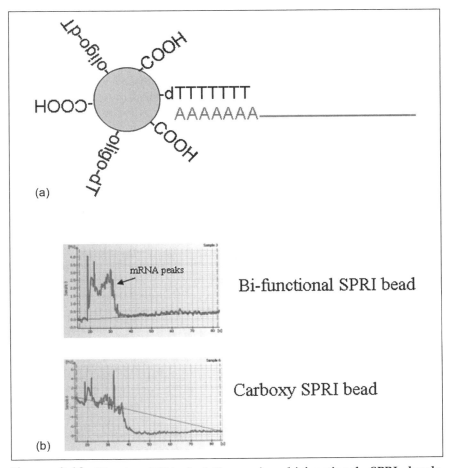

**Figure 9-19.** Direct mRNA isolation using bi-functional SPRI beads. (a) Schematic of oligo-dT/carboxy bead. (b) Total RNA was isolated from Hela-cell lysates using the SPRI RNAPrep kit with bi-functional carboxy oligo-dT beads or standard carboxy beads. Total RNA was eluted in a hybridization buffer that promotes poly-A messager RNA hybridization to oligo-dT, followed by wash and elution. mRNA was isolated only by the beads coupled to oligo-dT.

poly-A RNA. Following the wash and elution steps, mRNA is isolated (Figure 9-19).

Another bifunctional bead application that could have great utility for small interfering RNA (siRNA) gene transfections and other sensitive transfection-based assays is that of endotoxin removal from plasmid DNA. To close out this chapter, the following section describes this SPRI technique in further detail.

## Endotoxin Removal

Introduction of high-throughput DNA preparation methods has fueled the growth of large-scale sequencing efforts and has resulted in the generation of a vast collection of genomic and expressed gene sequences. Further characterization of these sequences in vitro may be applied by using high-throughput analysis of expressed genes for the evaluation of mammalian cell function (58). Moreover, gene therapy studies in vivo using plasmid and BAC DNA have been widely applied in animal models for both the characterization of disease states and for the evaluation of potential therapeutic intervention (41). Additionally, the increased understanding of genetic immunization using naked DNA vaccines has also held great promise as a novel therapeutic deliverable (32, 34). Unfortunately, such applications and therapies are very sensitive to contaminants typically present in nucleic acid preparations. All standard methods of plasmid DNA isolation from bacteria, including alkaline lysis, high pressure (French Press), boiling, and the use of lysozyme or detergents will induce the release of LPS, or endotoxin, from the outer membrane of the bacteria along with plasmid DNA. The LPS then forms micelles with physical characteristics (density, size, and charge distribution) similar to plasmid DNA, and as a consequence, it will be carried through the purification steps along with the plasmid DNA. In particular, contaminating endotoxins from the *E. coli* host typically used to prepare DNA molecules has been shown to induce apoptosis during culture of mammalian cells in vitro (30) as well as toxic shock, sepsis, and a variety of related clinical complications in vivo (15). The transfection efficiency of endotoxin-containing DNA in mammalian cells such as HeLa, Huh7, COS7, and LNH is reduced significantly compared to endotoxin-free DNA (55). A method for reliable and cost-effective preparation of endotoxin-free plasmid DNA (39), would be useful in transfection studies that utilize cell-based microarrays (4, 18, 35, 38, 52, 57, 58).

A number of peptides, proteins, and receptor motifs interact strongly with endotoxins. Some of these include lipopolysaccharide binding protein (LBP); bactericidal/permeability-increasing protein (BPI) (5); polymyxin and polymyxin analogs (28); amyloid P component (13); cationin protein 18 (12); MD-2 and Toll-like receptor (TLR) (49); TLR2 (44); CD14 (49); Bac7 (1, 34); a synthetic peptide derived from a protein found in bovine neutrophils (20); limulus factor-C and synthetic peptides derived from the Sushi3 domain thereof (33); and antibodies raised against the lipid A component of endotoxin (24). In principle, it should be possible to develop affinity adsorbents for endotoxin removal by immobilization of any of these proteins; however, most have not yet been investigated. One molecule that has been extensively used as an endotoxin absorbent is polymyxin B (PMXB), a cyclic polypeptide antibiotic. PMXB

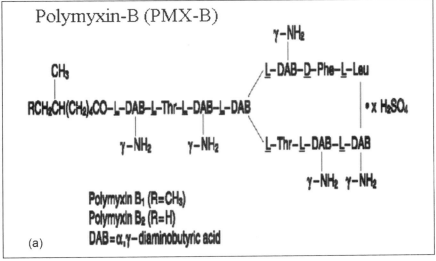

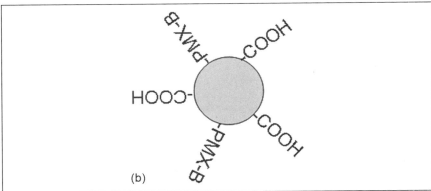

**Figure 9-20. Bi-functional SPRI beads for high throughput endotoxin removal.** (a) Chemical structure of polymyxin B (PMXB). (b) Schematic of bi-functional PMXB SPRI bead. (c) SPRI plasmid purification procedure modified to have an endotoxin removal (ERR) step. (See Plate 8.) (d) Results of endotoxin removal from plasmid using SPRI PMXB bi-functional beads and process outlined in (c). (See Plate 9.)

(Figure 9-20a) binds stoichiometrically to the lipid A moiety of endotoxin molecules, primarily through hydrophobic interactions (50).

PMXB has been linked to various supports, such as sepharose, for use in endotoxin removal (39). Production of affinity resins involve complex synthesis procedures and can thus be very costly. Studies of various affinity resins and resin formats with PMXB to remove endotoxins from plasmid DNA have exposed some problems, including significant DNA loss (39, 56). Nevertheless, PMXB affinity resins have been used to

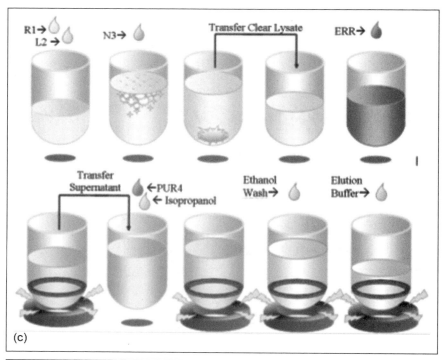

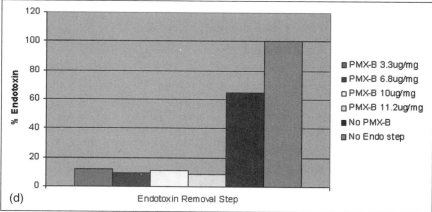

**Figure 9-20.** *Continued*

successfully remove endotoxins from DNA and protein-containing solutions. One report found DNA recovery following PMXB affinity purification to be greater than that of anion exchange chromatography (56).

Covalent coupling of the PMXB molecule to SPRI magnetic beads may improve the efficiency and efficacy of endotoxin removal and enable the development of a high-throughput, affordable, and easily automated, endotoxin-free DNA preparation method. To achieve this, Agencourt has developed a heterobifunctional magnetic particle that contains two distinct functional groups on the bead surface, each designed to target a different biochemical molecule. By covalently coupling PMXB to carboxylated SPRI beads, endotoxin can be selectively driven to the beads under low salt conditions and subsequently DNA driven to the same bead under SPRI conditions. Once separated and the supernatant removed, the DNA can be selectively eluted, leaving the endotoxin on the PMXB. This bifunctional bead allows for a single-step homogeneous assay for endotoxin removal and DNA purification (Figure 9-20b, c [color Plate 8], d [color Plate 9]).

## Summary

Nucleic acid preparation methods have developed over the years to be quick, reliable, and less dependent upon toxic reagents. Older methods of purification involving many manual steps and organic chemicals are still widely used for small sample numbers, but have given way to solid phase purifications that are more easily amenable to automation for higher throughput needs. SPRI chemistry has many advantages over other solid-support purification methods. It has demonstrated great utility and is widely used to provide efficient and cost effective nucleic acid purification to support genomics applications, such as large scale sequencing and genotyping. The requirement for multi-well nucleic acid purification continues to grow to support new applications such as molecular diagnostics, expression profiling, genotyping, transfection, and others. SPRI chemistry is now being applied to provide the same high quality sample preparation to support these growing applications.

## Material and Methods

For Figure 9-3, all kits were tested in 96-well format with 1/16 dilution of BigDye™ Terminator v3.1, 10 μL reactions, using the control pGem3zf template and M13F sequencing primer.

For each reaction combine:

- 0.5 μL BigDye Terminator v3.1
- 1.75 μL 5× BigDye 3.1 Buffer
- 1.0 μL pGem3zf control plasmid (at 200 ng/μL)
- 0.016 μL 200 μM M13F primer (5′ GTAAAACGACGGCCAGT)
- 6.73 μL Di-H$_2$O

The total volume equals 10 μL.

All elements were pooled in a reagent trough to make enough reaction mix for 300 samples. The mix was pipetted into three quadrants of 384-well PCR plate for ease of cycling.

Following cycling, all reactions were pooled back together into 1.5 mL aliquots (first pooled via 12-channel pipette into one row of a 96-well PCR plate, then pooled by a single channel into aliquots). The aliquots were frozen overnight for processing the next day.

When ready to perform the cleanup evaluation, 10 μL of sample was pipetted into three separate 96-well plates. CleanSEQ and Edge used a standard full-skirt 96-PCR plate, while the Promega magnet is only compatible with a semi-skirted 96-well PCR plate. In each plate, we were short 10 wells of sample (due to evaporation from cycling), so plates are actually 86 samples total.

- CleanSEQ: Requires 85% ethanol and 40 μL Di-$H_2O$ for elution. CleanSEQ is a product of Agencourt Biosciences, Inc. (Beverly, MA, USA).
- MagDTR™: Requires 100% ethanol for binding, 80% for washing and 20 μL Di-$H_2O$ for elution. MagDTR is from EdgeBiosystems, Inc. (Gaithersburg, MD, USA).
- Magnesil™ Green: Requires 90% ethanol for washing and 20 μL formamide for elution. Magnesil Green is from Promega (Madison, WI, USA).

All cleanups were eluted in 40 μL Di-$H_2O$.

Following elution, all plates were placed back on their respective magnets; 15 μL of sample was transferred to one 384-well plate for sequencing on the ABI 3700, and 15 μL of sample was transferred to a second 384-well plate for sequencing on the ABI 3730.

Control plasmid DNA, M13 primer, dye-terminators, ABI 3700 and ABI 3730 were from Applied Biosystems, Inc. (Foster City, CA, USA). The sequencing reactions were run on DNA analyzers according to manufacturer's specifications.

## References

1. About, S., King, I.B., About, J.S., et al. 2002. Buccal cell yield, quality and DNA collection costs: comparison of methods for large scale studies. *Cancer Epidemiol Biomarkers Prev* 11: 1130–1133.
2. Al-Soud, W.A., and Radstrom, P. 2001. Purification and characterization of PCR inhibitory components in blood cells. *J Clin Microbiol* 39: 485–493.
3. Baker, M.J. United States Patent #6,914,137-2005.

4. Baghdoyan, S., Roupioz, Y., Pitaval, A., et al. 2004. Quantitative analysis of highly parallel transfection in cell microarrays. *Nucleic Acids Res* 32: e77.
5. Beamer, L.J., Carroll, S.F., and Eisenberg, D. 1998. The BPI/LBP family of proteins: a structural analysis of conserved regions. *Protein Sci* 7: 906–914.
6. Birnboim, H.C. 1992. Extraction of high molecular weight RNA and DNA from cultured mammalian cells. *Methods Enzymol* 216: 154–160.
7. Chirgwin, J.M., Przybyla, A.E., MacDonald, R.J., et al. 1979. Isolation of biologically active ribonucleic acid from sources enriched in ribonuclease. *Biochemistry* 18: 5294–5299.
8. Chomczynski, P., Mackey, K., Drews, R., et al. 1997. DNAzol: a reagent for the rapid isolation of genomic DNA. *Biotechniques* 22: 550–553.
9. Chomczynski, P., and Sacchi, N. 1987. Single-step method of RNA isolation by acid guanidinium thiocyanate-phenol-chloroform extraction. *Anal Biochem* 1: 156–159.
10. Clements, L.D., Miller, B.S., and Streips, U.N. 2002. Comparative growth analysis of the facultative anaerobes *Bacillus subtilis, Bacillus licheniformis,* and *Escherichia coli. Syst Appl Microbiol* 25: 284–286.
11. David, R.G. 2005. High-throughput developments in fosmid DNA sequencing with SprintPrep technology at Agencourt Bioscience (poster presentation). Marco Island, FL: Advances in Genome Biology and Technology, February 2005.
12. de Haas, C.J., Haas, P.J., van Kessel, K.P., and van Strijp, J.A. 1998. Affinities of different proteins and peptides for lipopolysaccharide as determined by biosensor technology. *Biochem Biophys Res Commun* 252: 492–496.
13. de Haas, C.J., van der Zee, R., Benaissa-Trouw, B., et al. 1999. Lipopolysaccharide (LPS)-binding synthetic peptides derived from serum amyloid P component neutralize LPS. *Infect Immun* 67: 2790–2796.
14. DeAngelis, M.M., Wang, D.G., and Hawkins, T.L. 1995. Solid-phase reversible immobilization for the isolation of PCR products. *Nucleic Acids Res* 23: 4742–4743.
15. DiPiro, J.T. 1990. Pathophysiology and treatment of gram-negative sepsis. *Am J Hosp Pharm* 47: S6–S10.
16. Eberwine, J. United States Patent #5,514,545-1996.
17. Feng, T., Li, Z., Jiang, W., et al. 2002. Increased efficiency of cloning large DNA fragments using a lower copy number plasmid. *Biotechniques* 32: 992–996.
18. Friedman, A., and Perrimon, N. 2004. Genome-wide high-throughput screens in functional genomics. *Curr Opin Genet Dev* 14: 470–476.
19. Garcia Closas, M., Egan, K.M., Abruzzo, J., et al. 2001. Collection of genomic DNA from adults in epidemiological studies by buccal cytobrush and mouthwash. *Cancer Epidemiol Biomarkers Prev* 10: 687–696.
20. Ghiselli, R., Giacometti, A., Cirioni, O., et al. 2003. Neutralization of endotoxin in vitro and in vivo by Bac7(1-35), a proline-rich antibacterial peptide. *Shock* 19: 577–581.
21. Hawkins, T.L., McKernan, K.J., Jacotot, L.B., et al. 1997. A magnetic attraction to high-throughput genomics. *Science* 276: 1887–1889.
22. Hawkins, T.L. United States patents #5,898,071, #5705628-1998.

23. Heath, E.M., Morken, N.W., Campbell, K.A., et al. 2001. Use of buccal cells collected in mouthwash as a source of DNA for clinical testing. *Arch Pathol Lab Med* 125: 127–133.

24. Helmerhorst, E.J., Maaskant, J.J., and Appelmelk, B.J. 1998. Anti-lipid A monoclonal antibody centoxin (HA-1A) binds to a wide variety of hydrophobic ligands. *Infect Immun* 66: 870–873.

25. Holt, R.A., Subramanian, G.M., Halpern, A., et al. 2002. The genome sequence of the malaria mosquito *Anopheles gambiae*. *Science* 298: 129–149.

26. Huang, H., and Hopkins, P.B. 1993. DNA interstrand cross-linking by formaldehyde: nucleotide sequence preference and covalent structure of the predominant cross-linked formed in synthetic oligonucleotides. *J Am Chem Soc* 115: 9402–9408.

27. International Human Genome Sequencing Consortium. 2001. Initial sequencing and analysis of the human genome. *Nature* 409: 860–921.

28. Jacobs, D.M., and Morrison, D.C. 1977. Inhibition of the mitogenic response to lipopolysaccharide (LPS) in mouse spleen cells by polymyxin B. *J Immunol* 118: 21–27.

29. Kieleczawa, J. 2005. Controlled heat-denaturation of DNA plasmids. In: Kieleczawa, J., ed. *DNA Sequencing: Optimizing the Process and Analysis.* Sudbury, MA: Jones and Bartlett.

30. Kuwabara, T., and Imajoh-Ohmi, S. 2004. LPS-induced apoptosis is dependent upon mitochondrial dysfunction. *Apoptosis* 9: 467–474.

31. Lage, M.J., Leamon, H.J., Pejovic, T., et al. 2003. Whole genome analysis of genetic alterations in small DNAs used in hyperbranched strand displacement amplification and array CGH. *Genome Res* 13: 294–307.

32. Lewis, P.J., and Babiuk, L.A. 1999. DNA vaccines: a review. *Adv Virus Res* 54: 129–188.

33. Li, C., Ng, M.L., Zhu, Y., et al. 2003. Tandem repeats of Sushi3 peptide with enhanced LPS-binding and -neutralizing activities. *Protein Eng* 16: 629–635.

34. Liu, M.A. 2003. DNA vaccines: a review. *J Intern Med* 253: 402–410.

35. Maeda, I., Kohara, Y., Yamamoto, M., and Sugimoto, A. 2001. Large-scale analysis of gene function in *Caenorhabditis elegans* by high-throughput RNAi. *Curr Biol* 11: 171–176.

36. Marx, V. 2004. RNA quality: defining the good, the bad and the ugly. *Genomics Proteomics* 4: 14–21.

37. McKernan, K.J., McEwan, P.J., and Morris, W. United States Patent #6,534,262-2003.

38. Mishina, Y.M., Wilson, C.J., Bruett, L., et al. 2004. Multiplex GPCR assay in reverse transfection cell microarrays. *J Biomol Screen* 9: 196–207.

39. Montbriand, P.M., and Malone, R.W. 1996. Improved method for the removal of endotoxin from DNA. *J Biotechnol* 44: 43–46.

40. Morrow, J.R. 1996. Hydrolytic cleavage of RNA catalyzed by metal ion complexes. *Met Ions Biol Syst* 33: 561–592.

41. Nabel, G.J., Nabel, E.G., Yang, Z.Y., et al. 1993. Direct gene transfer with DNA-liposome complexes in melanoma: expression, biologic activity, and lack of toxicity in humans. *Proc Natl Acad Sci U S A* 90: 11307–11311.

42. Palumbi, S. 1996. Molecular systematics. In: Moritz, C., Hills, D.M., Moble, B.K., eds. *The Polymerase Chain Reaction*. Waltham, MA: Sinauer Associates.
43. Phillips, J., and Eberwine, J.H. 1996. Antisense RNA amplification: a linear amplification method for analyzing the mRNA population from single living cells. *Methods* 10: 283–288.
44. Sabroe, I., Jones, E.C., Usher, L.R., et al. 2002. Toll-like receptor TLR2 and TLR4 in human peripheral blood granulocytes: a critical role for monocytes in leukocyte lipopolysaccharide responses. *J Immunol* 168: 4701–4710.
45. Sambrook, J., and Russell, D.W. 2001. *Molecular Cloning: A Laboratory Manual*, 3rd ed, Cold Spring Harbor, NY: Cold Spring Harbor Laboratory Press.
46. Schena, M., Heller, R.A., Theriault, T.P., et al. 1999. Microarrays: biotechnology's discovery platform for functional genomics. *Trends Biotechnol* 17: 217–218.
47. Shimazu, R., Akashi, S., Ogata, H., et al. 1999. MD-2, a molecule that confers lipopolysaccharide responsiveness on Toll-like receptor 4. *J Exp Med* 189: 1777–1782.
48. Smith, C., Holmes, D.L., Simpson, D.J., et al. 2001. United States patent, #6,310,199-2001.
49. Soler-Rodriguez, A.M., Zhang, H., Lichenstein, H.S., et al. 2000. Neutrophil activation by bacterial lipoprotein versus lipopolysaccharide: differential requirements for serum and CD14. *J Immunol* 164: 2674–2683.
50. Srimal, S., Surolia, N., Balasubramanian, S., and Surolia, A. 1996. Titration calorimetric studies to elucidate the specificity of the interactions of polymyxin B with lipopolysaccharides and lipid A. *Biochem J* 315: 679–686.
51. Stulnig, T.M., and Amberger, A. 1994. Exposing contaminating phenol in nucleic acid preparations. *Biotechniques* 16: 402–404.
52. Vanhecke, D., and Janitz, M. 2004. High-throughput gene silencing using cell arrays. *Oncogene* 23: 8353–8358.
53. Venter, J.C., Adams, M.D., Myers, E.W., et al. 2001. The sequence of human genome. *Science* 291: 1304–1351.
54. Wang, J., Hu, S.R., Hamilton, K.R., et al. 2003. RNA amplification strategies for cDNA microarray experiments. *Biotechniques* 34: 394–400.
55. Weber, M., Moller, K., Welzeck, M., and Schorr, J. 1995. Short technical reports. Effects of lipopolysaccharide on transfection efficiency in eukaryotic cells. *Biotechniques* 19: 930–940.
56. Wicks, I.P., Howell, M.L., Hancock, T., et al. 1995. Bacterial lipopolysaccharide copurifies with plasmid DNA: implications for animal models and human gene therapy. *Hum Gene Ther* 6: 317–323.
57. Wu, R.Z., Bailey, S.N., and Sabatini, D.M. 2002. Cell-biological applications of transfected-cell microarrays. *Trends Cell Biol* 12: 485–488.
58. Ziauddin, J., and Sabatini, D.M. 2001. Microarrays of cells expressing defined cDNAs. *Nature* 411: 107–110.

# 10 The Sanger Institute High-Throughput Sequencing Pipeline

**Anthony P. West, Christopher M. Clee, and Jane Rogers**
*The Wellcome Trust Sanger Institute,*
*The Wellcome Trust Genome Campus,*
*Cambridgeshire, England, UK*

The Wellcome Trust Sanger Institute (WTSI) was established in 1993 to undertake the study of genomes, through large-scale sequencing and analysis. By February 2005, the WTSI had deposited in public databases over 2.5 billion base pairs of high quality sequence from organisms as diverse as bacteria (1), yeast (2), worms (3, 4) and human (5). Details of individual projects can be found on the WTSI Web site (http://www.sanger.ac.uk). Acceleration of sequencing to meet the goals of the Human Genome Project and the subsequent sustenance of the output was achieved through the introduction of large-scale automation supported by customized software development. This chapter describes the high-throughput approach to DNA sequencing implemented at the WTSI. It is not intended to provide a complete listing of available automation for the PCR and sequencing laboratory; the latter has been the subject of excellent reviews by Meldrum (6, 7).

The main factors driving the automation of large-scale sequencing were cost, space, flexibility, and the requirement for high quality data. At the WTSI, DNA sequencing was initially scaled up over the period from 1993 to 1998 to achieve a throughput of around 250,000 sequence reads per month. Sequence analysis was performed on Applied Biosystems 377[TM] and 373[TM] slab gel DNA sequencing instruments. Over 200 full time technicians were required to produce and load samples for sequence analysis on over 300 gels per day. With the introduction of basic automation of pipetting steps, using TomTec and Robbins dispensers, and filter

***DNA Sequencing II: Optimizing Preparation and Cleanup***
Edited by Jan Kieleczawa
©2006 Jones and Bartlett Publishers

plate technology to prepare DNA sequence templates in plasmid vectors, the throughput rate quadrupled through the years 1999 and 2000 to generate over a million reads per month for analysis on 143 Applied Biosystems 3700™ analyzers. At this point, the sequencing division occupied over 5000 square meters of total building space. By moving away from individual benchtop-based automation, to fewer high-throughput MiniTrak™ automation platforms, and incorporating the latest Applied Biosystems 3730xl™ automated DNA analyzers, space occupancy has been reduced by over 50%. Today, over one million reads per week are produced by fewer than 50 staff members. The quality of DNA sequence is generally high (typically less than 10% of the sequence fails quality control measures, which include rejection for sequencing and cloning vector content), and 96 sequence reads, each containing between 650 and 750 phred Q > 20 (8, 9) bases can be produced per instrument in 90 minutes. With the move to modular automation systems, we have also been able to reduce the use of both plasticware and sequencing reagent consumables. We now operate as one of the most cost-effective sequencing centers in the world. This could not have been achieved without the automation advances implemented over the last ten years.

Factors determining what and how much of a process to automate, are unique to each laboratory setting. At the WTSI, the adoption of automation to undertake processes that were initially carried out manually has evolved to incorporate "off the shelf" solutions as they have become available. This ethos has allowed the growth of a highly flexible system that can be expanded as necessary to minimize and prevent bottlenecks. The adoption of a modular system allows the pipeline modification to incorporate new reagents or materials (called the *phase in processes*) as they are adapted for automation. This provides flexibility for implementing new methodology. The strategy of investing development time on protocol simplification to allow the use of commercially available automation platforms, in preference to developing highly specialized "in-house" systems, was perceived to be relatively cost-effective. It has proved to be fast to implement and easy to support, because service support could be purchased as needed, spare parts are readily available, and multiple units could be purchased.

The main premise for any automation protocol is to keep to a minimum, processes that will leave a robot idle. Thus, non-liquid handling processes that have long incubation times, such as filtration or precipitation steps, are performed off-line. This approach maximizes the batch-processing throughput while keeping to a minimum, the number of automation platforms. An important consequence of implementing process automation that should be considered from the outset is the dependence that develops on suppliers of equipment and materials. Careful negotiation with all potential suppliers should be completed before purchase of equipment or consumables, such as plasticware, to ensure that suppliers can meet the demands of the process as changes in any part of the system can impact on the whole pipeline.

## Materials, Software, and Equipment

All 3730xl DNA analyzers, 3700 analyzers, 9700 thermocyclers, V3.1 BigDye™ terminator reaction mix, 3700/3730 10× running buffer, performance optimized polymers POP7™ and POP5™, and V3.1 BigDye terminator reaction mix 5× dilution buffer were purchased from Applied Biosystems Inc. (Foster City, CA, USA). Innova 4000 Shakers were purchased from New Brunswick Scientific (St. Albans, Hertfordshire, UK). The 384-well Masterblock™ was purchased form Greiner Bio-One (Stonehouse, Gloucestershire, UK). Diamond™ 384-well plates, Diamond EasyPeel™ foil seal, EasyPeel foil seals, ALPS™ 300 plate sealers and 100 µL 384-well storage plates were purchased from Abgene (Epsom, Surrey, UK). Velocity11 platelock sealers and easy peel sealing film were purchased from Velocity11 (Palo Alto, CA, USA). MiniTrak MPD V and VIII were purchased from Perkin-Elmer Life and Analytical Sciences (Monza, Milano, Italy). M.J. Research Tetrad™ thermocyclers, Hard-Shell™ 384 well plates, Microseal™ A and P+ were purchased from Bio-Rad Laboratories Ltd. (Hemel Hempstead, Hertfordshire, UK). The QSelect™ was purchased from Genetix Ltd. (New Milton, Hampshire, UK). Jouan centrifuges were purchased from Jouan, Inc. (Winchester, VA, USA). Eppendorf centrifuges were purchased from Eppendorf AG. (Hamburg, Germany). 384 Multidrop™ dispensers were purchased from Thermo Electron Corporation (Waltham, MA, USA). Circle Grow broth was purchased form QBiogene Inc. (Cambridge, Cambridgeshire, UK). Analytical grade 100% ethanol, analytical grade 100% isopropanol, 4 N sodium hydroxide (NaOH) solution, 30% (w/v) hydrogen peroxide ($H_2O_2$) solution, Decon-90™, sodium dodecyl sulfate (SDS), glucose, potassium acetate (KOAc), sodium acetate (NaAc), glacial acetic acid, and Whatman sequence clean-up blotting paper were purchased from VWR International Ltd. (Poole, Dorset, UK). Agarose and PicoGreen™ were purchased from Invitrogen Ltd. (Paisley, Scotland, UK). MilliQ™ water systems and Montage™ 384 PVDF filter plates were purchased from Millipore Ltd. (Watford, Hertfordshire, UK). Ribonuclease (RNase) A, ethylenediaminetetraacetic (EDTA), TRIZMA™ HCl (Tris), TRIZMA™ Base, TRIZMA™ acetate, magnesium chloride ($MgCl_2$), and Triton® X-100 were purchased from Sigma-Aldrich Company Ltd. (Gillingham, Dorset, UK). Virkon™ was purchased from Ascott Smallholding Supplies Ltd. (Ellesmere, Shropshire, UK). Oligonucleotides were purchased from Sigma Genosys (Haverhill, Cambridgeshire, UK). pUC18 vector was purchased from Cambrex Bio Science Wokingham Ltd. (Wokingham, Berkshire, UK). Foam seals were purchased from StarLab U.K. (Milton Keynes,

Hertfordshire, UK). Nanolid™ seals were obtained from Agencourt Bioscence Corporation (Beverley, MA, USA). Hybaid™ thermalcycler, was loaned by the Thermo Electron Corporation, (Waltham, MA, USA). Nanodrop was loaned by Innovadyne Technologies, Inc. (Chorley, Chesire, UK).

### Software

Phred and Phrap were obtained from Phil Green, Genome Sciences Department, University of Washington (Seattle, Washington, USA). Tomcat was obtained from The Apache Jakarta Project, The Apache Software Foundation (Forest Hill, MD, USA). JBoss was obtained from JBoss, Inc. (Atlanta, GA, USA). Platform™ Load Sharing Facility (LSF) was purchased from Platform Computing, Ltd. (Basingstoke, Berkshire, UK).

## Automation Pipelines

The Sanger Institute DNA sequence-processing pipeline has two main pathways to handle samples in a 384-well format and in a 96-/384-well format (Figure 10-1).

These pipelines are supported by stand-alone automation platforms (Figure 10-2) that run independently of each other. The modularity of the systems, and inclusion of buffers at every stage of the process, minimize down time in the pipelines in the event of system component failures.

Samples for genomic sequencing are generated by sub-cloning fragments of genomic DNA or large insert bacterial clones into plasmid vectors, such as pUC19, and propagation in *E. coli*. The sub-cloned DNA is retrieved either from colonies picked from agar plates using in-house picking robots, or from bacteria stored in glycerol by replication of glycerol storage plates using a QSelect™ robot. Cultures are grown in either a 96-well (1000–1500 μL) or 384-well (160 μL) format overnight, at 37°C, in orbital shaking incubators (Innova) and plasmid DNA is isolated using either in-house 96- or 384-tip pipetting robots or 96- or 384-tip MiniTrak automation platforms. The samples generated are processed for sequencing using 96- or 384-tip MiniTrak automation platforms. PCR products generated for sequencing are also processed with these robots. Sequencing is performed on M.J. Research Tetrad™ thermal cyclers and the resulting products are purified by ethanol precipitation and centrifugation (using Eppendorf 5810R centrifuges and Thermo 384 Multidrop dispensers for centrifugation and fluid dispensing, respectively). Cleaned sequenced products are analyzed using 3730xl or 3700 automated DNA analyzers.

The success of this pipeline is based not only on good automation protocols, but also on regular equipment maintenance. All pieces of equip-

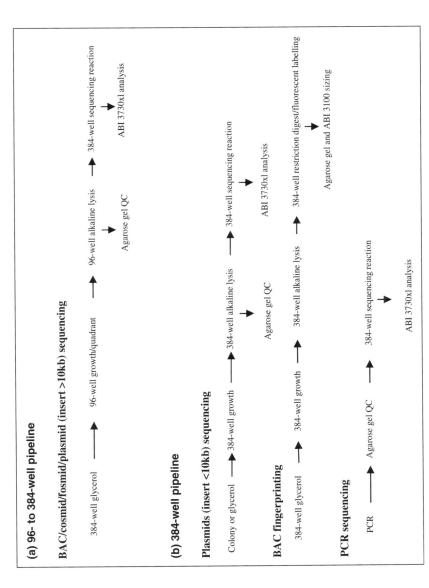

**Figure 10-1.** Sanger Institute high-throughput processing pipelines.

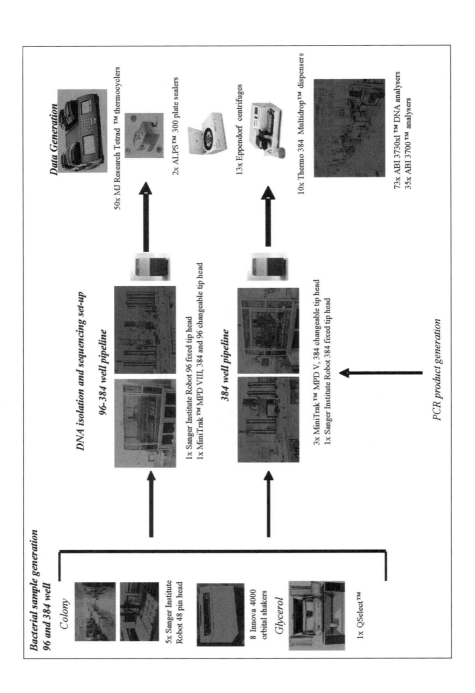

**Bacterial sample generation 96 and 384 well**

*Colony*

5x Sanger Institute Robot 48 pin head

8 Innova 4000 orbital shakers

*Glycerol*

1x QSelect™

**DNA isolation and sequencing set-up**

*96–384 well pipeline*

1x Sanger Institute Robot 96 fixed tip head
1x MiniTrak™ MPD VIII, 384 and 96 changeable tip head

*384 well pipeline*

3x MiniTrak™ MPD V, 384 changeable tip head
1x Sanger Institute Robot 384 fixed tip head

*PCR product generation*

**Data Generation**

50x MJ Research Tetrad ™ thermocyclers

2x ALPS™ 300 plate sealers

13x Eppendorf centrifuges

10x Thermo 384 Multidrop™ dispensers

73x ABI 3730xl ™ DNA analysers
35x ABI 3700 ™ analysers

ment in the Sanger Institute high throughput pipeline (PE MiniTrak automation platforms, Sanger Institute robotic platforms, Thermo 384 Multidrop dispensers, New Brunswick Laboratories Innova Shakers, AB 3730xl and 3700 analyzers, and centrifuges) are cleaned regularly, working parts are inspected, labware definitions are checked for drift, and dispensing heads are calibrated. Maintenance of clean liquid handling systems has been achieved by instituting a weekly cleaning routine consisting of soaking liquid handling parts with 0.5% (v/v) $H_2O_2$ (2h incubation), and 2% (v/v) Decon-90® detergent (15 min incubation), followed by thorough rinsing with MilliQ. Water tubing is regularly replaced to prevent contamination as this proved both quicker and more cost effective.

## Liquid Handling Automation Equipment

Liquid handling systems have to be robust in order to cope with prolonged use, and accurate in their dispensing characteristics. They must also be easy to program and to use. The automated liquid handling systems selected for the Sanger Institute sequencing pipeline exhibit all of these features.

### *Colony Picking Robots*

Automated systems designed and built in-house (Figure 10-3) were introduced in 1998 to provide a multitasking platform for picking and pipetting. The system is based on an XYZ robot with linear drives, chosen for its smooth action and ability to hold calibration. The platform bed and head are enclosed in a cabinet, which is kept under positive pressure by sucking air through dust and high-efficiency particulate air (HEPA) filters

◄ ─────────────────────────────────

**Figure 10-2.** Sanger Institute Automation pipeline equipment. The equipment currently used in the WTSI high-throughput sequencing pipeline. Samples enter the pipeline via colonies picked by in house developed robotic picking platforms or replication of glycerols using the Genetix QSelect (image courtesy Genetix). The samples are grown in Beckman Innova 4000 orbital shakers (image courtesy New Brunswick Scientific). Plasmid/BAC/PAC/fosmids are hand prepped using vacuum manifolds and purified on either Minitrak MPD V (384 tip heads), Minitrak MPD VIII (96- and 384-tip heads), or a Sanger Institute in-house automation platform (96- or 384-tip head), with centrifugation performed in Jouan GR422 (image courtesy of Thermo Electron Corp.). Sequencing reaction setup of purified PCR/plasmid/BAC/PAC/fosmid is performed on Minitrak MPD Vs and VIII. Labeled Sanger fragment generation is performed on M.J. Research Tetrad (image courtesy Bio-Rad) thermocycler with plates sealed by an ABgene ALPS 300s (image courtesy Abgene). Reaction cleanup is performed on Eppendorf 5810R centrifuges and Thermo 384 Multidrop dispensers (image courtesy Thermo Electron Corporation). Sequencing products are analyzed on the ABI 3700 and 3730xl DNA analyzers.

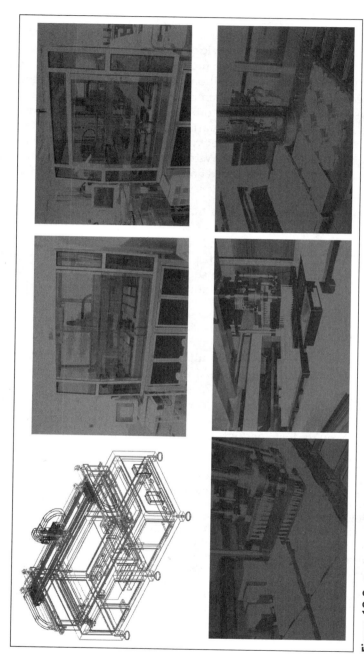

**Figure 10-3. In-house Sanger Institute automation.** The panels show the basic conceptual design of the in-house automation platform, an external view of the 96/48 head platform and 384 head platform enclosed within the cabinet, pipetting head and pin head with imaging.

and exhausting via the cabinet skirting. The picking system utilizes an air-driven, spring return 48-pin head, which was chosen as the best option to maximize throughput while keeping to a minimum any cross contamination. The head is kept sterile by routinely soaking it in disinfectant (Virkon™), followed by brief sonication to remove gross debris, soaking in 70% ethanol, and air drying between each picking cycle. Positional accuracy down to 1 micron is achieved by using optical encoding, and this ensures highly accurate picking, giving in excess of 99% growth following inoculation. The in-house pipetting platform (see section below) utilizes a 96 pin, fixed plastic tip head driven by a drive belt pulley-positive displacement piston configuration, which gives accurate dispensing down to 3 µL. The tips are cleaned using a constant running water bath and sonication system.

The QSelect™ is a recent addition to the automation pipeline at the Sanger Institute. The system has been introduced to improve the quality and efficiency of our library replication service. The QSelect is an ultra-high-throughput system for library replication and re-arraying. The system has capacity for up to 420 × 96-/384-well microplates per day. It has automated microplate stacking cassettes, bar code reading, and automated lid removal/replacement, allowing it to run for many hours while maintaining sample integrity. Rigorous pin sterilization via an ethanol wash bath and halogen pin dryer prevents cross-contamination.

## Liquid Handling Robots

Three years ago, the Sanger Institute developed a 384-well plasmid isolation protocol requiring the use of a high capacity 384-tip liquid handling system. The criteria for this system were that it should have a 384-changeable tip head with high precision dispensing (down to 1 µL with <10% coefficient of variance), no moving arms, fast tip head movement, the capability to continuously feed source and sample plates, and 24-hour callout servicing support. The system that met all these requirements and gave good quality results for the plasmid isolation protocol was the Perkin-Elmer Life and Analytical Sciences (PE) MiniTrak automation platform (Figure 10-4).

The MiniTrak automation platform is a robotic liquid handling system designed for high-throughput microplate handling. To facilitate high speed processing, the platform incorporates a bidirectional, linear conveyor to move plates through its various modules. The system design enables each module to function simultaneously, providing powerful parallel processing capabilities. The MiniTrak system is compatible with all Society for Biomolecular Screening (SBS) standard plate types, including 96-, 384-, 1536-, deep well, and thermal cycler plate formats. The system includes removable high capacity stacker cassettes for increased

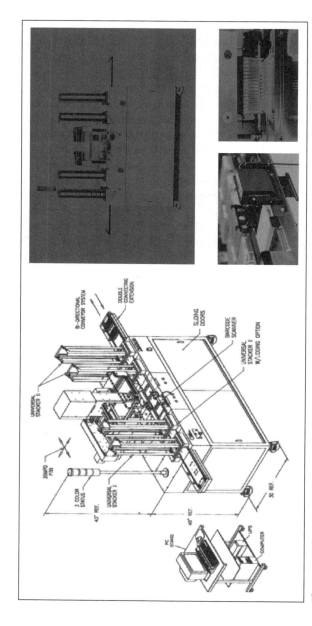

**Figure 10-4.** PE MiniTrak automation platform. The panels show the design of a MiniTrak MPD V with external views of the whole system, pipetting head, and linear track for moving plates. Images courtesy of PerkinElmer Inc.

walk-away time. The patented MultiPosition Dispense (MPD) module accurately positions a 96- or 384-channel disposable tip Dispense Head when aliquoting into 96-, 384-, or 1536-well plates. The system uses a proprietary dispensing method that eliminates tubing, pumps, and valves. Its design also permits movement of the dispense head between the conveyor and a multiposition deck located behind the conveyor. This allows additional reservoirs or cartridges (including chillers and bead or cell mixers) to be accessed, thus permitting reagent addition, tip wash, tip load, and a disposal chute for used tips. A unique tip attachment design ensures a secure seal without the need for O-rings, eliminating leaks and dropped tips.

The Thermo 384 Multidrop dispenser was introduced into the pipeline five years ago, facilitating rapid processing during the ethanol cleanup of sequencing reactions, which is used to remove unincorporated label. The Multidrop dispenser is an automated system designed for high-throughput screening and microvolume dispensing into 384- and 96-well plates. The system can switch between 96- and 384-well plates and dispense up to eight different liquids simultaneously. A high-precision peristaltic pump allows continuous dispensing directly from single or multiple reagent bottles. The dispensing cassette is detachable and can be autoclaved to ensure sterility.

## Protocol Development

As sequencing applications evolve, it is necessary to adapt the high throughput pipelines to accommodate new protocols and products. To date, we have established protocols for processing three main categories of substrate: (1) large insert vectors that include plasmids with inserts >10 kb, BACs, PACs, and fosmids; (2) small insert vectors, which include all plasmids with inserts <10 kb; and (3) PCR amplicons (Figure 10-5).

Standard protocols have been developed for DNA preparation and sequencing of each of the categories of substrate, which undergo continuous optimization. This enables the sequencing pipeline to handle multiple sample types, while limiting the number of protocols required. Wherever possible, a single protocol is employed for all samples entering one of the pipelines. Basic threshold parameters for quality and quantity are set, and must be achieved for the protocol to be used in the pipeline. Parameters have been set at an 80% quality pass (i.e., <20% of samples fail on quality assessment or because they contain cloning vector, sequencing vector, or bacterial host DNA) with an average phred Q20 (bases having a phred quality score of 20) of a 650-base read length. Protocol design for the sequencing pipeline is based on a continuous improvement cycle, whereby the consequences of any change to a step in the process, from

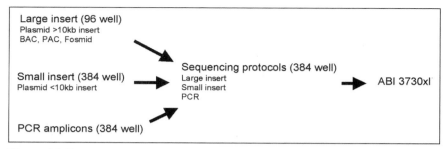

**Figure 10-5.** Protocol development pipelines.

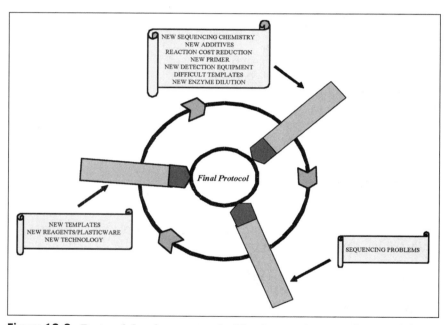

**Figure 10-6.** Protocol development cycle. The three main areas of protocol development are shown with the various parameters that affect these areas. Development areas are represented on a cycling process illustrating that each area is dependent on the previous process. Arrows indicate the direction of data flow with the circle initiated at sample isolation.

sample preparation to sequence output, is checked for effects on each of the subsequent steps until there is a complete revolution through the development circle (Figure 10-6). This ensures that modification of any individual step does not result in the breakdown of the pipeline and that both the quantity and quality of processed data are maintained.

Protocol changes are made in the areas of sample preparation and sequencing reaction setup to accommodate factors such as the introduction of new vectors/templates, different bacterial cell hosts, new reagents, plasticware, and technology. Changes in the template purification protocol, which could alter the quantity or quality of the DNA isolated, will have a "knock on effect" on the sequencing reaction setup, and may require changes to be made to the reaction mix dilution, reaction volume, or the thermal cycling protocol. The sequencing protocols may then require further modification if suboptimal results are obtained, and may even demand further modifications to be made to the sample preparation stage and the cycle to be repeated. The improvement cycle changes continually to accommodate the different sample types that are produced within the Institute.

Protocol development at the Sanger Institute is based on simplifying processes that are not dependent on any specific automation platform. In this respect, as noted above, protocols are under continuous review. Optimized processing or equipment improvements are implemented where possible. Examples of ongoing improvements and modifications are illustrated through developments of the 384-well small insert plasmid pipeline, and optimization of thermal cycling for sequencing.

## Plasmid Isolation

The genome sequencing projects undertaken at the WTSI require the generation of large numbers of plasmid samples containing inserts of between 4 and 6 kb of genomic DNA ($\sim$20 $\times$ 10$^6$/year). This is achieved using a 384-well plasmid isolation protocol in conjunction with the PE MiniTrak systems and in-house–developed robotics. The Sanger Institute differs from most genome centers in that it does not keep glycerol stocks of any of its sub-clones; therefore, for the requirement of "finishing," any purification technique must generate a minimum of 2 μg of DNA. The 384-well DNA isolation protocol (Figure 10-7) is based on alkaline lysis (10, 11) using modified buffers to improve automated handling. The robot adds the lysis components (Glucose, Tris, EDTA [GTE] pH 8 and NaOH, SDS lysis buffer) to the bacterial cells, and mixes the cells and solutions to fracture the cells and release the plasmid DNA. Potassium acetate is added to precipitate cellular debris, proteins, and high molecular weight DNA. The resulting solution is transferred to a 0.6-μm polyvinylidene fluoride (PVDF) filter (Millipore Montage 384), which acts as a clearing plate in retaining cell debris, etc., when the plate is centrifuged. The eluate is collected in a receiver plate containing 100% isopropanol, which precipitates the DNA. The resultant pellets are washed subsequently with 80%

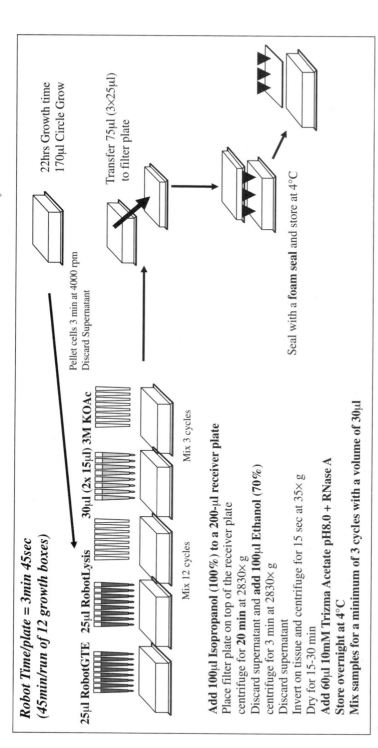

**Figure 10-7. Current 384-well plasmid isolation protocol.** The addition of Robot GTE (0.0225M Tris pH 8.0, 0.01M EDTA, 0.25% glucose, 0.06g/mL [w/v] RNase A), RobotLysis (0.2N NaOH, 0.7% [w/v] SDS, 0.1% [v/v] Triton X-100) and KOAc (3.6M KOAc, 14% [v/v] glacial acetic acid) and subsequent mixing and transfer to the filter plate were performed on a Minitrak MPD V. Bulk dispensing of CircleGrow, isopropanol, 70% (v/v) ethanol, and 10mM TRIZMA Acetate pH 8.0 + RNase A (0.016mg/mL [w/v]) were performed with a modified Thermo deep well 384-Multidrop dispenser. Centrifugation was performed in a Jouan GR422. A Greiner 384-well Masterblock (200 μL) was used for growth incubations and downstream processing. A Millipore Montage 384-well 0.6-μm PVDF filter was used for solution clearing.

ethanol. The whole process for one 384-well box takes 3 minutes 45 seconds of robot time followed by 30 minutes post-robot processing and 15 minutes for drying. Boxes are typically processed in batches of 12, with centrifugation in 3× four-bucket centrifuges. A single 12-plate run, therefore, takes approximately 45 minutes, followed by 1 hour 15 minutes post-robot time, which is equivalent to 1× plasmid DNA plate every 10 minutes for one run and 1× plasmid DNA plate every eight minutes in an 8-hour working day (Figure 10-8a).

We reviewed the protocol in an attempt to reduce automation handling time, and reagent volumes. This will improve batch processing during centrifugation and reduce storage space volume by using reduced volume plasticware. The new protocol is shown in Figure 10-9. By using smaller volumes, the number of movements a robot has to make to pipette a given volume can be reduced. With these modifications in place, processing time is reduced from 3 minutes 45 seconds, to 2 minutes 30 seconds per plate. Further efficiencies are achieved by using 100 µL 384-well receiver plates (ABgene) instead of 200 µL 384-well receiver plates (Greiner), which can be stacked in the centrifuges more efficiently, allowing simultaneous processing of 24× 384-well plates. A single 24-plate run now takes approximately 1 hour followed by 1 hour 15 minutes post-robot time, which is equivalent to 1× plasmid DNA plate every six minutes for one run and 1× plasmid DNA plate every four minutes in an eight hour working day (Figure 10-8b). Typical results derived from both protocols are shown in Figure 10-3. Both protocols produce high quality plasmid DNA with average total DNA concentration of approximately 2 to 3 µg/well (Figure 10-10; color Plate 11).

The sequencing reaction pipeline consists of a run of 12 plasmid DNA plates, which produces 24 sequencing plates (forward and reverse primed). On the MiniTrak platform, it takes approximately 45 minutes to process sequencing reaction run; in an eight hour working day, 10 runs can be performed that can generate 240 sequence plates per day (Figure 10-11). The plates are foil sealed using an ABgene ALPS 300 heat sealer and cycled for 2 hours 35 minutes on M.J. Research Tetrad thermal cyclers. The sequenced products are precipitated using an ethanol precipitation mix and collected by centrifugation, followed by inverted spins at low speed on to Whatman sequencing clean-up blotting paper to remove the supernatant from the 384-well plates (Figure 10-12). Ethanol solutions are dispensed using Thermo 384 Multidrop dispensers. Centrifugation is carried out in Eppendorf 5810R centrifuges, which have a maximum capacity of 12× 384-well plates. This process takes approximately 30 minutes (1× plate every 2.5 min); therefore, 240 plates can be processed with two Eppendorf centrifuges in five hours.

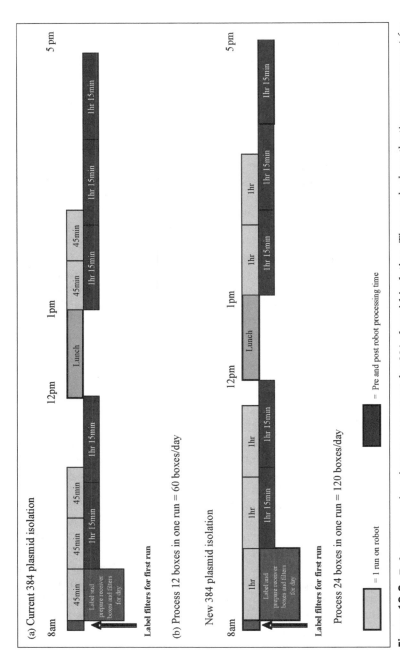

**Figure 10-8. Robot processing time management for 384 plasmid isolation.** The panels show the time management for an eight-hour period for processing the current (a) and new (b) 384 plasmid isolation protocol. The time periods in light blue represent 1 run on the robot with dark blue periods representing pre- and post-robot processing.

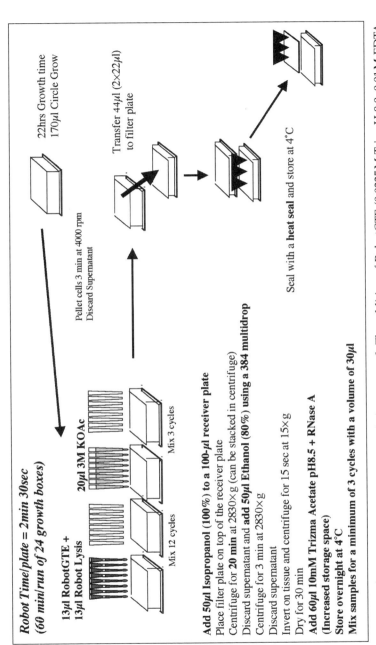

**Figure 10-9. New 384-well plasmid isolation protocol.** The addition of RobotGTE (0.0225 M Tris pH 8.0, 0.01 M EDTA, 0.25% glucose, 0.06 g/mL [w/v] RNase A), RobotLysis (0.2 N NaOH, 0.7% [w/v] SDS, 0.1% [v/v] Triton X-100) and KOAc (3.6 M KOAc, 14% [v/v] glacial acetic acid) and subsequent mixing and transfer to the filter plate were performed on a Minitrak MPD V. Bulk dispensing of CircleGrow, isopropanol, 80% (v/v) ethanol, and 10 mM TRIZMA Acetate pH8.0 + RNase A (0.016 mg/mL [w/v]) were performed with a modified Thermo deep well 384 Multidrop dispenser. Centrifugation was performed in a Jouan GR422. An ABgene EasyPeel was used to heat seal the plate using an ABgene ALPS 300. A Greiner 384 well Masterblock (200 μL) was used for growth incubations and an ABgene 384-well storage box (100 μL) was used for downstream processing. A Millipore Montage 384-well 0.6-μm PVDF filter was used for solution clearing.

185

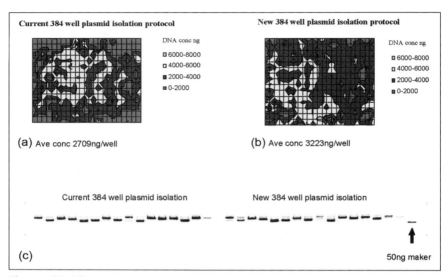

**Figure 10-10.** Comparison of plasmid yield and quality derived from the current and new 384-well plasmid isolation protocol. The comparison of the yield and quality of the plasmid DNA isolated from the 384 well protocols is described in Figures 10-7 and 10-9. Data profiles for the variance in total concentration per well obtained from isolation of the same sample (pUC19 with a known 4 kb insert) in every well are for the current isolation protocol (a; see Plate 11) and new isolation protocol (b). Plasmid concentration was determined using a modified PicoGreen (Invitrogen) assay using pUC18 as the standard. The quality of the plasmids isolated is shown in (c) for the current isolation protocol and new isolation protocol. 3 μL of sample, derived from the isolation of pUC19 with 2 to 4 kb or 4 to 6 kb inserts, was loaded on a 1.2% agarose gel with a pUC18 (50 ng) marker.

## Thermal Cycling

The success of a sequencing reaction and processivity of the pipeline is dependent not only on the reaction, but also the peripheral equipment used to perform the reaction, that is, the thermal cycler, the 384-well plate type and type of plate sealing during thermal cycling. Thermal cycling protocols can be developed on most thermal cyclers for successful results from any given dilution of V3.1 BigDye terminator reaction mix. The length of time these protocols can take, however, may significantly affect the workflow of a high-throughput pipeline. As noted previously, one requirement for a cost-effective, high-throughput sequencing pipeline is that samples are processed as rapidly as possible with the least amount of equipment. This requires a balance with regard to decreasing the sequencing reaction components to reduce costs, without increasing the processing time so that a standard level of performance is maintained.

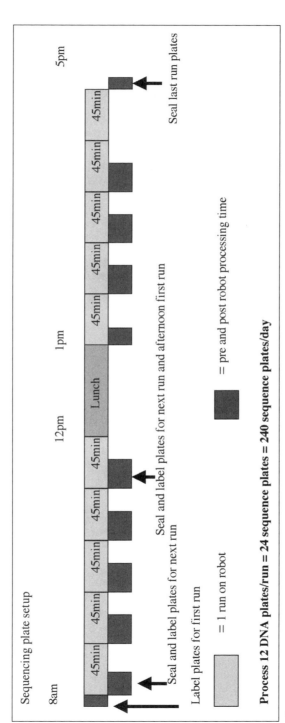

**Figure 10-11. Robot processing time management for sequencing setup.** Time management is for an eight-hour period for processing the sequencing setup protocol. The time periods in light gray represent one run on the robot, with dark gray periods representing pre- and post-robot processing.

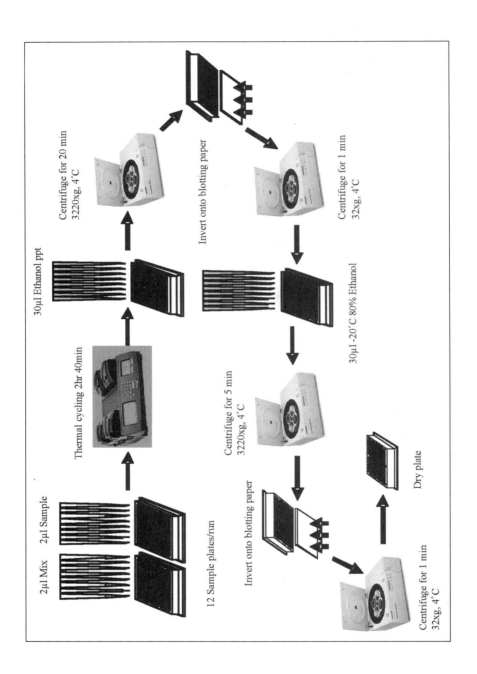

2 μl Mix     2 μl Sample

Thermal cycling 2hr 40min

30μl Ethanol ppt

Centrifuge for 20 min
3220xg, 4°C

Invert onto blotting paper

Centrifuge for 1 min
32xg, 4°C

30μl -20°C 80% Ethanol

Centrifuge for 5 min
3220xg, 4°C

Invert onto blotting paper

Dry plate

Centrifuge for 1 min
32xg, 4°C

12 Sample plates/run

188

Figure 10-13 illustrates the types of variation in cycling performance that result from alteration of cycling parameters, the use of different items of equipment, and varying reagent concentrations for the cycling process. In Figure 10-13a, the performance of three thermal cyclers (M.J. Research Tetrad, ABI 9700™, and Hybaid MBS™) is compared, running thermal cycling protocols for sequencing reactions assembled with two dilutions of sequencing reagents (Table 10-1). The results show significantly better performance characteristics for the M.J. Research and ABI thermal cyclers over the Hybaid thermal cycler in speed and quality of data produced.

Figure 10-13b shows the effect of varying the thermal cycler lid temperatures. V3.1 BigDye terminator mix reactions set up containing $0.0625\,\mu L$ reaction mix in total reaction volumes ranging from $400\,nL$ to $8\,\mu L$. The data show that for small volumes, $<1\,\mu L$, more consistent data quality can be achieved using a lid temperature of $110°C$, and using lid temperatures $<90°C$ for volumes $>1\,\mu L$ give better results.

Thermal cycling performance is dependent on the thermal properties of the plate seal and the efficiency of the seal during thermal cycling. Figure 10-13c shows a comparison of commercially available seals obtained from ABgene (EasyPeel foil heat seal), Agencourt (Nanolid™ plastic hedgehog type seal that inserts into the well to reduce the well volume) and M.J. Research (Microseal™ A, and Microseal P+ rubber pressure seals where the lip of the well cuts into it). The results show significant performance improvements for a 1/64 reaction performed with a lid temperature at $90°C$ when using the Microseal P+ membrane, which has been designed specifically for use with automated thermal cycler loading systems.

The use of different types of 384-well cycling plates, which have different thermal characteristics that are dependent on the plastic polymers used in the mould, can also affect the efficiency of thermal cycling. Figure 10-13d shows the comparison of data obtained from two 384-well thermal

---

**Figure 10-12. Ethanol precipitation sequencing reaction clean-up.** Sequencing reactions are setup on the Minitrak system, heat sealed with ABgene EasyPeel, and cycled using a MJ Research Tetrad thermocycler. Thirty microliters of the ethanol precipitation mix (79% (v/v) ethanol, 0.05 M NaAc) are added to the cycled sequencing reaction mix using Thermo 384 Multidrop dispensers. This is centrifuged at $3220\times g$ for 20 minutes and $4°C$ in Eppendorf 5810R centrifuges. The plates are then inverted on Whatman sequencing clean-up blotting paper and centrifuged for one minute at $32\times g$. Thirty microliters of 80% (v/v) ethanol at $-20°C$ is then added to each well and the plates centrifuged for five minutes at $3220\times g$ and $4°C$. The plates are inverted onto Whatman sequencing clean-up blotting paper and centrifuged for one minute at $32\times g$ and then dried. Thanks to Bio-Rad for use of MJTetrad image, and Eppendorf UK for those of 5180 centrifuges.

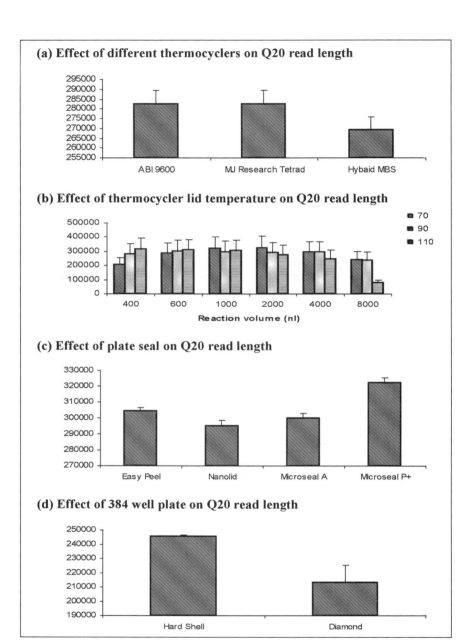

**Table 10-1.** Optimized thermal cycling protocols.

| Thermocycler | Cycles | Temperature (°C) | Time (sec) | Total Time |
|---|---|---|---|---|
| MJ Research | 1 | 96 | 30 | |
| Tetrad | 45 | 92 | 5 | |
| | | 50 | 5 | |
| | | 60 | 120 | |
| | 1 | 10 | forever | 2 h 35 min |
| ABI 9700 | 1 | 96 | 90 | |
| | 45 | 96 | 5 | |
| | | 50 | 10 | |
| | | 60 | 120 | |
| | 1 | 10 | forever | 2 h 50 min |
| Hybaid MBS | 1 | 96 | 90 | |
| | 55 | 96 | 5 | |
| | | 50 | 15 | |
| | | 60 | 120 | |
| | 1 | 10 | forever | 4 h 40 min |
| ABgene | 1 | 97 | 30 | |
| Diamond | 45 | 94 | 5 | |
| | | 48 | 5 | |
| | | 60 | 2 | |
| | 1 | 10 | forever | 2 h 35 min |

The cycling protocols represent the cycling conditions that gave maximum performance for the 1/64 or 1/128 sequencing reactions.

◄───────────────────────────────────

**Figure 10-13.** **Sequencing equipment optimization.** The data show the average phred Q20 bases/384 samples for the (a) effect of different thermocyclers ($n = 3$); (b) the effect of the thermocyler lid temperature ($n = 3$); (c) the effect of the plate seal ($n = 3$); and (d) the effect of the 384 plate ($n = 60$). Unless stated, the data are derived from: (i) 1/128 (0.0625 μL/reaction) V3.1 BigDye terminator mix reaction with 3 pmol primer (TGTAAAACGACGGCCAGT), 2 μL of the same sample at concentrations 2 to 50 ng (a, b) and, unless stated, the volume of the reaction was 4 μL (a); and (ii) 1/64 (0.125 μL/reaction) V3.1 BigDye terminator mix reaction (c, d) with 3 pmol primer (TGTAAAACGACGGCCAGT), varying samples and sample concentration in a total volume of 5 μL (2 μL reaction mix: 3 μL sample). Unless stated, an M.J. Research Tetrad thermocycler was used in conjunction with the M.J. Research HardShell 384-well cycling plate and ABgene EasyPeel foil seal and lid temperature set to 90°C. The ABI 9700 has a lid temperature of 103°C and the Hybaid MBS has a lid temperature of 110°C. The cycling protocols used are listed in Table 10-1. For all reactions, the M.J. Research Tetrad thermocycler cycling protocol was used except for the ABI 9700 (a; Table 10-1), Hybaid MBS (a; Table 10-1), and ABgene Diamond 384-well plate (d; Table 10-1). The sample was either pUC19 with a known 4 kb insert (a–c) or pUC19 with unknown inserts of 2 kb (d) ($n = 1 = 384$ samples = 1 plate; i.e., $n = 3 = [3 \times 384]$). Dispensing of volumes below 1 μL was performed using the Nanodrop dispenser from Innovadyne Technologies.

cycling plates designed for use in automated systems. Both plates were sealed with a foil heat seal (ABgene EasyPeel [HardShell] and Diamond EasyPeel) and the thermal cycling performed in an M.J. Research Tetrad™ thermal cycler. Slightly different thermal cycling protocol temperatures had to be used for the two types of plate to accommodate their different thermal properties (Table 10-1), but the results show that the HardShell 384-well plate in an M.J. Research Tetrad thermal cycler significantly outperformed the Diamond 384-well plate. Overall, these results illustrate the importance of careful evaluation and optimization of all of the components of a process in order to develop robust protocols that will deliver consistent, high quality data.

## Labeled Sample Detection

The central sequencing facility has a total of 73 of the 3730xl and 35 of the 3700 DNA analyzers, which run 24 hours per day, seven days per week. The facility is operated using a twin shift system, with staff working four twelve-hour days on and four days off. This means that an instrument failure is always picked up within twelve hours. Having the sequencers localized centrally has enabled the work force to become highly specialized in dealing with day-to-day issues and troubleshooting problems. The facility is also supported by on-site Applied Biosystems engineers and together they achieve a greater than 98% uptime on all of our equipment, including the older 3700 analyzers.

The instruments are set up to operate in one of two modes: to maximize throughput for PCR amplicons and inserts of less than 500 bp; or to maximize read length for genome/clone shotgun sequence assembly. The internal LIMS registers the type of sample being analyzed, based on its bar code, and supplies the run parameters accordingly (Figure 10-15). Samples are then separated and analyzed automatically using the appropriate run and analysis modules. Most shotgun sequence samples are processed on 50 cm arrays on either 3700 or 3730xl DNA analyzer instruments, to deliver average read-lengths in excess of 800 phred Q20 bases. Samples shorter than 500 bp are analyzed on 36 cm arrays using modules that can run about 40 runs per day.

Dye-labeled sequenced samples are normally held dehydrated at −20°C until they are ready to load. We have also found, however, that samples can be held in the dark at room temperature for several weeks. For loading, the samples are re-suspended in 0.1 mM EDTA made up with MilliQ water added using Thermo 384 Multidrop peristaltic dispensers. This solution was implemented to replace a 12% 2-pyrrolidinone–based loading solution used with V2.0 BigDye terminator mix, because it was found that the C and T dyes in the V3.0 BigDye terminator mix were susceptible to breakdown. Formamide is not a viable loading solution because it appears to induce sample breakdown (significant loss in signal over a 24-h period; data not shown) and it is also highly corrosive to the head of the Multidrop dispenser.

The run modules operated on 3730xl 36-cm arrays are based on the standard RapidSeq36_POP7 and StdSeq36_POP7 modules, with additional changes to allow the RapidSeq36 module to process 48 runs per day for shorter amplicons. We have modified the LongSeq50_POP7 module for our 50-cm array instruments. This allows us to process 15 runs per day with minimal loss of read length (Table 10-2). Sample sheets are delivered directly to the AB collection software from an Oracle database (described below).

The 3700 analyzers use POP5 polymer. We do not perform the recommended regeneration of arrays, but monitor data quality closely and regenerate if the data show signs of early loss of resolution or multiple blocked capillaries. Most of our 3700 arrays are used to collect data for over 10,000 runs before they are replaced. The main 3700 analyzer operational run module is based on the Seq2POP5DefaultModule (Table 10-2). By setting the sheath flow period to 1884 we can achieve a major saving on POP5 polymer and hence sample cost. Recent testing (data not shown) has demonstrated that the same sheath flow settings may be applied when using POP6 polymer. This has the added benefit, that signal intensity is increased significantly by the consequent compression of the sample plug entering the sheath flow. The 3700 analyzers are set up to achieve seven to eight runs per day with an average of 60 kb of phred Q20 bases per 96 lane run. The CapMinder tracking program (12) looks for instruments producing less than 30,000 phred Q20 bases over more than four runs, and flags failing instruments in a daily report. As left-to-right signal variation across the fluorescence cuvette on the 3700 analyzers is known to interfere with signal detection, we routinely run the Applied Biosystems signal variation assay on new arrays and run most of our instruments with individual cuvette temperature settings.

When sequenced, samples are loaded on to the 3700 analyzers in 96-well or 384-well microtiter plates, a tab-delimited file is downloaded to the sequencer personal computer (PC) using a bar code system, and then imported into the sequence collection software manually. The "Samplesheet-generator," which creates the tab-delimited file, adds on a pre-fix that tells the operator how to position the plates on the 3700 analyzer in order to assure the correct association of sequence data with samples.

## Sample Tracking

To be able to process over 100,000 sequenced samples per day and link the data to the multiple projects, it was essential to develop a sample

**Table 10-2.** Run modules in use on 3700 and 3730XL sequencers.

| Instrument Type | Array Length | Polymer Type | Rrun Time (sec) | Run Voltage (KV) | Data Delay (sec) | Sheat-flow Period | Sheat-flow Volume | Runs per Day | Q20 Bases per Day | Average Q20 readlength |
|---|---|---|---|---|---|---|---|---|---|---|
| 3700 | 50cM | POP5 | 7900 | 5700 | 300 | 1884 | 6000 | 8+ | $0.58 \times 10^6$ | 750bp |
| 3730 XL | 36cM | POP7 | 1060 | 13.2 | 80 | N/A | N/A | 40+ | $1.92 \times 10^6$ | 500bp |
| 3730 XL | 36cM | POP7 | 2450 | 8.5 | 120 | N/A | N/A | 24+ | $1.73 \times 10^6$ | 750bp |
| 3730 XL | 50cM | POP7 | 4500 | 9.6 | 300 | N/A | N/A | 15+ | $1.22 \times 10^6$ | 850bp |

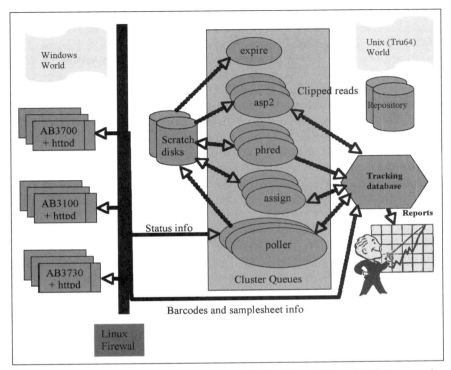

**Figure 10-14. A schematic of the CapMinder.** The relationship between the automated sequencers, the polling system, and Sanger Institute processing software are illustrated. Samples post-upload are checked for "ownership" (assignment), and "base called" using phred to estimate quality. They are then passed onto their project directory by Automated Sequence Preprocessing (ASP2) (13) software, ready for assembly using either phrap or phusion (14), before being passed on to a "Finisher" for human intervention. CapMinder is capable of tracking all of the interactions and generating reports that detail areas of potential concern to the sequencing facility managers, who check for both sample and instrument problems.

tracking process based on bar codes that can track samples and also be used to integrate data and generate project status reports through every stage of the sequencing pipeline. Bar code information is entered to the central database via graphical user interfaces (GUI). A polling program links the tracking database to a suite of custom applications (CapMinder), facilitating automatic transfer and processing of base-called sequence traces that are associated with specific samples.

CapMinder (Figure 10-14) was originally developed as a suite of Perl-based programs for integrating and managing a sequencing facility comprising multiple DNA sequencers and processing the data they produce.

It was designed to be modular, accommodating a variety of instrument models and analysis programs, scalable when more instruments or higher throughput are required, and extensible when new categories of data must be processed. By installing a small, dedicated Web server on each instrument, it is possible to work around incompatibilities between hardware platforms, operating systems, and programming languages. The instruments are polled every 10 minutes and their status recorded; if they aren't busy, they are queried for a list of unsent data and an attempt is made to upload the oldest set of sequence reads. The pollers can be multiplexed and automatically selected from non-overlapping subsets of instruments to manage. Dialogues between the poller and the Web servers are conducted using standard hyptertext transfer protocols (HTTPs). Results are transferred by creating either a temporary Zip file, in the case of the AB3700, or a "zipstream," in the case of the AB3730 instruments.

The Unified Data Collection (UDC) software, part of the new AB 3730xl DNA Analyzer, lends itself to tight integration into CapMinder (Figure 10-15; color Plate 10). Java Messaging and a Request/Response API make it easy to tap into the information flow inside the instrument. Installing a TomCat servlet container within the JBoss container manager bridges the Windows and Unix environments. This allows an instrument's state to be determined remotely, sample sheets to be fetched from a central database and loaded into UDC, and output data to be reliably uploaded as soon as it becomes available. UDC, and its Application Program Interface (API), has proved to be a powerful integration tool for the 3730xl DNA Analyzer. By collecting and processing raw data within minutes of its production, problems can be quickly detected and fed back to sequencing facility managers (15).

Uploaded run data are searched for embedded tracking database IDs (put there when the sample sheet was made from the plate bar code with the companion generator program). The samples are then re-named and the tracking database updated. All processing takes place on Platform LSF batch queues, which are customized to allow parallel processing without overloading the dedicated Unix server cluster. The initial assignment and re-naming task is invoked by the poller, but thereafter each run's route through the processing pipeline is controlled by a strategy (based on the analysis required) defined in the tracking database by the owners for the category of sequencing project (e.g., Shotgun/EST, small insert; BAC/PAC/fosmids, large insert, ExoSeq, PCR). Each stage invokes the next when it completes successfully, and terminates when the data have been delivered to their final destination or a non-recoverable error was encountered.

## Conclusion

This chapter is an overview of how the Sanger Institute approaches the implementation of automated high-throughput sequencing, from the

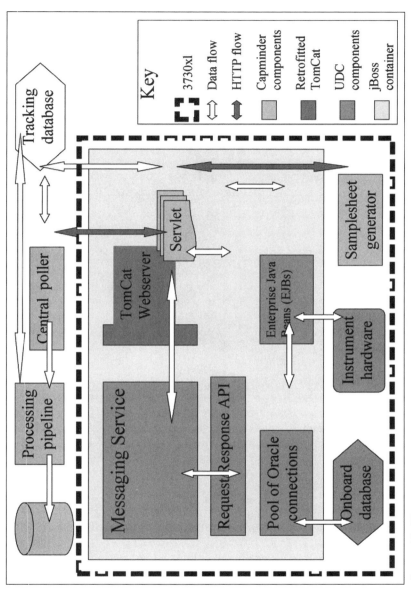

**Figure 10-15. Schematic showing a simplified ABI 3730xl UDC and CapMinder integration.** The relationships between the Sanger Institute database sample tracking software, and the driver software located on the ABI 3730xl sequencers are shown. Note that all interactions are Web based, utilizing our retrofitted copy of JBoss. Intermediary interactions (either poller-based or human-based) are shown by the black arrows, and denote "initiated" queries made between the user and the instrument. All other interactions are shown by the white arrows. (See Plate 10.)

197

development of laboratory protocols, selection of automation platforms and optimization of ancillary equipment performance, to the tracking of data. The system has been built to generate excellent data and to respond rapidly to changing scientific needs. The Sanger Institute will always maintain a balance between automation implementation and human intervention to keep it at the forefront of technology and science.

*Acknowledgments*
We thank Ms. Rachael Ainscough, Sequencing Operations Manager, for constructive comments on the manuscript, and Ms. Jane Garnett for help with the in-house–developed picking system. Thanks to John Attwood for all of his efforts in the creation, maintenance, and explanation of the Cap-minder suite of programs.

# References

1. Cole, S.T., Brosch, R., Parkhill, J., et al. 1998. Deciphering the biology of *Mycobacterium* tuberculosis from the complete genome sequence. *Nature* 393: 537–544.
2. Wood, V., Gwilliam, L.R., Rajandream, V.L., et al. 2002. The genome sequence of *Schizosaccharomyces pombe Nature* 415: 871–880.
3. The *C. elegans* Sequencing Consortium. 1998. *C. elegans*: sequence to biology. *Science* 282: 2012–2018.
4. Stein, D.L., Bao, Z., Blasiar, D., et al. 2003. The genome sequence of *Caenorhabditis briggsae:* a platform for comparative genomics. *PLoS Biol* 1: 166–192.
5. International Human Genome Sequencing Consortium. 2001. Initial sequencing and analysis of the human genome. *Nature* 409: 860–921.
6. Meldrum, D. 2000. Automation for genomics, part one: preparation for sequencing. *Genome Res* 10: 1081–1092.
7. Meldrum, D. 2000. Automation for genomics, part two: sequencers, microarrays, and future trends. *Genome Res* 10: 1288–1303.
8. Ewing, B., Hillier, L., Wendl, M., and Green, P. 1998. Basecalling of automated sequencer traces using Phred. I. Accuracy assessment. *Genome Res* 8: 175–185.
9. Ewing, B., and Green, P. 1998. Basecalling of automated sequencer traces using Phred. II. Error probabilities. *Genome Res* 8: 186–194.
10. Birnboim, H.C., and Doly, J. 1979. A rapid alkaline extraction procedure for screening recombinant plasmid DNA. *Nucleic Acid Res* 7: 1513–1523.
11. Ish-Horowicz, D., and Burke, J.F. 1981. Rapid and efficient cosmid cloning. *Nucleic Acid Res* 913: 2989–2998.
12. Clee, C.M., Attwood, J.A., Ainscough, R., et al. 2001. Monitoring high throughput sequencing at the Sanger Centre: the capmanager approach. Poster presentation. Marco Island, FL: *Advances in Genome Biology and Technology.*

13. Wendl, M.C., Dear, S., Hodgson, D., and Hillier, L. 1998. Automated sequence preprocessing in a large-scale sequencing environment. *Genome Res* 8: 975–984.

14. Mullikin, J.C., and Ning, Z. 2003. The Phusion Assembler. *Genome Res* 13: 81–90.

15. Attwood, J.A., Clee, C.M., Ainscough, R.A., and Mullikin, J.C. 2002. Integrating the Applied Biosystems 3730xl DNA Analyzer into the Sanger Institute's automated processing pipeline. Poster presentation. Boston: *Genome Sequencing and Annotation Conference 14.*

# Comparison of PCR Product Clean-Up Methods for High-Throughput Capillary Electrophoresis

**11**

## Darryl L. Irwin[1,2] and Keith R. Mitchelson[1]
[1]*Institute for Molecular Biosciences and*
[2]*Australian Genome Research Facility, The*
*University of Queensland, Brisbane, Australia*

The evolution from slab- to fluorescence-based capillary electrophoresis has revolutionized automated DNA fragment analysis, which has particular importance in high throughput diagnostics and completion of complex genotyping projects, reducing electrophoresis time, eliminating slab gel casting, and sample loading (7). However, analysis of polymerase chain reaction (PCR) products by capillary electrophoresis is often compromised by the presence of a high salt concentration. Typical PCR reactions contain KCl and $MgCl_2$, which cause variable electrokinetic injections and migration times because of chloride ions competing with the DNA for migration into the capillary (3, 8). In addition, the PCR mixtures contain unincorporated deoxyribonucleotide triphosphates (dNTPs), primers, and primer dimers, which also compete with the injection of the desired DNA fragments. These factors significantly reduce the amount of PCR product injected into the capillary and thus result in much lower than optimal signal strength. This is particularly problematic when the initial signal strength is low. Another variation seen in capillary electrokinetic injection is a bias toward the injection of smaller fragments in the presence of higher salt concentrations.

One method used to overcome these difficulties is modifying injection parameters. One such method of injection, termed *pressure injection*, eliminates the problems induced by the presence of salt; however, it can produce significant band broadening, which reduces resolution (2).

***DNA Sequencing II: Optimizing Preparation and Cleanup***
Edited by Jan Kieleczawa
©2006 Jones and Bartlett Publishers

Alternatively, the DNA concentration can be enriched before injection. In 1996, Devaney and Marino (2) reported using spin filters and membrane filters (both single or 96-well cartridges) to remove salts prior to capillary electrophoresis. These methods were effective and reproducible but lacked the capacity to be automated, which limited any high throughput capabilities. A number of other post-PCR clean-up techniques have also been suggested.

This chapter compares and contrasts five different clean-up methods that are routinely used prior to capillary electrophoresis. Multiplex PCR reactions are commonly used in our laboratory as this technique simultaneously analyzes multiple short tandem repeat (STR) alleles, determines dosage levels, and excludes contamination within a single reaction (4). The use of multiplex PCR confers several advantages when comparing clean-up methods. Multiplex reactions have a wide allele range allowing analysis of competition between large size and small sized products. The allele drop-out rate can be readily determined, and the various fluorophores can be compared for effects on injection efficiency and mobility. To the best of our knowledge, no formal comparison of these clean-up methods has been undertaken. Each clean-up method was analyzed for failed sizing fragment injection, size standard peak heights, total and target peak heights, primer removal, preferential injection of small fragment, and the number of STR alleles analyzed. These methods were also compared for automation versus manual processing, and overall cost. Further, conclusions are drawn as to the applicability of each method to single cell multiplexing where the PCR product is minimized.

## Materials and Methods

Multiplex fluorescent PCR of a number of well documented, highly conserved loci and STR regions was performed on extracted DNA from venous blood samples from 96 individuals. The highly conserved sex chromosome loci were AMEL and DYs14, and the STRs used were: D21S11, D13S631, D13S258, D18S51, D18S851, D13S317, and D18S391 (chromosome of origin being indicated by the number after the D). The multiplex alleles were balanced in terms of amplification efficiency and fluorescence output. Blood samples were collected in lithium/heparin anticoagulant from women aged 18 to 45 years. DNA was extracted using QIAamp DNA Blood Mini Kit (Qiagen Inc., Germantown, MD, USA) using the manufacturer's protocols. Multiplex reactions were carried out in 75 μL reaction volume resulting in sufficient product to simultaneously compare all clean-up methods. Each reaction contained 5.0 μL extracted DNA, forward and reverse primers, 1× PCR buffer with 1.5 mM $MgCl_2$ (Applied Biosystems, Inc., Foster City, CA, USA), 1.25 mM each dNTP

(Invitrogen, Melbourne, Australia) and 1.6 units Amplitaq Gold DNA polymerase (Applied Biosystems). PCR conditions were 94°C/15 minutes Hot Start activation followed by 39 cycles of 94°C/30 second denaturation, 59°C/45 second annealing, and 72°C/1 minute extension with a 72°C/10 minute final extension.

Following the cleanup, all samples were run through capillary electrophoresis. So that the samples could be compared adequately, a sizing standard labeled with the fluorescent dye Rox (purchased from GE Healthcare, formerly Amersham Biosciences, Piscataway, NJ, USA) was run in all wells. Each highly conserved loci tested as well as each STR loci was labeled with one of three fluorescent dyes (fluorophores). These dyes are 6-FAM, TET, and HEX (Qiagen, Inc.). Labeling with these fluorophores allows differentiation between samples when the fragments created are in the same size range; that is, where fragments from two different loci overlap on the electropherogram, the loci concerned are labeled with different fluorophores.

## Product Cleanup

The PCR products were processed using five various manual and automated methods. These methods are dilution, sephadex cleanup, magnetic bead cleanup, ammonium acetate/ethanol cleanup, and $MgSO_4$/ethanol cleanup. Common reagents were of the highest available purity (Sigma-Aldrich, St. Louis, MO, USA) unless otherwise stated.

### Dilution (Diln) (3)

Dilute 2 µL PCR product with 8 µL autoclaved filtered water, vortex, and spin down.

### Sephadex Cleanup (Seph) (1)

1. Following the manufacturer's protocols, use the Multiscreen Column Loader (Millipore, Billerica, MA, USA) to add the required amount of Sephadex G-75 Superfine (Sigma-Aldrich) to each well of a Millipore MAGV N22 filter plate (Millipore).
2. Add 300 µL of 0.2 µm filtered distilled water to each well of filter plate.
3. Leave the plate at room temperature for at least three hours to allow the Sephadex to hydrate.
4. Place an empty round-bottom microtiter plate (catalogue number 267245; Nalgene-Nunc International, New York, NY, USA) under

the filter plate, spin at $510\times g$ for 5 minutes in a plate centrifuge (Sigma, Osterode am Harz, Germany).
5. Place a new round-bottom microtiter plate under the filter plate and load 20 µL PCR product to the center of the Sephadex wells.
6. Spin in a Sigma 4K15 plate centrifuge at $510\times g$ for 5 minutes (as in point 4).
7. The cleaned sample is contained within the filtrate.

Comparisons were also made using a commercial 96-well Sephadex plate (CS; GE Healthcare) (1) using steps 4 to 7 of Sephadex clean-up method above.

### Magnetic Bead Cleanup (MagB) (6)

1. Wash 15 µL of magnetic beads (Magnasil Paramagnetic Particles [PMP]; Promega, Annandale, Australia) three times with 100 µL Milli-Q Water. This is achieved using the Paramagnetic Particle Concentrator (PMPC; Promega,) and the CRS Robotics system (CRS Robotics Corporation, Burlington, Canada) to fix the beads in place and the Hydra 96 (Matrix Technologies Corp., Hudson, NH, USA) to add and remove wash.
2. Add 40 µL of 2× binding buffer (20% polyethylene glycol [PEG] 8000, 2.5 M NaCl) to the beads.
3. Transfer 20 µL of PCR product to the bead/buffer mix, gently mix, and leave at room temperature for five minutes.
4. Concentrate the PMP-DNA complex on the PMPC for 30 seconds and aspirate the supernatant.
5. Wash PMP-DNA three times with 50 µL of 80% ethanol using Hydra/PMPC to remove residual salt and PEG.
6. Dry samples at 37°C for 15 minutes.
7. Resuspend samples in 10 µL formamide loading buffer (GE Healthcare), vortex, and pulse spin to collect clean the sample in the bottom of the well.
8. Incubate at 90°C for five minutes then place immediately on ice.

### Ammonium Acetate/Ethanol Cleanup (Aa/Eth) (1)

1. Add 2 µL of 7.5 M ammonium acetate and 60 µL 100% ethanol to 12 or 25 µL PCR product.
2. Vortex and then spin $2,862\times g$ for 20 minutes.
3. Blot on paper towel then spin plate upside down on paper towel $45\times g$ for one minute.
4. Add 100 µL of 70% ethanol.
5. Vortex and then spin $2,862\times g$ for 15 minutes.

6. Blot on paper towel and then spin the plate upside down on a paper towel 45× $g$ for one minute.
7. Add 20 µL water, vortex, and pulse spin to collect the clean sample in the bottom of the well.

### MgSO₄/Ethanol Cleanup (Mg/Eth)

1. Allow the PCR plate to come to room temperature.
2. Add 75 µL 0.2 mM MgSO₄ (BDH, Dorset, UK) in 70% ethanol (prepared in-house) to 12 or 25 µL PCR product at room temperature.
3. Mix thoroughly by vortexing.
4. Stand at room temperature for 15 minutes.
5. Spin in a plate centrifuge at room temperature for 15 minutes at 2,318× $g$.
6. Blot on a paper towel and then spin plate upside down on paper towel 179× $g$ for one minute.
7. Air dry on bench for five minutes.
8. Add 20 µL of water to resuspend the DNA, vortex and pulse spin to collect clean sample in the bottom of the well.

In all cases, post–clean-up processing involved adding 2 µL of the cleaned-up product to 3 µL loading buffer (GE Healthcare). Samples are then heated to 90°C for 60 seconds and then are placed immediately on ice. Analysis was completed using the Megabace 1000 capillary electrophoresis system with Genetic Profiler Version 1.5 software (GE Healthcare). Injection parameters were –3 kV for 45 seconds and run parameters were –10 kV for 75 minutes at 44°C.

## Results

Two representative electrophoretic profiles of multiplexes from two samples resulting from capillary electrophoresis are shown in Figure 11-1. These profiles demonstrate the wide range of alleles, various fluorophores used, and peak differences between individuals. Evaluation of clean-up methods was performed by comparing different color and base-pair–sized peak heights as well as total product availability. These values can be directly compared between clean-up methods because all samples originated from the same PCR (indicated as a single line in Figure 11-2), thus eliminating variation related to the PCR itself. Comparisons cannot be made directly between multiplex reactions (different lines in Figure 11-2), as they originate from different reactions; however, trends can be determined between different plates confirming reproducible results.

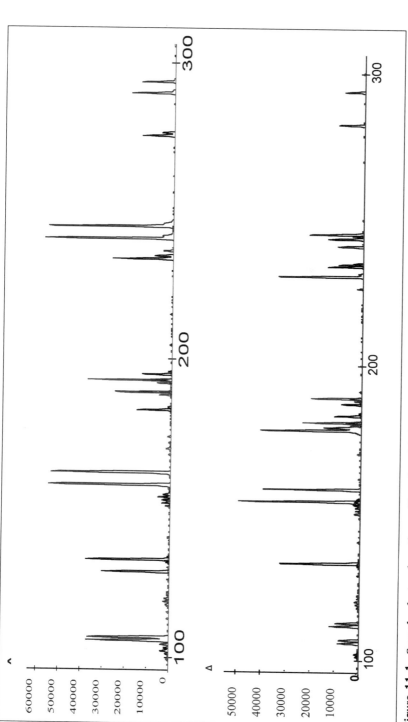

**Figure 11-1.** Sample electrophoretic profiles of multiplex of two individuals after capillary electrophoresis.

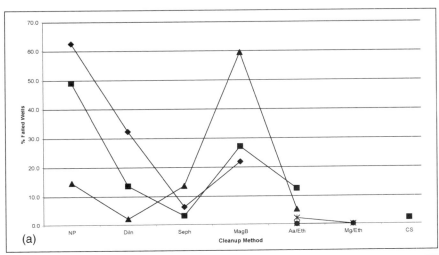

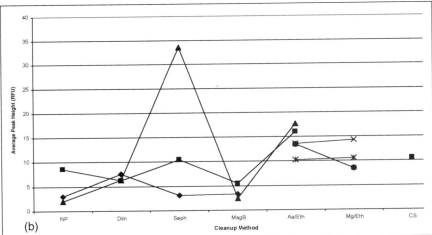

**Figure 11-2. Comparison of clean-up methods.** Evaluation of clean-up methods was performed by comparing different color and base-pair sized peak heights as well as total product availability. Multiplex PCR reaction product was split after amplification and used to evaluate a variety of clean-up protocols. These values can be directly compared between clean-up methods when samples originated from the same PCR (indicated as a single line and common plot point shapes in the panels), thus eliminating variation related to the PCR itself. Comparisons cannot be made directly between multiplex reactions (different lines in the panels) because they originate from different reactions; however, trends can be determined between different plates confirming reproducible results. Comparisons were made using the following parameters:

(a)  Failed sizing fragment injection. This compares the number of failed wells (% of total wells) because of an insufficient Rox sizing fragment injection.

*(continues)*

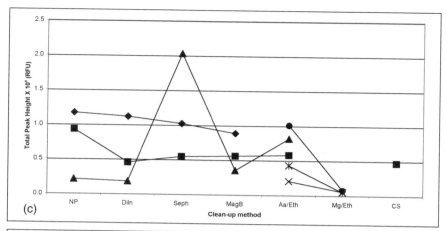

(c)

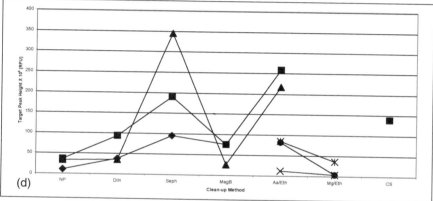

(d)

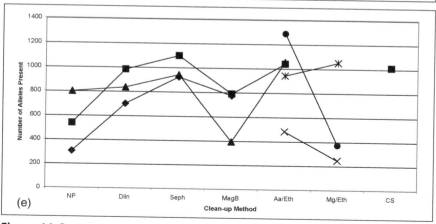

(e)

**Figure 11-2.** *Continued*
(b) Average Rox peak height. These data compare the average height (relative fluorescence units; RFU) of Rox peaks injected. A higher number indicates a better injection.

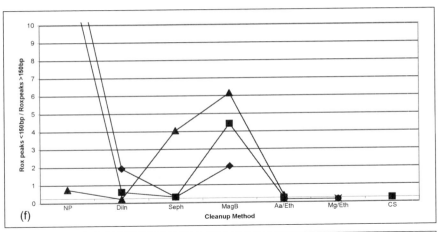

(f)

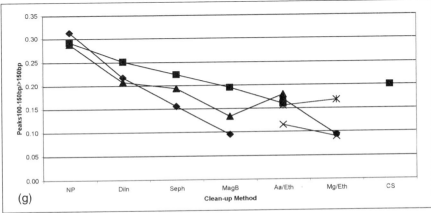

(g)

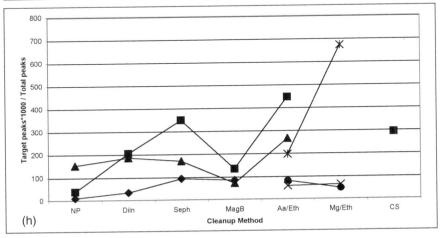

(h)

**Figure 11-2.** *Continued*
(c) Sum of the total peak heights. This compares the sum of the total peak heights (RFU) injected. A higher number indicates better injection.
(d) Sum of target peak heights. The sum of the target allele peak heights (RFU) injected is compared among the methods studied. A higher number indicates a better injection.

*(continues)*

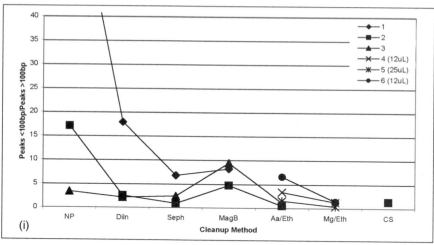

**Figure 11-2.** *Continued*
(e)   STR alleles detected. This compares the number of alleles detected. A higher number indicates a better injection.
(f)   Preferential injection of small Rox fragment. The preferential injection of small (<150 bp) Rox fragments is compared using the ratio of total Rox peak heights (RFU) <150 bp/total Rox peak heights (RFU) >150 bp. A lower number indicates a less preferential injection of smaller Rox fragments.
(g)   Preferential injection of small fragment. This compares the preferential injection of small (<150 bp) fragments using the ratio of total peak heights (RFU) <150 bp/total peak heights (RFU) >150 bp. A lower number indicates less preferential injection of smaller fragments.
(h)   Background peaks. This compares the level of background fluorescence using the ratio of sum target allele peaks (RFU) ×1000/sum total peaks (RFU). A lower number indicates lower background fluorescence.
(i)   Primer removal. The amount of primer removed by clean-up is compared using the ratio of sum primer peaks <100 bp (RFU)/Sum of peaks >100 bp (RFU). A lower number indicates a better primer removal.

To determine the level of salts present after clean-up, the numbers of failed wells were also compared (Figure 11-2a). This graph shows that a high level of failed sizing-fragment injection occurred with pure undiluted product and when the magnetic bead clean-up method was used. On the other hand, low levels of failed wells occurred with the ammonium acetate/ethanol, Sephadex, and magnesium sulphate/ethanol clean-up methods, while injection of diluted product produced a variable number of failed wells between plates.

The amount of Rox contained within the injection mixture should be standard between all plates and clean-up methods, because Rox is contained within the loading buffer. To determine the overall injection efficiency, the total heights of all Rox peaks injected from each plate were compared (Figure 11-2b). This graph shows that a consistently high level of Rox injection was achieved using the ammonium acetate/ethanol and magnesium sulphate/ethanol clean-up methods.

Figure 11-2c compares the sum of the total peak heights injected from each plate. The total peak heights were consistent across all clean-up methods except magnesium acetate/ethanol, which consistently showed reduced recovery. The sum of the total peak heights of the target STR peaks from each plate is shown in Figure 11-2d. This graph shows that low levels of target peaks occurred when injecting pure undiluted product and product that was cleaned with the magnetic bead or magnesium acetate/ethanol methods. A slight increase in total peak height occurred when the product was diluted, and the highest levels of target peaks were seen with ammonium acetate/ethanol and Sephadex clean-up methods. Comparison of the total number of alleles analyzed on each plate to determine the level of allele dropout, due to injection efficiency, are shown in Figure 11-2e. The number of alleles detected is lowest when injecting pure undiluted product, using the magnetic bead and magnesium sulfate/ethanol (12 µL initial volume) clean-up methods. There is a slight increase in the number of alleles detected when diluted product is injected and the highest numbers of alleles per plate were observed when using ammonium acetate/ethanol, Sephadex, and magnesium sulfate/ethanol (25 µL initial volume) clean-up methods.

To determine preferential injection of smaller peaks, Rox peak heights 100 to 150 bp were compared to those >150 bp (Figure 11-2f). The expected result of 0.25 (5/20) was determined due to the known peak sizes of the sizing standard contained in the loading buffer. A high preference (indicated by a higher number) toward injection of small fragments occurred when injecting pure undiluted product and when product from the magnetic bead cleanup was injected. This preference was reduced when injecting diluted product and the expected value of 0.25 was achieved with ammonium acetate/ethanol, Sephadex, and magnesium sulfate/ethanol clean-up methods. To further explore preferential injection, total peak heights 100 to 150 bp were compared to those >150 bp (Figure 11-2g). This graph shows a high preference toward smaller peak sizes when injecting pure undiluted product. When this product is diluted or cleaned using the Sephadex method, a reduction in this preference is observed. All other clean-up methods showed an overall low preference toward the injection of smaller-sized fragments with some variation.

To determine the levels of background and artefact peaks present after cleanup, the total peak heights were compared to the peak heights of the target alleles (Figure 11-2h). These results show that the highest levels of background (when compared to target peak heights) occur when injecting pure undiluted product and when the PCR product has been purified with the magnetic bead clean-up method. All other clean-up methods showed an overall lower background level with only a small amount of variation observed. The ammonium acetate/ethanol clean-up method was shown to have the lowest overall background ratio.

Another parameter to consider when comparing PCR clean-up methods is to determine the level of removal of very small peaks (primer dimer peaks). The ratio of peak heights for fragments <100 bp (primer front) to total peak heights is shown in Figure 11-2i. This panel demonstrates that the primer dimer was most efficiently removed with the ammonium acetate/ethanol, Sephadex, and magnesium sulphate/ ethanol clean-up methods. Moderate primer dimer removal occurred with the dilution and magnetic bead clean-up methods. As expected, the most inefficient removal of primer dimers and small fragments was observed where pure undiluted product was injected. This is demonstrated by the highest levels of primer front being observed in this case.

Table 11-1 shows the mean values across all three plates for each clean-up method. The numbers of failed wells for each method was compared to determine the level of salts present after cleanup. To determine injection efficiency, the total heights of all Rox peaks injected from each plate were compared. The amount of Rox contained within the injection mixture should be constant between all plates and clean-up methods, because Rox-labeled sizing peaks are contained within the loading buffer that is added post-cleanup. The sum of the total and target peak heights injected—calculated as a percentage of the sum of the crude product injection peak height—from each clean-up method can therefore be compared to determine injection efficiency.

The level of allele dropout due to injection efficiency was determined by comparing the total number of alleles analyzed, shown as a percentage of the total number of alleles analyzed when injecting crude product, for each clean-up method. To determine preferential injection of smaller peaks, Rox peak heights between 100 and 150 bp were compared to those >150 bp. For simplicity, the value of total peak heights <150 bp divided by total peak heights >150 bp has been assigned a value of 100 when no preferential injection has occurred. The preferential injection of peaks <150 bp results in values lower than 100 and preferential injection of peaks >150 bp results in values greater than 100. To further explore preferential injection, the sum of all peak heights from fragments with base pair size of 100 to 150 bp was divided by the sum of all the peak heights from a fragment with a base pair size >150 bp. This value was then calculated as a percentage of the value obtained when injecting crude product.

To determine the levels of background and artefact peaks present after cleanup, the sum of the total peak heights were compared to the sum of the peak heights of the target STRs. This value was then calculated as a percentage of the value obtained when injecting crude product. We also compared PCR clean-up methods to determine the level of removal of

**Table 11-1. Comparison of clean-up methods.**

A. Percentage of successful wells (lower values indicate more failed wells due to insufficient Rox size standard injection; higher values indicate higher reliability).
B. Total Rox size standard peak height (higher values indicate more efficient Rox injection).
C. Total peak height for all peaks (higher values indicate more efficient injection).
D. Total target peak height (higher values indicate more efficient injection).
E. Number of alleles per reaction (higher values indicate more alleles analyzed).
F. Preferential injection of small Rox fragment (lower value indicates a preference toward injection of fragment <150bp (i.e., preferential injection of primer dimer rather than PCR target product); target value is 100 (no preferential injection); higher value indicates preference toward injection of fragment >150bp).
G. Preferential injection of small peaks (lower values indicate preferential injection of small fragment and, therefore, higher values are better).
H. Background peak levels (higher value indicates lower background level and, therefore, more efficient).
I. Primer removal (higher value indicates more efficient primer removal).

| | Crude product | MagB | Diln | Seph | CS | Mg/Eth | Aa/Eth |
|---|---|---|---|---|---|---|---|
| A. Successful wells (%) | 58.0 | 63.9 | 84.0 | **92.4** | **97.9** | **100 (100)** | 93.4 (100) |
| B. Total Rox size standard peak height | 4499 | 3684 | 6687 | **15650** | 10513 | 10553 (11368) | **14583 (13449)** |
| C. Total peak height for all peaks (%) | 100 | 100 | 77 | **364** | 51 | 27 (26) | **221 (147)** |
| D. Total target peak height (%) | 100 | 335 | 240 | **829** | 390 | 290 (106) | **695 (600)** |
| E. Number of alleles per reaction (%) | 100 | **150** | **172** | **208** | **188** | **181 (57)** | **162 (142)** |
| F. Preferential injection of small Rox fragment | 2 | 6 | 28 | 16 | **107** | **167 (167)** | 98 (168) |
| G. Preferential injection of small peaks (%) | 100 | 495 | 417 | 1003 | **1220** | 506 (441) | **1664 (1500)** |
| H. Background peak levels (%) | 100 | 447 | 342 | **684** | **770** | 197 (129) | 674 (154) |
| I. Primer removal (%) | 100 | 428 | 412 | **1021** | **1257** | 429 (441) | **1858 (1500)** |

The best methods are in **bold**. Values within brackets indicate 12 µL initial volume.

very small peaks (primer dimer peaks). Peak heights <100 bp (primer front) were compared to the peak heights >100 bp. This value was then calculated as a percentage of the value obtained when injecting crude product.

## Discussion

Without any clean-up procedure, or at the very least some desalting, the analysis of pure undiluted PCR product after electro-injection is compromised. Figures 11-2a and e show that a high number of wells were not able to be analyzed because of failed Rox injection and the reduced number of alleles present in samples. This was due to the presence of high concentrations of salt, unincorporated dNTPs and primers as well as other reagents present during the PCR step that were not completely consumed by the PCR process.

Each of the different clean-up methods discussed in this chapter caused a reduction in both ion content and low molecular weight products. These methods were analyzed for: failed size standard injection, size standard peak height, total and target peak heights, primer removal, preferential injection of small fragment, and the number of STR alleles present (Figure 11-2). These methods were also evaluated for the suitability for automation versus manual processing and for the overall cost. Table 11-2 summarizes the clean-up methods comparing relative times, automation, cost and amount of product available for capillary electrophoresis analysis. Table 11-2 shows that a fivefold dilution of the PCR

**Table 11-2.** Comparison of clean-up methods.

| Method | Time (minutes) | Automation | Cost ($US/plate) | Analysis |
|--------|----------------|------------|------------------|----------|
| Neat Product | *** | *** | *** | * |
| | (0) | | (0) | |
| Dilution | *** | *** | *** | ** |
| | (5) | | (0) | |
| Aa/Eth | ** | ** | *** | *** |
| | (45) | | (0.10) | |
| Mg/Eth | ** | ** | *** | ** |
| | (45) | | (0.10) | |
| Seph | * | * | * | *** |
| | (180) | | (60) | |

Symbols are (***) highly preferable; (**) average; (*) not preferable.

product with distilled water is low in cost and has the full potential for automation, but results in some reduction in quality of the analyzed peaks. The dilution of the PCR product directly reduces the concentrations of salt, dNTPs, and primer, but also reduces the concentration of target DNA. This reduction in target DNA does not normally cause problems for analysis because DNA is in significant excess; capillary electrophoresis uses an electro-injection method (3) capable of overcoming this reduction in concentration. However, if the concentration of product is initially very low, as can be the case in single cell multiplexing (5), then the detection of alleles could be compromised. The dilution clean-up method increased the amount of primer injection, and preferentially injects small products compared to precipitation and filtering methods. The number of failed Rox injections increased in some plates compared to other methods. Overall, this cleanup was a cheap and readily automated method with only a small reduction in the efficiency of analysis and therefore applicable to multiplexes with a high level of product.

Sephadex gel filtration is a commonly used clean-up method prior to capillary electrophoresis. These columns (both single and 96/384 cartridges) can be purchased pre-hydrated or can be dispensed and hydrated manually. Our results indicate that the quality of the data after using this clean-up protocol is equivalent when using either commercial or in-house–prepared Sephadex plates. Sephadex successfully reduced the amount of contaminants such as salt, dNTPs, and primer dimer. However, a slight preferential injection of smaller peaks and incomplete primer removal resulted in preferential elution of smaller DNAs from the Sephadex. In general, this method was not as successful as the ammonium acetate/ethanol method. The Sephadex protocol cannot be fully automated and is expensive, costing approximately $60 per plate. A variety of filter plates, of varying cost, applicable to this process are available; however, these were not evaluated for clean-up efficiency due to the relatively high cost of this method. When using in-house–prepared plates, it is possible to re-use the filter plate to reduce cost, and in this case the cost of the Sephadex (about 3 grams per plate) would be $7 per plate.

Magnetic bead cleanup is fully automated and was performed using the CRS robotic platform. Magnetic bead cleanup only slightly reduced the problems of peak injection failure and preferential injection observed with analysis of pure undiluted PCR product. There is a trade-off between removal of all salt and loss of smaller DNA products during bead washing steps. The parameters for magnetic bead cleanup require further optimization for removal of contaminants prior to capillary electrophoresis. Furthermore, other magnetic bead purification methods are available and could be explored as they may be cheaper, perform better, or have different characteristics, for example, for retaining smaller fragments, but this was not pursued due to the relative high cost of the components.

Ammonium acetate/ethanol cleanup is also low in cost and able to be partially automated with no detectable reduction in quality of the analysis. Ammonium acetate/ethanol cleanup successfully reduced the concentration of salt and dNTPs interfering with injection as well as reducing the primer dimer concentrations. Additionally, the peak size of the target product and number of alleles remained the same, making it applicable for use where there is only a small amount of available PCR product. Preferential injection of smaller peaks was minimized and a strong Rox peak height was observed across all plates. This method was also successfully applied to 12 μL reaction volumes with no apparent compromise being made on analysis. This method can be partially automated and the components required are relatively inexpensive.

Magnesium acetate/ethanol cleanup also successfully reduced the preferential injection of small products and reduced primer dimer concentration as well as nonspecific background peaks. Although the overall target peak heights (excluding Rox) decreased, this did not reduce the number of alleles analyzed when the initial reaction volume was 25 μL because of an abundance of starting DNA fragments. However, when applied to smaller (12 μL) reaction volumes, there was a significant amount of allele dropout that was most likely because of the peak heights dropping below the threshold detected by our current instrumentation. Where very small amounts of DNA are produced during PCR, the magnesium acetate/ethanol cleanup is not the best method to use.

Comparisons between all five clean-up methods conclude that two (dilution and ammonium acetate/ethanol) are clearly superior in reducing salt, dNTPs, and primer concentrations without compromising analysis quality of the product. Fivefold dilution of the PCR product with distilled water is quick, low cost, and able to be fully automated, with only a small reduction in quality of the analyzed peaks. Ammonium acetate/ethanol cleanup is also low cost and able to be partially automated with no detectable reduction in quality of the analysis.

## Acknowledgments

The authors thank *BioTechniques, The International Journal of Life Science Methods* for permission to use figures and tables from a published article entitled, Irwin, D.L., Mitchelson, K.R., and Findlay, I. 2003. PCR product cleanup methods for capillary electrophoresis. *BioTechniques* 34: 932–936.

The authors also thank Dr. Ivan Biros, Dr. Ian Findlay, and Mr. Brendan Mulcahy for technical assistance, and Ms. Rachel MacKay for critical review of the manuscript.

## References

1. Desalting. Piscataway, NJ: GE Healthcare. Available at http://www.chromatography.amershambiosciences.com/aptrix/upp00919.nsf/Content/LabSep_Prod%5CSelGuides%5CMedia%5CDesalting. Accessed June 2005.

2. Devaney, J.M., and Marino, M.A. 2001. Purification methods for preparing polymerase chain reaction products for capillary electrophoresis analysis. *Methods Mol Biol* 162: 43–49.

3. Devaney, J.M., Marino, M.A., Williams, P.C., et al. 1996. The evaluation of fast purification methods for preparing polymerase chain reaction (PCR) products for capillary electrophoresis analysis. *Appl Theor Electrophor* 6: 11–14.

4. Findlay, I., Mathews, P.L., and Quirke, P. 1998. Multiple genetic diagnoses from single cells using multiplex PCR: Reliability and allele dropout. *Prenat Diagn* 18: 1413–1421.

5. Findlay, I., Matthews, P.L., Mulcahy, B.K., et al. 2001. Using MF-PCR to diagnose multiple defects from single cells: implications for PGD. *Mol Cell Endocrinol* 183: S5–S12.

6. Promega, DNA & RNA Purification. Annadale, Australia: Promega. Available at http://www.promega.com/applications/dna_rna/. Accessed June 2005.

7. Silva, W.A., Costa, M.C.R., Valente, V., et al. 2001. PCR template preparation for capillary DNA sequencing. *BioTechniques* 30: 537–543.

8. Williams, P.E., Marino, M.A., Del Rio, S.A., et al. 1994. Analysis of DNA restriction fragments and polymerase chain-reaction products by capillary electrophoresis. *J Chromatogr A* 680: 525–540.

# 12

# Evaluation of Methods for Cleanup of DNA Sequencing Reactions

### Jan Kieleczawa and Katarzyna Bajson

*Biological Technologies Department, Wyeth Research, Cambridge, MA, USA*

DNA sequencing technology made a number of breakthroughs in sequence quality, throughput, and the read length in recent years (9, 14, 23). However, the two critical factors in garnering the optimal sequencing results with a minimal interfering background remain the quality of the template and the effective removal of the dye-terminators and other ingredients of the sequencing mix (see also Chapter 11) (3, 10). This is especially critical after the introduction of capillary electrophoresis technology as a primary separation medium (3, 11, 13), where samples are loaded using electrokinetic injection (3, 15, 16) and any charged molecules will interfere with the efficient intake of fluorescently labeled DNA fragments.

Since the inception of the fluorescent sequencing technology, the method of choice for the removal of unincorporated dye-terminators, dye-primers, and other contaminants was alcohol precipitation (4; and older versions of this protocol). There are a number of variations of this method, and the advice from Applied Biosystems (4) is to tailor it to a specific instrument platform and the version of the BigDye™ used for the DNA sequencing. In general, this method is labor-intensive, time-consuming, and not easily amenable for automation; it also occasionally produces the so-called "T-blobs" (22; J. Kieleczawa, unpublished observation). Despite these obvious drawbacks, it is still used in many small and big sequencing laboratories, as it is very cost-effective, produces very clean fragments, and recovers most of the DNA fragments, as measured, for example, by the amount of fluorescent signal detected by an automated DNA sequencer (discussed below). The variations of the alcohol include ethanol, isopropanol (4) or even n-butanol (19) in the presence or absence of different additives like ammonium or sodium acetate, EDTA, liquid acrylamide, or phenol/chloroform (3, 19, 22).

***DNA Sequencing II: Optimizing Preparation and Cleanup***
Edited by Jan Kieleczawa
©2006 Jones and Bartlett Publishers

A very popular method of removal of sequencing contaminants is by gel filtration using single or multiple (most often in a 96- or 384-well format) filtration cartridges. These cartridges can be purchased ready-made or they can be prepared in-house. Advantages of these methods are that they are easily handled manually, are reliable and less labor intensive compared to alcohol precipitation, and that the liquid handling portion of the purification protocols can be automated using robotic workstations. However, they can add from $0.1 to $1 per sample to the cost of sequencing depending on the source of the filtration device.

Another, although not very often used, clean-up method is by digestion of dNTPs with shrimp alkaline phosphatase (SAP) followed by ethanol precipitation (22). This method is limited to clean-ups of the ABI PRISM dRhodamine and ABI PRISM FS rhodamine terminators, but generally it is not recommended for BigDye terminators (22). Although it yields very pure sequencing products, it is time consuming because it combines at least two 30- to 45-minute steps and has the same limitations with respect of automation as pure alcohol precipitation. The use of SAP can add about $0.1 per sample to the sequencing process.

In recent years, the method that has gained popular widespread use to remove excess dye-terminators and other contaminants is magnetic separation. Although a number of variations of this technology exists (6, 20, 21), the solid-phase reversible immobilization (SPRI) technology is most widely used, as it is very easily automatable and produces good quality sequencing data (7, 8, 18). Depending on the size and the source of the kit, the added cost can vary from $0.1 to $0.5 per sample.

Although the above-mentioned methods for purification of sequencing reaction have been used for many years, to our best knowledge there is no comprehensive publication that provides a side-by-side comparison of these methods. This chapter evaluated three categories of methods used for the clean-up of DNA sequencing reactions: filtration, alcohol precipitation, and magnetic separation. The primary criteria for evaluations of each method are quality of read length expressed in terms of phred score $Q \geq 20$ (5) and the fluorescent signal strengths produced by an automated DNA sequencer. As each method was performed manually, we also provide a somewhat arbitrary criterion: "ease-of-use." Furthermore, the first correctly called base after the 3' end of M13 forward primer and the error distribution in a few discrete ranges are provided. Finally, for five clean-up procedures, we evaluated Taq dilution limits showing data for read length and signal strength.

## Materials and Methods

The BigDye terminator mix and DNA standard (pGem3zf) were purchased from Applied Biosystems, Inc. (Foster City, CA, USA). Five times

BigDye dilution buffer (100 mM Tris, pH 8.9, 10 mM $MgCl_2$) was prepared in house from reagents of the highest purity.

Throughout the course of all testing, we made sure that the same lots of the Taq mix, 5× dilution buffer and Sephadex G-50™ superfine (catalog #17-0041-01; GE Healthcare [formerly Amersham Biosciences], Piscataway, NJ, USA) were used. All experiments were performed within a two-month period on the same capillary array as to minimize the variations arising from the aging of the array.

The clean-up protocols tested in this chapter can be divided into the following categories:

A. Using pre-made filtration 96-well cartridges:
   1. Dye-terminator removal kit (catalog #AB-0943a; ABgene®, Rochester, NY, USA).
   2. SEQueaky Kleen™ $H_2O$ Dye-terminator removal kit (catalog #732-6530; BioRad Laboratories, Hercules, CA, USA).
   3. Performa® DTR V3 96-Well Short Plate Kit (catalog #89939; EdgeBiosystems, Gaithersburg, MD, USA).
   4. GenCLEAN Dye-terminator removal 96-well kit (catalog #K1010; Genetix USA, Inc., Boston, MA, USA).
   5. SigmaSpin 50™ Post-Reaction Clean-up Plate (catalog #S-4559; Sigma-Aldrich, St. Louis, MO, USA).
B. Using filter plates filled in-house with G-50 Sephadex beads:
   6. V96-well plate, 0.2 μm, PVDF (catalog #3504; Corning, Inc., Corning, NY, USA).
   7. 96-Well plate, 0.66 Glass Fiber, 1.2 μm PES (catalog #3511; Corning).
   8. EdgeBiosystem plate from the Performa® kit after discarding the original content and thorough washing with water (see method # 3).
   9. Innovative plate (catalog #F20011; MAT 0.45 μm PVDF 400 μL long-drip; Innovative Microplate, Chicopee, MA, USA).
   10. MAHVN45 96-well filter plate (catalog #MAHVN45; Millipore).
   11. AcroPrep96 filter plate, 0.45 μm GHP membrane 350 μL (catalog #5030; Pall Sciences, East Hills, NY, USA).
   12. AcroPrep96 filter plate, 0.45 μm GHP membrane, 1 mL (PN S5054; Pall Sciences).
   13. MPF-011, long-drip, hydrophilic PVDF, 400 μL, 0.45 μm filter plate (Phenix Research Products, Candler, NC, USA).
   14. MPF-046, short-drip, hydrophilic PVDF, 300 μL filter plate (Phenix Research Products).
   15. The 96-well dye terminator removal UNIFILTER®, 800 μL, long-drip filter plate (catalog #S7700-2801; Whatman, Inc., Clifton, NJ, USA).

C. Magnetic separation methods:
   16. CleanSEQ dye-terminator removal kit (catalog #000121;
       Agencourt Biosciences Inc., Beverly, MA, USA).
   17. MagDTR™ dye terminator removal kit (catalog #41744;
       EdgeBiosystems).
   18. Wizard® Magnesil™ Sequencing reaction cleanup system
       (catalog #A1830; Promega, Madison, WI, USA).
D. Alcohol precipitation methods:
   19. Ethanol precipitation was performed as described in (11;
       pp. 4–9), except that centrifugation was carried out at room
       temperature.
   20. Isopropanol precipitation was performed as described in (21;
       pp. 23–25).

## DNA Sequencing

The master mix (for 96 reaction) consisting of 25 µL of control pGem3zf
DNA at 0.2 mg/mL was mixed with 100 µL of 5 µM M13F primer (5′-
TGTAAAACGACGGCCAGT) and 575 µL of 10 mM Tris/0.01 mM EDTA,
pH 8.0 (=TEsl), and 7 µL of this mix was aliquoted into 0.2 mL Thermo-12
Strip tubes (ABGene). Twenty-four tubes were left on ice and the other 24
tubes were subjected to a controlled heat-denaturation step (5 min at 98°C)
as described (12). Upon heat denaturation, tubes were placed on ice and
3 µL of diluted (0.8 mL of dye mix plus 0.4 mL of 5× dilution buffer) BigDye
V3.1 dye terminator was added to all samples. Tubes were briefly spun
down to collect volumes at the bottoms of tubes, gently vortexed, and
then subjected to cycle sequencing in MJ-225 thermal cycler (MJ Research,
Waltham, MA, USA). The cycling conditions were as follows: [(10
sec/96°C)(5 sec/50°C)(2 min/60°C)] × 40. The final dilution of dye mix
was fourfold in a 10 µL reaction. The protocol where samples were not
heat-denatured corresponds to a standard ABI-like DNA sequencing protocol described in (4). We refer to the sequencing protocol with heat-denaturation step as "modified." Upon completion of sequencing cycles, plates
were spun down and 20 µL of de-ionized water was added to all samples.
For plates listed in series A, the sequencing reactions were purified using
the manufacturer's recommended protocols. Plates in series B were prepared using the gel-loading tool (catalog #MACL09645; Millipore) to dispense Sephadex G-50 beads into tested filter plates. Three hundred
microliters of water were added and plates were left at room temperature
for at least two hours prior to their use. Excess water was removed by
centrifugation in an Eppendorf 5810 model centrifuge (Brinkmann, Westbury, NY, USA) just before transferring the sample. Sequenced samples
were transferred to prepared filter plates and purified reactions were
recovered into a 96-well collection plate by centrifugation. In both steps,

the spinning conditions were 2100 rpm (887× *g*) for five minutes at room temperature. Purified samples were heat-denatured for two minutes at 90° to 95°C and the data were collected on an ABI3730 DNA genetic analyzer (Applied Biosystems) using conditions suggested by the manufacturer. The next day, the same plate was rerun on ABI 3100 genetic analyzer (Applied Biosystems) under conditions suggested by the manufacturer. The ABI3100 was equipped with an 80-cm array. For each chromatogram read length, using Q ≥ 20 (5), and the fluorescent signal intensities were recorded. The average and standard deviation for 24 data points for each treatment were calculated using an Excel spreadsheet or hand-held calculator. To calculate the first correctly called base and error distribution, each chromatogram was assembled against a pGem3zf vector using the Sequencher V4.2 analysis program (Gene Codes, Inc., ann Arbor, MI, USA) and the appropriate data were recorded in Excel spreadsheet for statistical analysis.

Detailed data are presented for an ABI3730 genetic analyzer and only general conclusions will be given for data derived from an ABI3100.

For Taq dilution limits experiments, the aliquots of undiluted BigDye™ dye-terminator V3.1 were diluted with 5× dilution buffer to maintain the original magnesium concentration and buffer strength, but not the nucleotide concentration. The dilution range was from 4× (2 μL of undiluted Taq mix in a 10 μL reaction volume) to 128× (0.0625 μL of undiluted Taq mix in 10 μL reaction volume). These experiments were carried out at three different DNA concentrations (20, 50, and 200 ng of pGem3zf) using both standard and modified sequencing conditions. Each data point is the average of six independent samples.

## Result and Discussion

### Clean-Up Procedures

The data are discussed separately for each category of clean-up method.

### Pre-Made Clean-Up Plates

In Table 12-1 read length and signal strength data are shown for pre-made plates. When the standard ABI protocol was used for the sequencing, the best results were produced using the EdgeBiosystem and Abgene plates. Twelve percent shorter reads were obtained with the Genetix plate and significantly shorter reads (33%–43%) were produced using either the BioRad or Sigma plates. With modified sequencing protocols, the longest reads were produced using plates #1 and #3, but all other plates gave data within 7% of the best results. With a standard sequencing protocol, there

**Table 12-1.** Read length, signal strengths, and ease-of-use of pre-made clean-up plates.

| Treatment ⇒ Plate Description ⇓ | Standard DNA Sequencing Protocol No Heat-Denaturation | | Modified DNA Sequencing Protocol With Heat-Denaturation | | Ease-of-Use | Cost/Sample ($) |
|---|---|---|---|---|---|---|
| | RL | SS | RL | SS | | |
| ABGene plate (#1) | 1005.3 ± 30.6 | 241.2 ± 127.0 | 1013.9 ± 16.9 | 748.9 ± 373.5 | ●●●◗ | 0.67–0.72 |
| BioRad plate (#2) | 583.5 ± 104.0 | 51.6 ± 12.5 | 956.1 ± 26.2 | 274.3 ± 133.0 | ●●●◗ | 0.70 |
| EdgeBiosystems plate (#3) | 1031.7 ± 9.2 | 216.1 ± 106.8 | 1032.4 ± 10.3 | 577.8 ± 238.7 | ●●●● | 0.53–0.83 |
| Genetix plate (#4) | 910.4 ± 75.8 | 84.9 ± 32.8 | 967.0 ± 39.2 | 209.0 ± 87.1 | ●●● | 0.39 |
| Sigma plate (#5) | 685.5 ± 135.9 | 60.1 ± 13.2 | 958.4 ± 37.9 | 178.5 ± 73.7 | ●●◗ | 0.74 |

Abbreviations are: RL, read length; SS, signal strength. The number of symbols (●) denotes the ease of using the plate. More dots indicates that the method is easier to use. The data obtained with the modified protocol were much more consistent compared to data obtained with a standard sequencing protocol, and the ratio of standard deviation over the average read length ranged from 1% to 4%; the same ratio for the standard sequencing protocol ranged from 1% to 19.8%.

224

was a fourfold difference in signal strength between the plate #3 and plate #2, which contributes somewhat to the difference in read lengths. The fluorescent signal of 60 units, under these experimental conditions, is only about 1.5 to 2 times higher than the signal produced with no DNA (blank lane). For a modified protocol, signal intensity is about two- to threefold higher compared to that produced by standard protocol and there is a smaller, if insignificant, difference in the overall read length produced by all five clean-up plates.

The geometry and the long-drip feature made the plate from Edge-Biosystems easiest to handle (score 5). The other three plates (#1, #2, and #5) had dimensions that made it difficult to fit in an Eppendorf 5810 centrifuge. In addition, because of the depth of wells, it was inconvenient to transfer the reactions to the middle of the filtration beds. The plate #4 required an additional preparation step prior to its use. For these reasons, all of these plates received a somewhat lower score in the "ease-of-use" category. The cost of cleaning of one sample was quite similar between all plates and depended on the size of kit (although price of plate from Genetix was significantly lower). It is possible that through special arrangements with vendors or quantity price breaks, the price of clean-up kits may be lower.

### In-House–Prepared Clean-Up Plates

The plates tested in this category represent popular (but by no means all) filtration 96-well devices available on the market. Table 12-2 summarizes the data in a similar fashion to that in Table 12-1. The longest reads were obtained using plates from EdgeBiosystems (#8) and from Innovative Microplate (#9). However, all other plates were within 11% of the best read lengths data. Of interest is the high signal strength produced by a Pall plate (#11) even when using standard sequencing protocol. The data suggest that this plate could be particularly useful when sequencing samples with extremely low concentrations of DNA. When the modified sequencing protocol was used, the longest reads were again produced with plates #8 and #9, but all other plates provided data that were within 9% of the two best clean-up plates. The differences in read lengths and signal strengths are partially due to the various types of membranes used in different filter plates. On the other hand, the plates were prepared and treated in the same way, indicating that these differences may not arise from optimal centrifugation parameters. In the "ease-of-use" category, plates #8 and #9 (which upon visual inspection look identical) proved to be easiest to use. Their long-drip design feature eliminated the need for a very careful alignment with the collection plate and any possibility for cross-contamination. The Whatman plate (#15) also has a long-drip design but it is much deeper (1 mL) and, hence, we had difficulty with handling

**Table 12-2.** Read length, signal strengths and ease-of-use in in-house type of clean-up plates.

| Treatment ⇒ Plate Description ⇓ | Standard DNA Sequencing Protocol No Heat-Denaturation | | Modified DNA Sequencing Protocol With Heat-Denaturation | | Ease-of-Use | Cost/Sample ($) |
|---|---|---|---|---|---|---|
| | RL | SS | RL | SS | | |
| Corning plate (#6) | 910.2 ± 24.4 | 94.3 ± 25.5 | 964.0 ± 20.5 | 232.1 ± 91.2 | ●●● | 0.20 |
| Corning plate (#7) | 944.3 ± 13.7 | 154.6 ± 27.1 | 964.9 ± 13.3 | 308.1 ± 49.9 | ●●● | 0.20 |
| EdgeBiosystems plate (#8) | 1020.4 ± 16.0 | 157.4 ± 47.6 | 1034.9 ± 14.1 | 350.7 ± 144.2 | ●●●● | 0.07* |
| Innovative plate (#9) | 999.8 ± 78.1 | 117.6 ± 50.6 | 1035.9 ± 10.4 | 426.6 ± 187.4 | ●●●● | 0.16 |
| Millipore plate (#10) | 901.3 ± 48.5 | 78.0 ± 36.8 | 979.5 ± 15.0 | 258.6 ± 100.7 | ●●● | 0.18 |
| Pall plate (#11) | 942.0 ± 14.3 | 390.1 ± 51.3 | 945.3 ± 11.7 | 952.5 ± 150.4 | ●●● | 0.21 |
| Pall plate (#12) | 959.7 ± 42.9 | 88.7 ± 24.3 | 1021.4 ± 15.9 | 200.8 ± 84.6 | ●●● | 0.19 |
| Phenix plate (#13) | 906.8 ± 43.9 | 86.7 ± 38.9 | 991.9 ± 21.1 | 299.1 ± 145.7 | ●●● | 0.19 |
| Phenix plate (#14) | 924.4 ± 39.5 | 86.8 ± 35.3 | 997.6 ± 14.3 | 276.0 ± 109.4 | ●●● | 0.18 |
| Whatman plate (#15) | 920.2 ± 35.5 | 89.2 ± 26.1 | 966.5 ± 35.8 | 213.8 ± 93.9 | ●●◗ | 0.25 |

Abbreviations are RL, read length; SS, signal strength. The number of symbols (●) denotes the ease of using the plate that was prepared in-house. Note that the data obtained with the modified protocol was more consistent compared to data obtained with a standard sequencing protocol, and ratio of standard deviation over the average read length ranged from 1% to 3.6%. The same ratio for the standard sequencing protocol ranged from 1.5% to 8%. It is worth noting that, in general, the data obtained with in-house filter plates were more consistent compared to pre-made plates. The * next to plate #8 in the Cost/Sample column indicates just the cost of Sephadex, as the filter plate itself was re-used plate #3 from Table 12.1.

during the centrifugation and sample transferring steps. Purification cost per sample was very similar for all plates and ranged from $0.16 to $0.25 (including the cost of Sephadex G-50 at $0.07/well).

## Purification Using Magnetic Beads

Three commercially available kits were tested in this category. The read lengths produced with the magnetic beads kits from Agencourt and Edge-Biosystems were almost identical and very reproducible. On the other hand, the Promega kit gave significantly shorter reads even when using modified sequencing protocol (Table 12-3). The performance of the Promega kit improved as the amount of the sequenced DNA increased above 50 ng per reaction (data not shown). Compared to filtration methods, magnetic beads protocols are much more labor-intensive, cumbersome, and on average twice as expensive; in a manual format it does not offer any advantages. However, out of all five clean-up protocols, they were the easiest to automate. Currently, they are implemented in number of medium and large DNA sequencing centers and, in fact, the magnetic clean-up protocol is the integral part of a fully automated DNA sequencing workstation, Parallab 350™ (Brooks Automation, Inc., Chelmsford, MA, USA). Further information on this platform is available on the company's Web site (http://www.brooks.com/pages/2627_parallab_350.cfm).

The cost per purification using magnetic beads can vary quite significantly from about $0.16 to $0.45, depending on the size of the kit and manufacturer.

## Alcohol Precipitation Clean-Up Procedures

Only two alcohol precipitation procedures were evaluated in this category. For both the standard and modified protocols, read lengths were similar to those produced by other clean-up procedures (Table 12-4). On the other hand, the signal strengths were among the highest out of all clean-up protocols, particularly with ethanol precipitation. Again, this may allow for much lower (10- to 20-fold) use of template or significantly higher dilution of Taq mix, which is the decisive factor in cost of a sequencing reaction. The cost of cleanup is at least an order of magnitude lower compared to other clean-up procedures. However, the precipitation methods are very tedious, time-consuming, and difficult or impossible to automate. This method is successfully used in some large sequencing centers (see Chapter 10) primarily because it is cost-effective and allows for a higher dilution of dye-terminators, thus significantly reducing overall cost. This method can be particularly attractive if labor costs are relatively low.

Table 12-3. Read length, signal strengths and ease-of-use using magnetic beads and alcohol purification methods.

| Treatment ⇒ Clean-up Method ⇓ | Standard DNA Sequencing Protocol No Heat-Denaturation | | Modified DNA Sequencing Protocol With Heat-Denaturation | | Ease-of-Use | Cost/Sample ($) |
|---|---|---|---|---|---|---|
| | RL | SS | RL | SS | | |
| **Magnetic Beads** | | | | | | |
| Agencourt (#16) | 966.8 ± 34.3 | 381.8 ± 165.5 | 995.0 ± 14.4 | 1371.4 ± 674.6 | ●●● | 0.3–0.4 |
| EdgeBiosystems (#17) | 974.6 ± 33.4 | 254.9 ± 138.6 | 1009.7 ± 19.6 | 670.1 ± 366.4 | ●●● | 0.16–0.24 |
| Promega (#18) | 636.9 ± 183.0 | 38.7 ± 16.1 | 868.3 ± 147.5 | 207.9 ± 236.1 | ●● | 0.45 |
| **Ethanol** | | | | | | |
| Precipitation (#19) | 957.3 ± 83.3 | 585.0 ± 275.6 | 973.5 ± 42.3 | 2304.9 ± 1225.4 | ●● | ~0.02 |
| **Isopropanol** | | | | | | |
| Precipitation (#20) | 926.8 ± 47.0 | 216.4 ± 143.8 | 951.4 ± 18.0 | 924.1 ± 495.0 | ● | ~0.01 |

Abbreviations are RL, read length; SS, signal strength. The number of symbols (●) denotes the ease of using the protocols for the magnetic beads and alcohol purification methods. The data obtained with the modified sequencing protocol was more consistent compared to data obtained with a standard sequencing protocol, and the ratio of standard deviation over the average read length ranged from 1.4% to 4.3% (excluding Promega's kit). The same ratio for the standard sequencing protocol ranged from 3.4% to 8.7% (excluding Promega's kit).

**Table 12-4.** Stability of read lengths over time in the sequencing samples purified using three different filter clean-up procedures.

| Time | Read Lengths & Signal Strengths | Plate from EdgeBiosystems filled with Sephadex G-50 (#)* | | SigmaSpin 50™ plate (#5)* | | Whatman Unifilter plate filled with Sephadex G-50 (#15)** | |
|---|---|---|---|---|---|---|---|
| | | No HD | +HD | No HD | +HD | No HD | +HD |
| Time 0 | RL | 1008.1 ± 17.7 | 1039.3 ± 11.8 | 820.4 ± 287.5 | 1016.8 ± 26.4 | 1018.3 ± 13.3 | 1033.5 ± 11.0 |
| | SS | 98.8 ± 30.7 | 288.2 ± 110.1 | 65.0 ± 25.9 | 174.6 ± 75.2 | 122.4 ± 38.6 | 344.3 ± 132.5 |
| Time 1 | RL | 971.6 ± 50.1 | 1018.3 ± 11.9 | 475.7 ± 246.9 | 895.7 ± 130.3 | 968.0 ± 28.9 | 983.2 ± 24.9 |
| | SS | 79.9 ± 22.9 | 233.1 ± 94.4 | 44.7 ± 14.6 | 111.4 ± 49.8 | 100.6 ± 36.0 | 282.6 ± 122.7 |
| Time 2 | RL | 970.1 ± 64.9 | 1013.7 ± 12.8 | 83.5 ± 159.4 | 414.6 ± 192.7 | 700.8 ± 92.3 | 791.3 ± 62.5 |
| | SS | 86.3 ± 28.3 | 256.6 ± 103.2 | 34.2 ± 9.4 | 63.6 ± 22.1 | 25.8 ± 9.4 | 61.6 ± 19.9 |
| Time 3 | RL | 985.3 ± 18.7 | 1014.4 ± 12.0 | 500.3 ± 266.5 | 944.5 ± 44.9 | NT | NT |
| | SS | 88.5 ± 24.7 | 225.1 ± 85.5 | 44.8 ± 11.1 | 120.7 ± 53.6 | NT | NT |
| Time 4 | RL | 975.5 ± 18.3 | 1007.4 ± 12.8 | 47.3 ± 147.4 | 430.6 ± 225.9 | NT | NT |
| | SS | 91.3 ± 25.3 | 239.2 ± 93.2 | 32.6 ± 7.8 | 66.4 ± 25.5 | NT | NT |

Abbreviations are RL, read lengths; SS, signal strengths; HD, heat denaturation; NT, not tested. Please refer to the text for an explanation of times 3 and 4.

* The times 1 and 2 for these plates were 5.5 and 31 hours, respectively.

** The times 1 and 2 for this plate were 16 and 24 hours, respectively.

**Table 12-5.** Comparison of read lengths and signal strength between data obtained from an ABI3730 and ABI3100.

| ABI Machine ⇒ | ABI3730 | | ABI3100 | |
|---|---|---|---|---|
| Sequencing Protocol ⇒<br>Readout Parameter ⇓ | Standard protocol | Modified protocol | Standard protocol | Modified protocol |
| Read length | 904.9 ± 123.5 | 981.0 ± 39.4 | 804.7 ± 84.7 | 890.0 ± 50.6 |
| Difference in read length (Modified – Standard) | | 77 | | 85 |
| Signal strength | 173.7 ± 140.8 | 549.2 ± 524.2 | 59.3 ± 42.9 | 157.3 ± 110.4 |
| Ratio $\frac{\text{Modified}}{\text{Standard}}$ | | 3.0 | | 2.7 |

## Stability of the Read Lengths and Signal Strength Over Time

As indicated in the Material and Methods section, after the clean-up step, all plates were run on an ABI3730 DNA analyzer and subsequently (next day) on an ABI3100 DNA analyzer. In most cases, under our experimental conditions the reads on ABI3100 were shorter by about 100 bases compared to reads on ABI3730 (Table 12-5). In an experiment where we ran samples first on the ABI3100 and then on the ABI3730, the reverse was true (data not shown). However, when clean-up plates from Sigma (#5) were tested, no data on the ABI3100 were produced. To further investigate this loss of data, we performed a time-course experiment with two control clean-up plates (#8 and #15) and the Sigma plate (#5). These data are presented in Table 12-4. In one set of experiments, the reactions were purified using these plates and immediately run on an ABI3730. The same plates were subsequently rerun at the times specified in Table 12-4. In the other set of experiments, the reactions were purified using clean-ups #5 and #8, and the plates were left at room temperature for nine hours and then run on an ABI3730 (the same plate was again rerun after 31 hours from the time of purification). For clean-up procedure #8, even after 31 hours from the clean-up time, the decrease in read length (and signal strength) was on the order of 2% to 4%, depending upon which sequencing protocol was used. A somewhat larger decrease in read length was observed with clean-up #15 (25%–30%); however, with clean-up #5, 60%

to 90% of read length was lost after 31 hours and 12% to 40% was lost just after 5.5 hours, regardless of the order of treatment. The reason for such rapid loss of the read lengths using SigmaSpin50™ kit is not known, but is most likely related to the type of beads and membrane used in this filtering device. Some loss of read lengths was reported after prolonged storage of sequencing reactions (23); however, the loss was related to the storage of reactions in formamide, which tends to break down into acrylic acid, especially in water solutions.

## Repetitive Use of the Same Filter Plate

One of the ways to decrease the cost of reaction purification using filtering devices is to reuse the same filter plate after a thorough cleaning after each use. As indicated in Table 12-2, the cost for a single clean-up using in-house–prepared filter plates is about $0.20, of which two thirds constitute the cost of a filter plate. If the same plate can be reused several times with no loss of read lengths (and signal strength) and without any cross-contamination, then the overall cost of a cleanup can be reduced to only the cost of the G-50 beads. Figure 12-1a shows that the read length stayed constant when the same plate was reused for up to ten times. Similarly, the signal strength remained almost unchanged (Figure 12-1b) throughout the experiment, indicating that the DNA is not being retained on the filter membrane and a thorough water wash is sufficient to remove any possible particles.

## Effects of Amount of DNA on Read Length and Signal Strength

For four clean-up methods (Agencourt/EdgeBiosystems magnetic beads and Innovative/EdgeBiosystems filter plates), the effects of various concentrations of DNA on the $Q \geq 20$ read lengths and signal strength were tested. Figure 12-2a shows the read-length data for the DNA amount (1250 bp PCR fragment) in the range of 0.1 to 100 ng per reaction. For very low concentrations—below 1 ng—both magnetic methods performed significantly better than the filter methods. Above this concentration, each method performed equally well. However, there seemed to be no appreciable differences between two magnetic methods or two filter plates. The signal strength was significantly higher for both magnetic bead methods (Figure 12-2b), but this did not impact the read length when the average signal was above 100 units. In principle though, it is possible that a higher dilution of dye-terminator mix can be used with magnetic clean-up methods, provided that the nucleotide concentration is not limiting for full extension of sequencing products.

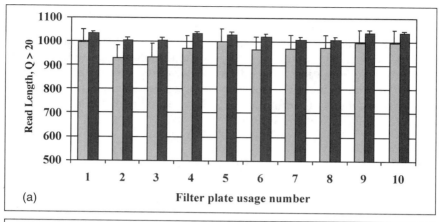

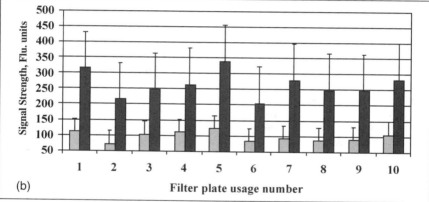

**Figure 12-1.** Repetetive use of the same in-house–prepared filter plate for the purification of sequencing reactions. (a) When the same filter plates were re-used up to ten times, the read length remains constant. (b) Similarly, when the same plates were used up to ten times, the signal strength remains nearly constant. Symbols are (light bars) read lengths for the standard sequencing protocol; (■) reads for the modified sequencing protocols.

## First Correctly Called Base and Error Distribution

The first correctly called base and the error distribution in four discrete ranges are shown in Table 12-6. For most filtration methods, when a standard DNA sequencing protocol is used, the first correctly called base was about 42 nucleotides away from the 3′ end of M13 primer. When a heat-denaturation step was added to a sequencing protocol (modified), the first correctly called base was about 40 nucleotides away from the end of the primer. The magnetic bead method (#7) from EdgeBiosystems produced

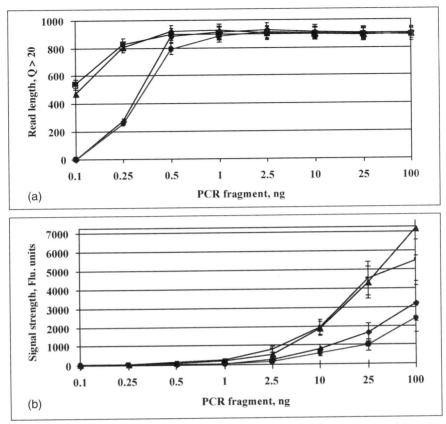

**Figure12-2. Testing of four different purification protocols.** (a) Read lengths versus the amount of DNA. (b) Signal strength versus the amount of DNA. Symbols are (■) magnetic beads purification using Agencourt's kit; (▲) magnetic beads purification using Edge Biosystems kit; (◆) pre-made filter plate from EdgeBiosystems; (●) Innovative filter plate and G-50 Sephadex.

DNA fragments that were the closest to the primer, especially in combination with the modified sequencing protocol. On the other hand, the magnetic procedure from Promega (#18) retained shorter fragments and the first correctly identified base was about 80 nucleotides away form the primer. The performance of this method improved as the amount of the DNA used for sequencing increased above 200 ng in sequencing reactions (data not shown). Each tested method varied with respect of number of misreads compared to the control sequence of pGem3zf DNA. None of the methods was able to produce zero errors in the first 50 bases. This was primarily due to the KB caller's (KB is a default base caller on an ABI3730 DNA genetic analyzer) inability to frequently call the double "A"

**Table 12-6.** The first correctly called base and the error distribution produced when using different DNA sequencing reaction clean-up methods.

| Method # Name | Sequencing Protocol | Position of First Base Called | Number of Errors in the Range | | | | |
|---|---|---|---|---|---|---|---|
| | | | 1–50 | 51–500 | 501–850 | 851–1000 | |
| #1 ABGene | Standard | 42.1 ± 1.5 | 0.7 ± 0.5 | 3.4 ± 2.8 | 0.2 ± 0.6 | 2.0 ± 2.5 | |
| | Modified | 39.4 ± 1.1 | 1.0 ± 0.0 | 1.3 ± 1.1 | 0.0 ± 0.0 | 1.2 ± 0.5 | |
| #2 BioRad | Standard | 43.1 ± 2.6 | 6.7 ± 4.4 | 7.4 ± 5.4 | 9.4 ± 7.9 | 22.3 ± 7.0 | |
| | Modified | 41.0 ± 1.8 | 0.5 ± 0.5 | 0.0 ± 0.0 | 0.0 ± 0.0 | 1.9 ± 1.4 | |
| #3 EdgeBiosystems | Standard | 42.3 ± 1.2 | 0.7 ± 0.5 | 0.0 ± 0.0 | 0.0 ± 0.0 | 1.4 ± 0.8 | |
| | Modified | 39.3 ± 1.2 | 1.0 ± 0.2 | 0.0 ± 0.0 | 0.0 ± 0.0 | 1.0 ± 0.5 | |
| #4 Genetix | Standard | 43.2 ± 1.4 | 1.0 ± 1.0 | 0.3 ± 0.9 | 1.0 ± 1.5 | 4.4 ± 2.6 | |
| | Modified | 40.9 ± 1.4 | 0.8 ± 0.4 | 0.0 ± 0.0 | 0.2 ± 0.5 | 2.9 ± 2.3 | |
| #5 Sigma | Standard* | 42.2 ± 1.9 | 2.3 ± 3.4 | 4.4 ± 4.1 | 10.8 ± 6.0 | 18.2 ± 5.8 | |
| | Modified | 41.0 ± 1.6 | 0.6 ± 0.5 | 0.6 ± 0.7 | 0.5 ± 0.9 | 4.1 ± 1.7 | |
| #6 Corning 1 | Standard | 43.5 ± 1.4 | 0.5 ± 0.5 | 2.4 ± 1.5 | 0.6 ± 0.8 | 4.3 ± 2.4 | |
| | Modified | 42.0 ± 1.2 | 0.8 ± 0.4 | 1.0 ± 0.6 | 0.0 ± 0.0 | 1.6 ± 1.1 | |
| #7 Corning 2 | Standard | 45.3 ± 4.7 | 2.6 ± 1.8 | 0.0 ± 0.0 | 1.7 ± 1.4 | 17.7 ± 5.2 | |
| | Modified | 44.3 ± 1.4 | 1.7 ± 1.1 | 0.2 ± 0.6 | 1.0 ± 1.0 | 11.9 ± 4.5 | |
| #8 EdgeBiosystems | Standard | 44.2 ± 1.9 | 0.8 ± 0.4 | 0.0 ± 0.0 | 0.1 ± 0.3 | 1.8 ± 1.2 | |
| | Modified | 42.2 ± 1.5 | 1.0 ± 0.2 | 0.0 ± 0.0 | 0.0 ± 0.0 | 1.1 ± 0.7 | |
| #9 Innovative | Standard | 42.6 ± 1.1 | 0.6 ± 0.5 | 1.3 ± 1.0 | 0.0 ± 0.0 | 2.1 ± 1.4 | |
| | Modified | 39.6 ± 0.9 | 0.8 ± 0.4 | 0.0 ± 0.0 | 0.0 ± 0.0 | 0.9 ± 0.5 | |
| #10 Millipore | Standard | 44.0 ± 1.6 | 0.6 ± 0.5 | 3.2 ± 1.3 | 0.6 ± 0.9 | 5.1 ± 3.1 | |
| | Modified | 40.5 ± 1.1 | 0.9 ± 0.3 | 0.5 ± 0.8 | 0.0 ± 0.0 | 1.2 ± 0.9 | |

| | | | | | |
|---|---|---|---|---|---|
| #11 | Standard | 41.6 ± 1.2 | 1.5 ± 1.2 | 3.3 ± 3.4 | 1.1 ± 1.0 | 11.7 ± 3.9 |
| Pall 1 | Modified | 40.0 ± 1.3 | 1.2 ± 0.7 | 0.6 ± 0.9 | 0.6 ± 1.2 | 13.8 ± 4.2 |
| #12 | Standard | 43.5 ± 1.8 | 0.9 ± 1.4 | 3.0 ± 2.1 | 0.5 ± 0.8 | 4.8 ± 3.5 |
| Pall 2 | Modified | 41.1 ± 1.1 | 0.7 ± 0.4 | 0.3 ± 0.5 | 0.1 ± 0.3 | 2.0 ± 1.2 |
| #13 | Standard | 43.5 ± 1.9 | 0.4 ± 0.5 | 0.2 ± 0.4 | 1.6 ± 2.0 | 4.8 ± 3.7 |
| Phenix 1 | Modified | 41.0 ± 1.1 | 0.8 ± 0.4 | 0.0 ± 0.0 | 0.0 ± 0.0 | 1.3 ± 1.3 |
| #14 | Standard | 45.2 ± 1.9 | 0.7 ± 0.6 | 0.0 ± 0.0 | 0.7 ± 1.1 | 5.5 ± 3.6 |
| Phenix 2 | Modified | 41.6 ± 1.4 | 0.8 ± 0.4 | 0.0 ± 0.0 | 0.0 ± 0.0 | 1.2 ± 1.1 |
| #15 | Standard | 44.9 ± 1.5 | 0.6 ± 0.6 | 0.2 ± 0.5 | 0.8 ± 0.8 | 6.4 ± 3.8 |
| Whatman | Modified | 42.9 ± 1.1 | 0.8 ± 0.4 | 0.0 ± 0.0 | 0.1 ± 0.5 | 1.8 ± 1.5 |
| #16 | Standard | 40.8 ± 1.0 | 0.8 ± 0.4 | 2.3 ± 3.7 | 0.1 ± 0.3 | 2.2 ± 1.6 |
| Agencourt | Modified | 39.5 ± 1.1 | 1.0 ± 0.0 | 0.3 ± 0.6 | 0.0 ± 0.0 | 1.7 ± 1.2 |
| #17 | Standard | 37.9 ± 1.9 | 0.3 ± 0.6 | 0.3 ± 0.8 | 0.0 ± 0.0 | 1.7 ± 1.8 |
| EdgeBiosystems | Modified | 35.5 ± 2.6 | 0.7 ± 0.8 | 0.1 ± 0.3 | 0.0 ± 0.0 | 1.4 ± 0.8 |
| #18 | Standard* | 84.5 ± 43.7 | 18.9 ± 5.6 | 3.1 ± 2.4 | N/A | N/A |
| Promega | Modified* | 45.9 ± 8.3 | 6.0 ± 4.8 | 0.6 ± 1.4 | 2.1 ± 3.2 | 7.8 ± 8.4 |
| #19 | Standard | 37.6 ± 1.3 | 0.9 ± 0.7 | 0.0 ± 0.0 | 0.0 ± 0.0 | 3.6 ± 3.1 |
| Ethanol | Modified | 37.4 ± 1.4 | 0.9 ± 0.3 | 0.0 ± 0.0 | 0.0 ± 0.0 | 3.4 ± 2.8 |
| #20 | Standard | 42.1 ± 1.3 | 0.8 ± 0.4 | 0.5 ± 0.9 | 0.4 ± 0.5 | 5.8 ± 2.5 |
| Isopropanol | Modified | 39.5 ± 1.2 | 1.0 ± 0.0 | 0.0 ± 0.0 | 0.0 ± 0.0 | 3.8 ± 2.5 |
| Average of all clean-up | Standard | 42.6 ± 2.1 | 1.2 ± 1.4 | 1.7 ± 2.0 | 0.5 ± 0.5 | 4.2 ± 2.6 |
| methods | Modified | 40.4 ± 1.9 | 0.9 ± 0.2 | 0.3 ± 0.4 | 0.1 ± 0.3 | 1.9 ± 1.0 |

For the calculation of averages for all clean-up procedures (the last row in the Table) some methods were excluded because of the very poor and inconsistent results (*symbols indicate these methods).

at the very beginning of the pGem3zf sequence. An error rate higher than one was primarily due to the "blobs" that occurred around bases 55 to 60, and sometimes around bases 90 to 100. Four clean-up methods (#3, #8, #14, #19) resulted in no misreading errors in the range of 51 to 850, regardless whether the standard or modified sequencing protocol was used. However, the inclusion of a heat-denaturation step significantly improved the quality of the data. Based on all of the data presented in Table 12-6, when the heat-denaturation step is part of a sequencing protocol, there were 4.4 fewer misread errors in the range of 1 to 1000 bases compared to the standard sequencing protocol data, and the first correctly called base was by about two nucleotides closer to the end of M13 primer.

## Dye-Terminator BigDye Dilution Limits

Figure 12-3(a–c) shows the Q ≥ 20 read lengths for various Taq dilutions at 20, 50, and 200 ng of DNA using five different sequencing reaction clean-up protocols. Under these experimental conditions, regardless of clean-up protocol, the read lengths dropped significantly when the dilution factor was above 32×, with an optimal read length achieved at Taq dilution of 16×: at 128× Taq dilution, the read length was still around 150 to 200 bases. Signal strength was not a reliable predictor of long quality read lengths (Figure 12-3d). For all clean-up procedures, even at 96× Taq dilution, the signal was 3 to 10 times above the background (under normal sequencing conditions such signal results in optimal reads), yet the read lengths were only 20% to 30% of optimal. At 128× Taq dilution, the signal was still above the background level (1.5- to 5-fold), but this does not produce good quality data.

## Conclusions

This chapter compared 20 different clean-up procedures according to three categories of analysis. Each method has its distinct advantages and disadvantages, so a particular method can be tailored to the specific needs and conditions of a sequencing laboratory. If cost is the guiding principle for selecting of a clean-up procedure, then alcohol precipitation should be the method of choice. In a sequencing laboratory with a throughput of five to ten 96-well plates/day, the filtration option is very attractive, especially when plates can be prepared in-house and filter device re-used many times. The cost of single purification therefore can be reduced to a cost of Sephadex beads ($0.07 or less, depending on the size of Sephadex package), which translates to a five- to tenfold and three- to sixfold lower

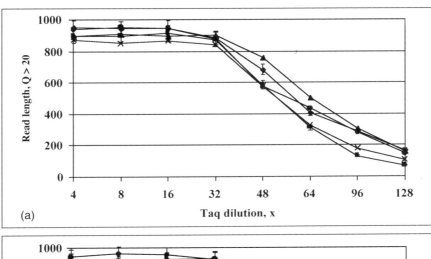

(a)

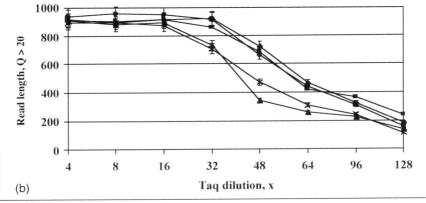

(b)

**Figure 12-3.** **Effects of Taq dilution on Q ≥ 20 read length and signal strength.**
Read lengths at 20 (a) at 50 (b) and at 200 ng (c) of DNA/reaction are shown. The
average read length is slightly shorter at higher Taq dilutions as the amount of
DNA in the sequencing reaction increases. This can be attributed to faster exhaustion of nucleotide pool available for incorporation during the synthesis. (d) Signal
strength at 50 ng of DNA per reaction. The overall pattern is similar for the other
two concentrations with values lower or higher at 20 or 200 ng of DNA per reaction. Note that the signal strength does not linearly correlate with the changing
amount of dye terminator mix in the sequencing reaction. Symbols are: (■) magnetic beads purification using Agencourt's kit; (▲) magnetic beads purification
using Edge Biosystems kit; (◆) pre-made filter plate from EdgeBiosystems; (●)
Innovative filter plate and G-50 Sephadex; (×) values for the ethanol precipitation
protocol.

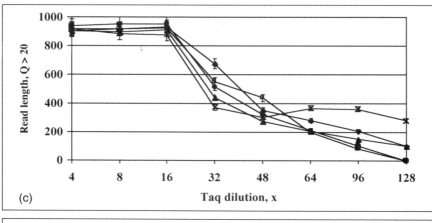

(c)

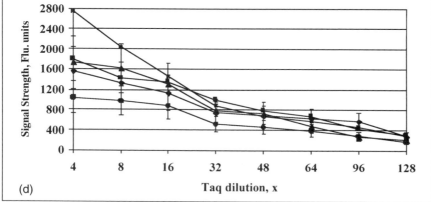

(d)

**Figure 12-3.** *Continued*

purification cost compared to pre-packaged filtering devices and magnetic methods, respectively. The three- to fourfold higher cost compared
to alcohol precipitations is compensated by slightly longer reads and
decreased amount of hands-on involvement.

In all methods, less variation (as measured by standard deviation)
was observed when a modified DNA sequencing protocol was used. In
addition, read length increased on average by 77 bases and the signal
strength tripled with a modified sequencing protocol. Table 12-5 summarizes the data for both ABI3730 and ABI3100 genetic analyzers. In general,
the reads from ABI3100 were shorter, but this was mainly due to the time
lapse between the purification and the run of samples on this instrument.
As pointed out in the text, the read lengths decreased after prolonged
storage of purified DNA reactions, which was highly method-dependent;

Table 12-5 presents only the summary data. Finally, Table 12-6 summarizes the data regarding the first correctly called base and the error distribution for all compared clean-up protocols. Once again, the incorporation of the heat-denaturation step as a part of sequencing protocol resulted in longer and cleaner data. There are a number of advantages of using a modified over a standard DNA sequencing protocol (12). Based on the data presented in this study, we can identify two additional advantages: the significantly fewer number of sequencing errors and the first correctly called base is closer to a sequencing primer. The critical cost driver of doing DNA sequencing is the price of dye-terminator dye mix and many laboratories save money by diluting the original mix. To maintain an optimal ratio of nucleotides, some laboratories perform sequencing reactions in relatively small volumes (500 nL to 1 µL) and in specialized plasticware, which is possible with the aid of precise liquid handling robots. Here, we demonstrate that simply with the adjustment of magnesium and buffer to the original values, it is possible to use 1/16 to 1/32 amount of the original Taq mix without sacrificing the quality of reads and without using expensive laboratory robots.

## References

1. *ABI PRISM™ BigDye™ Terminator Cycle Sequencing Ready Reaction Kit.* 1998. Protocol. P/N 4303237, Rev B. Foster City, CA: Applied Biosystems.
2. *Automated DNA Sequencing. Chemistry Guide.* 2000. P/N 4305080B. Foster City, CA: Applied Biosystems.
3. *ABI Prism® 3100 Genetic Analyzer. Sequencing Chemistry Guide.* 2001. Foster City, CA: Applied Biosystems.
4. *ABI PRISM® BigDye™ Terminator v3.1 Cycle Sequencing Kit.* 2002. Protocol. Part number 4337035 Rev. A. Foster City, CA: Applied Biosystems.
5. Ewing, B.G., Hillier, L., Wendl, M.C., and Green, P. 1998. Base-calling of automated sequencer traces using Phred. II. Error probabilities. *Genome Res* 8: 186–194.
6. Fanga, B.M., Dahlberg, O.J., Deggerdal, A.H., et al. 1999. Automated system for purification of dye-terminator sequencing products eliminates up-stream purification of templates. *BioTechniques* 26: 980–983.
7. Hawkins, T.L., O'Connor, T., Roy, A., and Santillan, C. 1994. DNA purification and isolation using solid-phase. *Nucleic Acid Res* 22: 4543–4544.
8. Hawkins, T.L., McKernan, K.J., Jacotot, L.B., et al. 1997. A magnetic attraction to high-throughput genomics. *Science* 276: 1887–1899.
9. International Human Genome Sequencing Consortium. 2001. Initial sequencing and analysis of the human genome. *Nature* 409: 860–921.
10. Irwin, D.I., Mitchelson, K.R., and Findlay, I. 2003. PCR product cleanup methods for capillary electrophoresis. *BioTechniques* 34: 932–936.
11. Karger, A.E. 1996. Separation of sequencing fragments using an automated capillary electrophoresis instrument. *Electrophoresis* 17: 144–151.

12. Kieleczawa, J. 2005. Controlled heat-denaturation of DNA plasmids. In: Kieleczawa, J., ed. *DNA Sequencing: Optimizing the Process and Analysis.* Sudbury, MA: Jones and Bartlett; 1–10.
13. Marziali, A., and Akeson, M. 2001. New DNA sequencing methods. *Annu Rev Biomed Eng* 3: 195–223.
14. McKernan, K. 2005. Future of DNA sequencing: towards an affordable genome. In: Kieleczawa, J., ed. *DNA Sequencing: Optimizing the Process and Analysis.* Sudbury, MA: Jones and Bartlett; 177–196.
15. Ruiz-Martinez, M.C., Salas-Solano, O., Carrilho, E., et al. 1998. A sample purification method for rugged and high-performance DNA sequencing by capillary electrophoresis using replaceable polymer solutions. A. Development of the cleanup protocol. *Anal Chem* 70: 1516–1527.
16. Salas-Solano, O., Ruiz-Martinez, M.C., Carrilho, E., et al. 1998. A sample purification method for rugged and high-performance DNA sequencing by capillary electrophoresis using replaceable polymer solutions. B. Quantitative determination of the role of sample matrix components on sequencing analysis. *Anal Chem* 70: 1528–1535.
17. Sambrook, J., and Russell, D.W. 2001. *Molecular Cloning,* 3rd ed. Cold Spring Harbor, NY: Cold Spring Harbor Laboratory Press.
18. Sawakami-Kobayashi, K., Segawa, O., Hornes, E., et al. 2003. Multipurpose robot for automated cycle sequencing. *BioTechniques* 34: 634–637.
19. Tillett, D., and Neilan, B.A. 1999. n-Butanol purification of dye terminator sequencing reactions. *BioTechniques* 26: 606–610.
20. Tong, X., and Smith, L.M. 1992. Solid-phase method for purification of DNA sequencing reactions. *Anal Chem* 64: 2672–2677.
21. Tong, X., and Smith, L.M. 1993. Solid-phase purification in automated DNA sequencing. *J DNA Sequenc Map* 4: 151–162.
22. *User Bulletin. Precipitation Methods To Remove Residual Dye Terminators From Sequencing Reactions.* P/N 4304655 Rev. A. 1998. Foster City, CA: Applied Biosystems.
23. Venter, J.C., Adams, M.D., Myers, E.W., et al. 2001. The sequence of human genome. *Science* 291: 1304–1351.

# 13 Thermostable DNA Polymerases for a Wide Spectrum of Applications: Comparison of a Robust Hybrid TopoTaq to Other Enzymes

Andrey R. Pavlov, Nadejda V. Pavlova,
Sergei A. Kozyavkin, and Alexei I. Slesarev
*Fidelity Systems, Inc., Gaithersburg, MD, USA*

Thermostable DNA polymerases play a central role in current methods for DNA amplification and sequencing. Since their introduction, DNA polymerases from thermophilic organisms brought fundamental changes to modern biotechnology. Numerous thermostable DNA polymerases from natural sources were identified, sequenced, cloned, and expressed to optimize the DNA amplification (17). Various blends of thermophilic DNA polymerases and mixtures of the polymerases with other proteins are used to accomplish specific purposes of DNA synthesis. Nonetheless, natural DNA polymerases and their combinations do not satisfy all needs in DNA production. Diverse techniques that use DNA amplification often require special characteristics that could not be easily achieved with natural polymerases. Consequently, the search for enzymes with better properties and efforts to modify existing enzymes continues.

The engineering of new enzymes traditionally was performed through the use of point-directed mutagenesis of catalytic domains to eliminate $3' \rightarrow 5'$ exonuclease activity, improve ddNTP utilization for DNA sequencing, or alter fidelity of amplification (1, 5, 11, 15, 20, 23). The removal of the entire $5' \rightarrow 3'$ nuclease domains was often performed to abolish the nuclease activity.

A different approach was recently applied to construct new proteins by fusing catalytic domains of thermostable DNA polymerases and DNA binding domains of other thermostable enzymes (6, 13, 16, 21, 22).

*DNA Sequencing II: Isolation and Preparation*
Edited by Jan Kieleczawa
©2006 Jones and Bartlett Publishers

Specifically, we attached nonspecific DNA binding domains, such as Helix-hairpin-Helix (HhH) domains of DNA topoisomerase V of *Methanopyrus kandleri* (3), to catalytic domains of *Thermus aquaticus* (*Taq*) and *Pyrococcus furiosus* (*Pfu*) DNA polymerases. This method resulted in a significant improvement of performance of the enzymes operating in an abnormal environment. The efficacy of this engineering approach was affirmed using the *Taq* and *Pfu* DNA polymerases, which belong to different structural families. We produced active chimeric DNA polymerases that maintained high processivity at high levels of salts (16) and other inhibitors of DNA synthesis, such as phenol, blood, and DNA intercalating dyes (18).

In the *Taq*-HhH chimeric polymerase (TopoTaq) and its variants, the gains in processivity, thermostability, and specificity result in shortened extension times, and more robust and higher yield amplification. The performance of TopoTaq is enhanced by the built-in "hot start" properties and by the presence of DNA topoisomerase activity that facilitates DNA strand separation.

This chapter presents several examples demonstrating specific properties of TopoTaq and its successful use for DNA amplification. We also compare some of the properties of this enzyme to other known DNA polymerases.

## Materials and Methods

Chimeric polymerases TopoTaq™, TopoTaqBL™, TopoTaqSq™, Genomic DNA Amplification Kit (GH100), all primers (unless indicated otherwise), and Fimers were manufactured by Fidelity Systems, Inc. (Gaithersburg, MD, USA). The DNA polymerase activity was determined in a standard primer extension assay (16) with $1\,\mu M$ fluorescent primer-template junction (PTJ) duplex substrate at 70°C. One unit of TopoTaq DNA polymerase is equivalent to one unit of DNA polymerase activity determined in a DNA acid precipitation assay.

AmpliTaq DNA polymerase, Stoffel fragment of DNA polymerase 10× PCR Buffer II, and BigDye™ terminator V2 and V3 (BDT v2 and BDT v3) kits were products of Applied Biosystems, Inc. (Foster City, CA, USA). M13mp18(+) single-stranded DNA and ALF M13 universal fluorescent primer were obtained from Amersham Biosciences (Piscataway, NJ, USA). Primers for amplification of regions of human gene *SLC5A1* were kindly provided by Dr. Natalia K. Abuladze (Division of Nephrology, David Geffen School of Medicine at UCLA, Los Angeles, CA, USA). Human genomic DNA, dUTP, and 7-deaza-dGTP were purchased from Roche Diagnostics Corp. (Indianapolis, IN, USA). QIAquick PCR purification kit was purchased from Qiagen Inc. (Valencia, CA, USA). pET21d expression vector was obtained from Novagen, Inc. (Madison, WI, USA). Restriction enzymes were purchased from New England Biolabs (Beverly, MA, USA) and used in common cloning techniques. dATP, dTTP, dGTP, and dCTP were obtained from Fermentas Life Sciences (Hanover, MD, USA), SeaKem Agarose was purchased from FMC BioProducts (Rockland, ME, USA), and all other chemicals were of highest available reagent grade.

All DNA polymerase activity assays, PCR, and cycle sequencing reactions were carried out in Peltier Thermal Cycler PTC-225 (M.J. Research, Watertown, MA, USA). The sequencing reactions were stopped with 20 µL of 20 mM Na$_2$EDTA. The samples were purified by centrifugation through Sephadex G-50 spin columns, dried down, resuspended in loading buffer and analyzed on an ABI Prism 377 DNA sequencer (Applied Biosystems, Inc.) using manufacturer's recommended protocol.

## Examples of the Enzyme Activities

### Example 1: Comparison of Strand Displacement Activity for TopoTaq, AmpliTaq, and Stoffel Fragment

To evaluate the strand displacement ability of DNA polymerases, a fluorescent duplex substrate (strand displacement fluorescent substrate; SDFS) was formed. A 26-nt primer was synthesized that anneals to M13 single-stranded (ss) DNA (positions 5791–5766) and recreates an *Mse*I restriction site. A double-stranded (ds) part of M13 ssDNA was cut with *Mse*I after annealing of the primer, *Mse*I was inactivated by heating (65°C, 15 min), and the resulting DNA substrate was annealed to a fluorescent ALF M13 Universal Primer (positions 6313–6290) and a 42-nt long Stop Primer, complimentary to the template 20 nt downstream (positions 6270–6229) of the attached labeled primer (Figure 13-1a). DNA polymerase reaction mixtures were formulated essentially as earlier described (16), except the mixtures contained 16 nM SDFS, and the primer extensions were carried out for three minutes at 60°C with equal amounts of each DNA polymerase used. The DNA polymerases formed fluorescent extension products with the substrate, having the maximal length of 533 nt that could be resolved on a sequencing gel, as shown in Figure 13-1b.

### Example 2: Dye Terminator Cycle Sequencing Using TopoTaqSq

TopoTaqSq polymerase contains a mixture of engineered chimeric DNA polymerases with ThermoFidelase designed to be added to BigDye terminator kits for robust cycle sequencing through the G-C–rich area in DNA templates. Sequencing reaction mixtures (total volume 10 µL) contained 4 µL BDT v3 kit, 2 µL 5× TopoTaq amplification buffer with 10 mM MgCl$_2$, 1 µL TopoTaqSq (1.2 U/mL), 1 µL plasmid DNA pUC18 (200 ng/µL) with a cloned stop sequence from bacterial genomic DNA *Bifidobacterium lactis* NCC362, or intact *E. coli* cells transformed with the

```
            ALF M13 Universal Primer                              Stop Primer
FL*- cgacgttgtaaaacgacggccagt → extension       aggtcgactctagaggatccccgggtaccgagctcgaattcg → ... 450
     ||||||||||||||||||||||||                    ||||||||||||||||||||||||||||||||||||||||||||
6313 gctgcaacatttgctgccggtcacggttcaacggtacggacgtccagctgagatcctcctaggggcccatggctcgagcttaagc     6229

          gagtccactattaaagaacgtggact  26
          ||||||||||||||||||||||||||
5791 ...ctcaggtgataatttcttgcacctga 5766

                                      Total length = 533 nt

                      ↑ MSE I
```

(a)

**Figure 13-1. Comparison of strand displacement activities for DNA polymerases.** (a) The strand displacement fluorescent substrate (SDFS) used in the strand displacement primer extension experiments. (b) A urea-polyacrylamide gel picture of the extension products. Strand displacement synthesis of DNA was carried out with 0.125 units of each AmpliTaq DNA polymerase (1), Stoffel fragment (2), or TopoTaq (3) polymerase, respectively. The extension products were analyzed on an ABI Prism 377 DNA sequencer as described in the Materials and Methods section.

plasmid (prepared from 20 μL overnight culture), and 1 μL pUC18 Fimer. The tubes with the reaction mixtures were placed in the thermal cycler and cycle sequencing was performed as follows:

1. Initial template denaturing for three minutes at 100°C.
2. 400 cycles (5 sec at 95°C/30 sec at 55°C/90 sec at 60°C).
3. Holding at 4°C until ready for purification.

Figure 13-2 displays the sequencing traces obtained with the purified plasmid template (panel b) and the *E. coli* cells (panel c). The trace obtained with the kit alone under the same cycling conditions is shown in Figure 13-2a.

### Example 3: PCR Amplification from G-C–Rich Templates with TopoTaq DNA Polymerase

The enhanced strand displacement ability of TopoTaq DNA polymerase allows for consistent amplification of G-C–rich templates and can produce up to 12 kb long products from G-C–rich plasmid and genomic DNAs. An example of PCR conditions is presented below.

| | |
|---|---|
| • Plasmid DNA (5 ng) or bacterial genomic DNA (20 ng) | 1 μL |
| • Primer mixture (10 μM each) | 1 μL |
| • 2× Amplification buffer with 6 mM MgCl$_2$ | 10 μL |
| • dNTP mixture (10 mM each) | 1 μL |
| • TopoTaq DNA polymerase | 1 U |
| • Deionized water | to 20 μL |

G-C–rich (67%–70% G-C) DNA targets from *Methanopyrus kandleri* AV19 were cloned into plasmid *pet21d* (Novagen) between restriction sites NcoI and HindIII, as described in (16), and amplified with T7 promoter and T7 terminator primers: GAAATTAATACGACTCACTATAGGG and GCTAGTTATTGCTCAGCGG. In addition, G-C–rich regions were amplified directly from archaeal genomic DNA of *M. kandleri* AV19 using specific primers with annealing temperatures 64° to 67°C.

A three-step cycling PCR protocol was employed for this amplification:

1. Initial template denaturing for 40 seconds at 94°C.
2. 30 Cycles (30 sec at 94°C/30 sec at annealing temperature/1 min/kb of amplified target at 72°C) (50°C for plasmid DNA and 60°C for genomic DNA, respectively).
3. Final extension at 72°C for 6 minutes. Holding at 4°C until ready for purification.

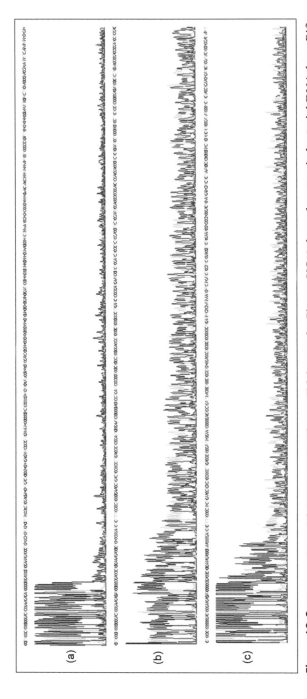

**Figure 13-2.** Sequencing with BDT v3 Kit, ThermoFidelase and a Fimer pUC18 through a stop in bacterial DNA from *Bifidobacterium lactis* NCC362 cloned into pUC18. (a) Purified plasmid DNA without additions. (b) Purified plasmid DNA with the addition of 0.67U TopoTaqSq. (c) Direct sequencing from *E. coli* cells with addition of TopoTaqSq.

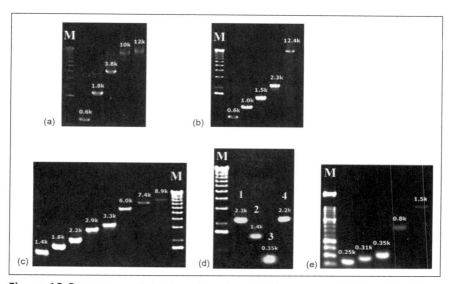

**Figure 13-3.** **PCR amplification of various DNA targets with TopoTaq DNA polymerase.** G-C–rich DNA regions of *M. kandleri* AV19 amplified from templates cloned into plasmid *pET21d* (a) or directly from archaeal genomic DNA (b). Examples of PCR amplification of exons of a unique gene SLC5A1 with TopoTaq from purified human genomic DNA (c), human genomic DNA amplified with kit GH100 (d), and with TopoTaqBL from fresh blood (e) are shown.

The products were resolved on 1% agarose gels shown in Figure 13-3 (panels a and b show the plasmid and genomic DNA targets, respectively) after staining with ethidium bromide.

### Example 4: PCR from Human Genomic DNA

TopoTaq can be successfully used for high-yield amplification of targets from human genomic DNA. However, the optimal elongation temperature in PCR is lower than that for plasmid or bacterial targets, or than that commonly used in PCR protocols with other thermostable DNA polymerases. The example shows amplification of regions of human genomic DNA containing a unique gene, *SLC5A1*, with multiple exons.

Amplification of regions of human genomic DNA containing the *SLC5A1* gene with multiple exons

- Human genomic DNA (50 ng)                                    1 μL
- Primer mixture (10 μM each)                                  1 μL
- 2× Amplification buffer with 6 mM $MgCl_2$                   10 μL

| | |
|---|---|
| • dNTP mixture (10 mM each) | 1 μL |
| • TopoTaq DNA polymerase | 1 U |
| • Deionized water | to 20 μL |

1. Initial template denaturing for one minute at 95°C.
2. 30 Cycles (30 sec at 95°C / 30 sec at annealing temperature [58°C] / 1 min/kb of amplified target at 66°C).
3. Final extension at 72°C for six minutes. Holding at 4°C until ready for purification.

Depending on template purity, as little as 20 ng of human genomic DNA could be used to achieve successful amplifications. The products were applied to a 1% agarose gel electrophoresis and stained with ethidium bromide (Figure 13-3c).

Chimeric DNA polymerases built with TopoTaq technology are highly resistant to variety of inhibitors (16, 18) and especially suited for PCR amplification in the presence of a substantial amount of blood. We have optimized the TopoTaqBL DNA polymerase for amplification of DNA templates from whole fresh or frozen blood with the following simple protocol:

| | |
|---|---|
| • Fresh or frozen blood | 1 μL |
| • Primer mixture (10 μM each) | 1 μL |
| • 2× Amplification buffer with 6 mM MgCl₂ | 10 μL |
| • dNTP mixture (10 mM each) | 1 μL |
| • TopoTaqBL DNA polymerase | 1 U |
| • Deionized water | to 20 μL |

We carried out the reaction with the cycling protocol as above, except the initial denaturing was carried out for 5 minutes and the elongation temperature in the cycles was 68°C. Figure 13-3e shows amplified exons of the gene *SLC5A1* directly from fresh blood with TopoTaqBL DNA polymerase after staining with ethydium bromide in 1% agarose gel.

## Example 5: Specific PCR Using DNA Amplified with Random Hexamer Primers

Random-priming isothermal amplification techniques are based on use of the bacteriophage Phi29 DNA polymerase having remarkable processivity and strand displacement ability (4). These techniques have resulted in a number of protocols for nonspecific strand displacement amplification of circular and linear molecules of DNA.

While amplification of circular plasmid DNAs, which proceeds through mechanisms of rolling circle amplification (8, 10, 14), is used for generation of DNA templates for high throughput operations (9), amplification of long linear genomic DNAs (7, 12) is less studied and used.

However, it is the amplification of genomic DNA that is especially attractive for successful work with unique DNAs that are usually not abundant. Although PCR protocols usually require very small amounts of templates, a preliminary amplification of DNA can be very useful in such cases. On the other hand, any foreign DNA contamination in DNA samples can be co-amplified with the target DNA template; a successful specific PCR can assure the presence of the target template in the amplified DNA.

We amplified human genomic DNA with Phi29 DNA polymerase and then used the amplified DNA for specific PCR of regions containing a unique gene *SLC5A1* with multiple exons.

### Protocol for Amplification of Genomic DNA with Genomic DNA Amplification Kit

1. Thoroughly mix 1 μL of material to be amplified (human genomic DNA, 200 ng/μL) and 1 μL denaturation solution and incubate the mixture for three minutes at room temperature. Stop the denaturation by adding 2 μL of neutralization buffer.
2. Combine 12.5 μL of 4× amplification buffer and 10 units of Phi29 polymerase; add $H_2O$ to 46 μL. Add the mix to the denatured genomic DNA to the total volume 50 μL.
3. Incubate for 16 hours at 30°C.
4. Inactivate the enzyme (10 min at 65°C).

The protocol produced amplified human DNA template for about 100 PCR reactions, as shown below.

### PCR Amplification Protocol with TopoTaq DNA Polymerase

- Amplified genomic DNA                              0.5 μL
- Primer mixture (10 μM each)                        1 μL
- 2× Amplification buffer with 6 mM $MgCl_2$         10 μL
- dNTP mixture (5 mM each)                           2 μL
- TopoTaq DNA polymerase                             1 U
- Deionized water                                    to 20 μL

A three-step cycling PCR was performed:

1. Initial template denaturing for one minute at 95°C.
2. 30 cycles (30 sec at 95°C/30 sec at annealing temperature [50°C]/ 2 min at 68°C).

3. Final extension at 72°C for six minutes. Holding at 4°C until ready for purification.

The products of PCR amplification are shown in Figure 13-3d. Lanes 1 to 3 demonstrate bands for pre-amplified template, and lane 4 shows the product of PCR amplification with 100 ng original human genomic DNA, as a template. No products could be observed (not shown) under these conditions with 2 ng human genomic DNA (the amount of the original DNA introduced to PCR with the amplified template).

## *Example 6: Direct PCR DNA Amplification from Bacterial Cultures with TopoTaq DNA Polymerase*

TopoTaq DNA polymerase offers a robust PCR amplification of both genomic and plasmid DNA targets directly from bacterial cultures. To perform PCR amplification from plasmid targets, DNA sequences were cloned into a low-copy plasmid, pet21d, and amplified with T7-promoter (GAAATTAATACGACTCACTATAGGG) and T7-terminator (GCTAGT-TATTGCTCAGCGG) primers.

To amplify 16S rRNA targets from genomic DNA templates, we synthesized four universal primers that specifically anneal to conservative regions encoding bacterial 16S RNAs:

- 16SU1—TGTGTACAAGGCCCGGGAACGTATTCAC
- 16SU2—CCGCAAGGTTAAAACTCAAATGAATTGAC
- 16SU3—AGCCGCGGTAATACGGAGGGTGCAAG
- 16SU4—TGATCATGGCTCAGATTGAACGCTGG

Amplification with primer pairs 16SU1–16SU2, 16SU1–16SU3, and 16SU1–16SU4 produced 0.5, 0.85, and 1.38 kb long fragments, respectively.

The PCR products from the plasmid and genomic templates could be obtained both from bacterial glycerol stocks and from individual colonies on agar plates without the need for DNA isolation and employing similar cycling protocols.

## *PCR Amplification From Colonies*

- Cells* (1 colony suspended in 15 µL deionized water)    1 µL
- Primer mixture (10 µM each)    1 µL
- 2× Amplification buffer with 6 mM MgCl$_2$    10 µL
- dNTP mixture (10 mM each)    1 µL

*Thoroughly resuspended cells were added to premixed PCR reagents.

- TopoTaq DNA polymerase                                    1 U
- Deionized water                                           to 20 µL

## PCR Amplification From Glycerol Stocks

- Cells* (Glycerol stock)                                   1 µL
- Primer mixture (10 µM each)                               1 µL
- 2× Amplification buffer with 6 mM MgCl₂                   10 µL
- dNTP mixture (10 mM each)                                 1 µL
- TopoTaq DNA polymerase                                    1 U
- Deionized water                                           to 20 µL

The PCR tubes with the reaction mixtures were placed in a thermal cycler, and three-step cycling was performed as follows:

1. Initial template denaturing for 15 seconds at 100°C.
2. 30 cycles (15 sec at 100°C/30 sec at annealing temperature*/2 min at 68°C) (*50°C for plasmid DNA and 55°C for genomic DNA, respectively.)
3. Final extension at 72°C for six minutes. Holding at 4°C until ready for purification.

Figure 13-4 demonstrates PCR amplification products obtained both from bacterial colonies spread on agar plates using common protocols (plasmid DNA target in *E. coli* strain BL21, genomic DNA targets in *Escherichia coli, Klebsiella pneumoniae, Lactobacillus bulgaricus, Lactobacillus acidophilus,* and *Lactobacillus casei*), and from glycerol stocks (plasmid DNA target in *E. coli,* genomic DNA targets in *Escherichia coli, Klebsiella pneumoniae,* and various strains of *Lactobacillus,* including *bulgaricus, acidophilus, casei, curvatus, gasseri, reuteri, sakei, delbrueckii, brevis, kefiri, plantarum*). Additionally, products of DNA amplification from cultures of *Streptococcus thermophilus, Leuconostoc mesenteroides, Oenococcus oeni,* and *Pediococcus pentosaceous* are shown.

### Example 7: Production of Modified DNA with TopoTaq DNA Polymerase by PCR Amplification

High efficiency of the TopoTaq DNA polymerase allows producing modified DNA molecules by PCR amplification using regular templates and specific deoxyribonucleotides. Because the hybrid polymerase has a catalytic domain of the *Taq* polymerase (16) (DNA polymerase of structural family A (reviewed in reference 17), it is not inhibited by

---

*Thoroughly resuspended cells were added to premixed PCR reagents.

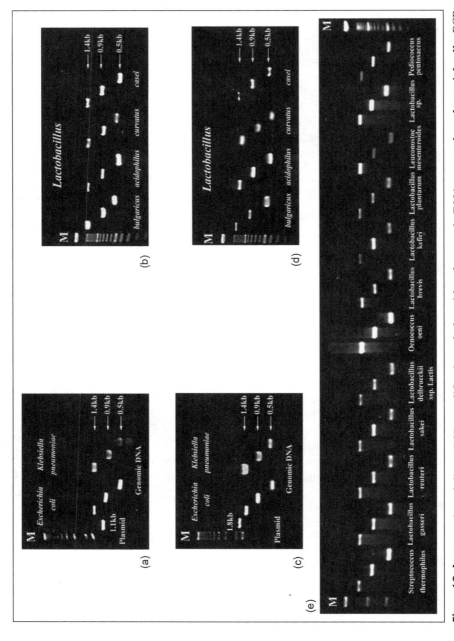

**Figure13-4. Examples of direct PCR amplification of plasmid and genomic DNA targets from bacterial cells.** PCR products were obtained with TopoTaq DNA polymerase from individual colonies on agar plates (a and b) and from bacterial glycerol stocks (c–e) without DNA isolation steps.

252

dU-containing templates. Thus, TopoTaq efficiently synthesizes DNA molecules, in which U replaces T. For synthesis of dU-modified DNA, we used a standard PCR amplification protocol in which dNTP mixture contained dUTP instead of dTTP. In control PCR amplifications, the mixture contained regular deoxyribonucleotides.

- Plasmid DNA (5 ng)                                        1 µL
- Primer mixture (10 µM each)                               1 µL
- 2× Amplification buffer with 6 mM MgCl₂                   10 µL
- dNTP mixture (10 mM each)*                                1 µL
- TopoTaq DNA polymerase                                    1 U
- Deionized water                                           to 20 µL

Two targets cloned into plasmids were amplified: a 1.8 kb G-C–rich (60%) insert from *Methanopyrus kandleri* archaeal DNA, cloned into plasmid *pet21d*, and a 4.5 kb A-T–rich (80%) fragment of DNA from chromosome 10 of *Plasmodium falciparum* containing microsatellite DNA sequences cloned into plasmid pFKB50 (obtained from The Institute for Genomic Research, Rockville, MD).

The PCR tubes with the reaction mixtures were placed in a thermal cycler to perform a three-step cycling PCR:

1. Initial template denaturing for two minutes at 94°C.
2. 30 cycles (30 sec at 94°C / 30 sec at annealing temperature** / elongation*** at 72°C).
3. Final extension at 72°C for six minutes. Holding at 4°C until ready for purification.

Figure 13-5a shows products obtained in the amplification and indicates that TopoTaq incorporates equally efficiently dUTP and dTTP nucleotides. The products were purified from the mixture using QIAquick PCR purification kit and sequenced. Figure 13-5a also shows examples of chromatograms for PCR products amplified with dU or dT.

Although with lesser efficiency, the TopoTaq is also able to replace dGTP with 7-deaza-dGTP (dzGTP) in PCR (16). We used the following protocol to amplify a region in genomic DNA of *Bifidobacterium lactis*, NCC362, which causes strong stops in sequencing reactions using genomic DNA and primers adjacent to the stop region:

- Genomic DNA (150 ng)                                      1 µL
- Primer mixture (10 µM each)                               1 µL

---

*dATP, dGTP, dCTP, and dUTP.
**50° and 54°C for *M. kandleri* and *P. falciparum* targets, respectively.
***30 sec and 3 min for *M. kandleri* and *P. falciparum* targets, respectively.

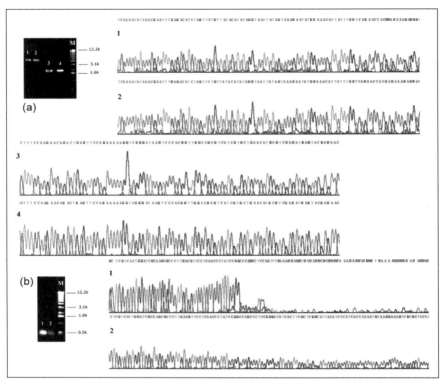

**Figure 13-5.** Synthesis of modified DNA using PCR amplification with TopoTaq. (a) dT-(1) and dU-(2) PCR products from a cloned *P. falciparum* template, and dT-(3) and dU-(4) PCR products from cloned *M. kandleri* DNA. (b) Effects of replacement dGTP (1) by 7-deaza-dGTP (2) in a PCR amplification from a genomic *B. lactis* DNA template. Numbers at sequencing traces on the panels correspond to the numbers on agarose gels. In each panel, the numbers of sequencing traces correspond to the numbers of traces on agarose gels.

- 2× Amplification buffer with 6 mM MgCl₂      10 μL
- dNTP mixture (5 mM each)*      2 μL
- TopoTaq DNA polymerase      1 U
- Deionized water      to 20 μL

    A three-step cycling PCR was performed in the thermal cycler:

1. Initial template denaturing for one minute at 95°C.
2. 30 cycles (30 sec at 95°C/1 min at annealing temperature [61°C]/ 2 min at 72°C).

*dATP, dTTP, dCTP, and dzGTP.

**3.** Final extension at 72°C for six minutes. Holding at 4°C until ready for purification.

The products obtained in the amplification (Figure 13-5b) were purified with the QIAquick PCR purification kit according the manufacturer's instructions and then sequenced. Sequencing of PCR products amplified with dzGTP resulted in clean sequencing data (Figure 13-5b; trace 2), as TopoTaq is able to efficiently replace dG by dzG in difficult regions of the template and decrease the stability of DNA secondary structures. In contrast, trace 1 in Figure 13-5b shows the sequencing reaction with the same fragment amplified with dGTP, as a control, with stops at the specific position on the template.

## Discussion

Existing thermostable DNA polymerases have very useful applications in biotechnology, but the search for new enzymes with better properties continue, and so do attempts to modify characteristics of known enzymes. Some DNA amplification technologies often pose conflicting requirements for DNA synthesis (reviewed in reference 17).

A method to improve performance of the proteins in laboratory environments (which most of the time differ from a natural environment) revolves around joining the DNA binding domains and catalytic domains of DNA processing enzymes. The most spectacular results so far were achieved by the attachment of nonspecific DNA binding domains from other thermostable enzymes (such as Helix-hairpin-Helix [HhH] domains of DNA topoisomerase V of *Methanopyrus kandleri* (3) or DNA binding domains from sequence nonspecific dsDNA binding protein Sso7d from *Sulfolobus solfataricus*) to catalytic domains of DNA polymerases (16, 18, 22).

Based on our previous results (2, 3, 16, 18), we designed our commercial TopoTaq DNA polymerase to have enhanced thermostability, better strand displacement, and increased resistance to inhibitors, which made it suitable for robust performance in a variety of DNA amplification techniques. The enzyme is efficient and performs different tasks using similar protocols. The difference between the various protocols is determined by the nature of templates and the conditions of primer annealing, for example, a longer time for the initial denaturation in protocols for direct amplification from blood or some bacterial cells is required because of the slow release of DNA from cells. However, the high thermostability of the polymerase allows for prolonged incubation at high temperatures. It is important to note that amplification protocols with the TopoTaq polymerase are generally more stringent in their requirement to use correct primer annealing temperatures (Tm). In practice, Tm depends both on the primer concentration and the effective template concentration (which is usually unknown, since it depends on the availability of the annealing sites on template) as well as the composition of amplification media (19). However, for the majority of applications, using common programs to estimate Tm should suffice for successful amplification with TopoTaq line of DNA polymerases.

## Acknowledgments

This work was supported in part by Department of Energy and National Cancer Institute (NIH) grants DE-FG02-99ER83009 and 2R44CA101566-02 (to S.A.K. and A.I.S.). We thank Ms. Anna Pavlova for help with the preparation of this manuscript.

## References

1. Arezi, B., Hansen, C.J., and Hogrefe, H.H. 2002. Efficient and high fidelity incorporation of dye-terminators by a novel archaeal DNA polymerase mutant. *J Mol Biol* 322: 719–729.
2. Belova, G.I., Prasad, R., Kozyavkin, S.A., et al. 2001. A type IB topoisomerase with DNA repair activities. *Proc Natl Acad Sci U S A* 98: 6015–6020.
3. Belova, G.I., Prasad, R., Nazimov, I.V., et al. 2002. The domain organization and properties of individual domains of DNA topoisomerase V, a type 1B topoisomerase with DNA repair activities. *J Biol Chem* 277: 4959–4965.
4. Blanco, L., Bernad, A., Lazaro, J., et al. 1989. Highly efficient DNA synthesis by the phage phi 29 DNA polymerase. Symmetrical mode of DNA replication. *J Biol Chem* 264: 8935–8940.
5. Bohlke, K., Pisani, F.M., Vorgias, C.E., et al. 2000. PCR performance of the B-type DNA polymerase from the thermophilic euryarchaeon *Thermococcus aggregans* improved by mutations in the Y-GG/A motif. *Nucleic Acids Res* 28: 3910–3917.
6. Davidson, J.F., Fox, R., Harris, D.D., et al. 2003. Insertion of the T3 DNA polymerase thioredoxin binding domain enhances the processivity and fidelity of Taq DNA polymerase. *Nucleic Acids Res* 31: 4702–4709.
7. Dean, F.B., Hosono, S., Fang, L., et al. 2002. Comprehensive human genome amplification using multiple displacement amplification. *Proc Natl Acad Sci U S A* 99: 5261–5266.
8. Dean, F.B., Nelson, J.R., Giesler, T.L., et al. 2001. Rapid amplification of plasmid and phage DNA using Phi29 DNA polymerase and multiply-primed rolling circle amplification. *Genome Res* 11: 1095–1099.
9. Detter, J.C., Jett, J.M., Lucas, S.M., et al. 2002. Isothermal strand-displacement amplification applications for high-throughput genomics. *Genomics* 80: 691–698.

10. Detter, J.C., Nelson, J.R., and Richardson, P.M. 2004. Phi29 DNA polymerase based circle amplification of templates for DNA sequencing. In: Demidov, V.V., and Broude, N.E., eds. *DNA Amplification: Current Technologies and Applications*. Norfolk, U.K.: Horizon Bioscience; 245–266.

11. Evans, S.J., Fogg, M.J., Mamone, A., et al. 2000. Improving dideoxynucleotide-triphosphate utilisation by the hyper-thermophilic DNA polymerase from the archaeon *Pyrococcus furiosus*. *Nucleic Acids Res* 28: 1059–1066.

12. Lage, J.M., Leamon, J.H., Pejovic, T., et al. 2003. Whole genome analysis of genetic alterations in small DNA samples using hyperbranched strand displacement amplification and array-CGH. *Genome Res* 13: 294–307.

13. Motz, M., Kober, I., Girardot, C., et al. 2002. Elucidation of an archaeal replication protein network to generate enhanced PCR enzymes. *J Biol Chem* 277: 16179–16188.

14. Nelson, J.R., Cai, Y.C., Giesler, T.L., et al. 2002. TempliPhi, phi29 DNA polymerase based rolling circle amplification of templates for DNA sequencing. *BioTechniques* 44–47.

15. Patel, P.H., and Loeb, L.A. 2000. Multiple amino acid substitutions allow DNA polymerases to synthesize RNA. *J Biol Chem* 275: 40266–40272.

16. Pavlov, A.R., Belova, G.I., Kozyavkin, S.A., et al. 2002. Helix-hairpin-helix motifs confer salt resistance and processivity on chimeric DNA polymerases. *Proc Natl Acad Sci U S A* 99: 13510–13515.

17. Pavlov, A.R., Pavlova, N.V., Kozyavkin, S.A., et al. 2004. Recent developments in the optimization of thermostable DNA polymerases for efficient applications. *Trends Biotechnol* 22: 253–260.

18. Pavlov, A.R., Pavlova, N.V., Kozyavkin, S.A., et al. 2004. Thermostable chimeric DNA polymerases with high resistance to inhibitors. In: Demidov, V.V., and Broude, N.E., eds. *DNA Amplification: Current Technologies and Applications*. Norfolk, U.K.: Horizon Bioscience; 3–20.

19. Rychlik, W., Spencer, W.J., and Rhoads, R.E. 1990. Optimization of the annealing temperature for DNA amplification in vitro. *Nucleic Acids Res* 18: 6409–6412.

20. Tabor, S., and Richardson, C.C. 1995. A single residue in DNA polymerases of the *Escherichia coli* DNA polymerase I family is critical for distinguishing between deoxy- and dideoxyribonucleotides. *Proc Natl Acad Sci U S A* 92: 6339–6343.

21. Villbrandt, B., Sobek, H., Frey, B., et al. 2000. Domain exchange: chimeras of *Thermus aquaticus* DNA polymerase, *Escherichia coli* DNA polymerase I and *Thermotoga neapolitana* DNA polymerase. *Protein Eng* 13: 645–654.

22. Wang, Y., Prosen, D.E., Mei, L., et al. 2004. A novel strategy to engineer DNA polymerases for enhanced processivity and improved performance in vitro. *Nucleic Acids Res* 32: 1197–1207.

23. Yang, S.-W., Astatke, M., Potter, J., et al. 2002. Mutant *Thermotoga neapolitana* DNA polymerase I: altered catalytic properties for non-templated nucleotide addition and incorporation of correct nucleotides. *Nucleic Acids Res* 30: 4314–4320.

# 14 Prolonged Storage of Plasmid DNAs Under Different Conditions: Effects on Plasmid Integrity, Spectral Characteristics, and DNA Sequence Quality

Jan Kieleczawa and Paul Wu
*Wyeth Research, Cambridge, Massachusetts, USA*

The understanding of DNA stability during a prolonged storage under different conditions is important for several reasons. For one, the usage of naked DNAs for gene therapy gains in popularity as it is viewed as an alternative to viral and liposomal gene transfer technologies (10, 11, 14, 17). To fulfill its potential, the plasmid needs to be in as intact a form as possible for prolonged periods of time (4). When stored in aqueous solutions, DNA is thought to degrade in a two-step process of depurination and β-elimination (6, 12). Unfavorable environmental conditions significantly impact the rates of these degrading processes (6).

Secondly, high quality plasmid DNA is essential for many biological applications, including transformation, restriction analysis, PCR, and DNA sequencing. The quality of a plasmid DNA preparation is highly dependent on the method used and subsequent storage conditions (17). If a good purification method is used, then the plasmid DNA is stable for many years as measured, for example, by transformation efficiency and agarose gel electrophoresis (16). On the other hand, lower quality purification methods may lead to a faster degradation of plasmid DNA. At least three factors could contribute to the acceleration of these processes. First, the residual endonuclease I activity will degrade DNA. The example of such activity was observed when Qiagen preparation method was used to isolate plasmid DNA from *E. coli* D1210 cell line; after the overnight storage at 4°C in 10 mM Tris-Cl, pH 8.0, almost all of the plasmid DNA was degraded. On the other hand, when the same plasmid was isolated using EdgeBiosystems preparation method, the DNA remained largely intact (J.K., unpublished results). Second, trace levels of transition metal ions leads to the formation of hydroxyl radicals, which over time may produce oxidized bases and strand breaks (3, 7). Third, prolonged exposure to an acidic environment during plasmid preparation may lead to depurination, which eventually causes hydrolysis of DNA molecules (6). Therefore, the preparation protocols and storage conditions that minimize the effects of the above-mentioned factors would lead to more stable plasmid DNAs.

This chapter examines DNA sequencing patterns, spectral characteristics, and band integrity for two plasmid DNAs stored for up to two months under several different conditions. We also examined the effect of BigDye™ dye-terminator mix stored for several days at various temperatures on read length and signal strength. In addition (though this is not related to the main topic of this chapter), various combinations of dye-terminators and 5× dilution buffers were evaluated to find the optimal pair. We believe that data presented in this chapter will be very practical for those who are worried about leaving DNA samples at elevated temperatures, either by accident or when a shipment of samples takes several days to arrive. Our data show that leaving sequencing mix even at elevated temperatures for several days does not lead to the destruction (or decrease) of the activity of this enzyme, or to an increase in number of misreads compared to an enzyme mix stored under recommended conditions.

## Materials

Unmodified pGem3zf vector was purchased from Promega (Promega Corporation, Madison, WI, USA) and plasmid p1467 containing 630 base-pair insert in pDONR221 vector (pDONR221; Invitrogen, Carlsbad, CA, USA) was prepared at Wyeth. The pGem3zf will be referred to as DNA #1 and p1467 as DNA #2 throughout the remaining of this chapter. Agarose was purchased from Invitrogen.

The BigDye terminator mix v3.1 was purchased from Applied Biosystems (Foster City, CA, USA). Five times BigDye dilution buffer (100 mM Tris, pH 8.9, 10 mM $MgCl_2$) was prepared in house from reagents of highest purity. All other reagents were of highest available purity.

## Methods

Two plasmid DNAs were transformed into Mach1 T1R competent cells (Invitrogen) and plated onto agar plates with either ampicillin (100 μg/mL) or kanamycin (50 μg/mL). DNAs from an overnight 100 μL culture were prepared using Maxi preparation kit (Marilgen Biosciences Inc., Ijamsville, MD, USA). The DNAs were resuspended at the concentrations of 0.92 (pGem3zf) and 0.96 (p1467) mg/mL in 10 mM Tris-Cl, 0.01 mM EDTA, pH 8.0 (*TEsl*). Immediately after the plasmid preparation, the following aliquots were distributed to 0.5 mL tubes for various storage conditions:

1. DNA sequencing analysis. To 0.71 μL of DNA #1 or 0.68 μL of DNA #2 and add either water or TEsl to a final volume of 78 μL. The final amount for sequencing is 50 ng of DNA/reaction. (A more detailed description is in the DNA sequencing section below.)
2. Agarose gel analysis. Take 0.43 μL of DNA #1 or 0.42 μL of DNA #2 and add water or TEsl to final volume of 20 μL. This will result in 200 ng of DNA/lane. (A more detailed description is in the agarose gel electrophoresis section below.)
3. Spectral analysis. To 95 μL of water or TEsl, add 5 μL of either DNA #1 or #2 and distribute 7 μL for each storage condition. This will assure five to six spectral measurements from the same sample. (A more detailed description is in the spectral analysis section below.)

After all aliquots were distributed, the tubes were placed at room temperature (22°–23°C), at 4°C, at −20°C and at −80°C. In all cases, the samples were protected from light. For time zero (t = 0) day controls, DNA sequencing and spectral analysis were carried out on the day of the plasmid preparation. For agarose gel analysis, aliquots were flash-frozen on dry ice and placed in a −80°C freezer. Similarly, after one week, two weeks, one month, and two months, respective aliquots were subjected to DNA sequencing and spectral analyses, but samples for agarose analysis were treated in the same fashion as for t = 0 controls until samples for all time points were collected.

### DNA Sequencing

Two protocols were used for DNA sequencing. Protocol #1 (referred henceforth as *standard*) corresponds to an ABI-like protocol described earlier (1). Briefly, to 78 μL of DNA mix that is aliquoted for sequencing, add 13 μL of 5 μM M13 forward primer 5'-TGTAAAACGACGGCCAGT; the same primer is used for both plasmid DNAs. Vortex thoroughly and split 7 μL of this mix into 12 separate 0.2 mL tubes. Add 3 μL of 1.5 diluted BigDye v3.1 terminator mix (0.8 mL of undiluted Taq mix + 0.4 mL of the ABI 5' dye-terminator buffer = 4 × final Taq dilution), and start the cycle sequencing protocol (5 sec at 50°C/2 min at 60°C/10 sec at 96°C) × 40.

Protocol #2 (referred henceforth as *modified*). To 78 μL of DNA mix aliquoted for sequencing, add 13 μL of 5 μM M13 forward primer as above. Vortex and split 7 μL of this mix into 12 separate 0.2 mL tubes. Heat-denature the samples at 98°C for 2.5 minutes for the DNAs stored in water or for 5 minutes for the DNAs stored in TEsl (8). Place the tubes on ice, add Taq dye-terminator mix as above, and cycle as in Protocol #1.

For experiments where read length and signal strength were evaluated after storing dye-terminator mix for several days at various temperatures, the aliquots of diluted (as above) BigDye v3.1 mix were stored for 1, 3, 6, and 8 days at 4°C room temperature and at 37°C. DNA sequencing was carried out using both Protocols #1 and #2.

For experiments where two different versions of dye-terminators were mix-and-matched with two different 5× dilution buffers, the same final dilution factor was used.

Following the cycling regime, which was the same for all experiments reported in this study, excess unincorporated dye-terminators was removed as described earlier (9). Purified samples were heat-denatured for two minutes at 90° to 95°C, and the data were collected on an ABI3730 DNA genetic analyzer (purchased from Applied Biosystems, Foster City, CA, USA) using conditions suggested by the manufacturer.

For each individual sample, read length (using Q ≥ 20 phred value, 5) and signal strength were recorded. The values presented in all of the tables are the average of 12 data points unless stated otherwise.

### *Agarose Gel Electrophoresis*

After all samples were collected (2 months from the start of the experiment), each original 20 μL volume was split into two 10-μL aliquots (so each aliquot contained 200 ng DNA). One aliquot was heat-denatured for 2.5 minutes (samples stored in water) or 5 minutes (samples stored in TEsl), and the second aliquot was left untreated. To all samples, 1 μL of 10× DNA loading buffer was added and electrophoresis was run for 1.5 to 2 hours as described earlier (2, 13). The digital image of an agarose gel was then taken using the EDAS 290 system (Eastman Kodak Company, Rochester, NY, USA).

### Spectral Analysis

One-microliter aliquots of DNA sample were placed in the ND-1000 Nanodrop™ spectrophotometer (purchased from Nanodrop Technologies, Wilmington, DE, USA) and absorbancies at 230 nm, 260 nm, and 280 nm as well as ratios 260/280 and 260/230, and concentrations (ng/µL) were recorded. Each data point is an average of five to six measurements.

## Results and Discussion

Table 14-1 shows the comprehensive spectral data for all storage conditions. In general, data for DNA stored in TEsl were less variable compared to the DNA stored in water and, not surprisingly, the spectral values for DNAs stored at −80°C were the most consistent. There are number of higher values (DNA concentrations after one week of storage at 4°C and at −20°C, and all ratios for both DNAs stored in TEsl for 1 month that may not be consistent with the overall data. The cause(s) for this is unclear at the moment, and more studies (more plasmid DNAs and different preparation methods) are needed to investigate the observed phenomenon further. Chapter 16 contains a detailed explanation of all spectral ratios and their changes.

There was very little effect on the sequence read lengths when DNAs were stored in TEsl for up to two months regardless of the storage temperature (Tables 14-2 and 14-3). Changes in signal strength follow the established pattern with signal strengths being two- to three-fold higher after heat-denaturation step was added to a sequencing protocol (8).

When DNAs were stored in water, a small decrease (3%–8%) in read lengths was seen only after one month of storage at room temperature when using the standard sequencing protocol. After two months of storage under the same conditions, read lengths fell to 35% to 50% of the original values. More dramatic changes were observed when the modified sequencing protocol was used. The 17% to 23% and 49% to 62% shorter reads were observed after one and two months of storage in water at room temperature, respectively. These observations could be correlated to agarose gel data shown in Figure 14-1. After one month of storage (room temperature in water) there was a significant redistribution of DNA forms toward a nicked form and the DNA was completely degraded after two months. This was further underscored when DNA samples were heat-denatured; only DNAs stored for one and two months were completely converted to a degraded form that later had a

**Table 14-1.** Spectral data for all samples stored under indicated conditions.

| Time | DNA | Solution | 23°C/RT | | | 4°C | | |
|------|-----|----------|---------|---------|------|---------|---------|------|
| | | | 260/280 | 260/230 | Conc. | 260/280 | 260/230 | Conc. |
| 0 Days | 1 | H₂O | 1.73 ± 0.03 | 2.10 ± 0.04 | 46.05 ± 1.11 | NA | NA | NA |
| | | TEsl | 1.85 ± 0.06 | 2.10 ± 0.04 | 45.50 ± 1.23 | NA | NA | NA |
| | 2 | H₂O | 1.74 ± 0.05 | 2.06 ± 0.04 | 48.40 ± 1.13 | NA | NA | NA |
| | | TEsl | 1.78 ± 0.02 | 1.79 ± 0.03 | 47.82 ± 0.92 | NA | NA | NA |
| 7 Days | 1 | H₂O | 1.90 ± 0.07 | 2.31 ± 0.06 | 44.80 ± 0.63 | 1.91 ± 0.04 | 2.28 ± 0.03 | 57.52 ± 0.63 |
| | | TEsl | 1.97 ± 0.03 | 2.28 ± 0.07 | 47.10 ± 0.98 | 1.98 ± 0.06 | 2.26 ± 0.01 | 54.25 ± 0.65 |
| | 2 | H₂O | 1.88 ± 0.06 | 2.28 ± 0.07 | 50.34 ± 1.96 | 1.87 ± 0.02 | 2.25 ± 0.03 | 62.58 ± 2.43 |
| | | TEsl | 1.89 ± 0.05 | 1.97 ± 0.06 | 50.15 ± 1.63 | 1.92 ± 0.03 | 1.77 ± 0.06 | 52.55 ± 3.17 |
| 14 Days | 1 | H₂O | 1.81 ± 0.03 | 2.15 ± 0.06 | 47.24 ± 0.64 | 1.74 ± 0.05 | 1.78 ± 0.03 | 49.44 ± 1.53 |
| | | TEsl | 1.91 ± 0.05 | 2.36 ± 0.05 | 47.67 ± 0.47 | 1.84 ± 0.04 | 2.25 ± 0.07 | 48.73 ± 1.96 |
| | 2 | H₂O | 1.75 ± 0.05 | 1.99 ± 0.03 | 51.92 ± 1.13 | 1.78 ± 0.07 | 1.99 ± 0.02 | 49.28 ± 3.24 |
| | | TEsl | 1.89 ± 0.04 | 2.07 ± 0.01 | 50.83 ± 0.74 | 1.86 ± 0.02 | 1.85 ± 0.04 | 54.00 ± 4.59 |
| 28 Days | 1 | H₂O | 1.78 ± 0.01 | 2.01 ± 0.02 | 52.33 ± 1.17 | 1.78 ± 0.05 | 2.13 ± 0.06 | 47.13 ± 2.66 |
| | | TEsl | 2.09 ± 0.03 | 4.67 ± 0.16 | 49.24 ± 1.40 | 2.18 ± 0.08 | 6.08 ± 0.72 | 43.28 ± 1.50 |
| | 2 | H₂O | 1.73 ± 0.10 | 1.62 ± 0.04 | 57.19 ± 0.79 | 1.74 ± 0.01 | 2.05 ± 0.04 | 51.72 ± 1.90 |
| | | TEsl | 2.11 ± 0.02 | 4.01 ± 0.08 | 49.44 ± 0.46 | 2.06 ± 0.01 | 3.31 ± 0.18 | 47.09 ± 1.04 |
| 56 Days | 1 | H₂O | 1.73 ± 0.10 | 1.70 ± 0.08 | 61.40 ± 3.03 | 1.78 ± 0.16 | 2.11 ± 0.30 | 53.33 ± 5.12 |
| | | TEsl | 1.87 ± 0.04 | 2.19 ± 0.03 | 51.42 ± 1.24 | 1.85 ± 0.04 | 2.31 ± 0.04 | 49.71 ± 0.45 |
| | 2 | H₂O | 1.81 ± 0.07 | 2.02 ± 0.03 | 63.43 ± 0.86 | 1.78 ± 0.22 | 2.14 ± 0.01 | 51.13 ± 2.40 |
| | | TEsl | 1.77 ± 0.03 | 1.74 ± 0.02 | 60.38 ± 0.66 | 1.87 ± 0.04 | 1.89 ± 0.06 | 54.25 ± 0.42 |

Abbreviations are DNA #1, pGem3zf; DNA #2, p1467 plasmid; TEsl, Tris-Cl, 0.01 mM EDTA, pH 8.0; RL, read length; SS, signal strength; RT, room temperature; No HD, sequencing protocol without heat-denaturation; + HD, sequencing protocol with heat-denaturation; NA, not applicable. For time = 0 days (measurements performed on the day of DNA preparations) only one set of data is presented and is placed under RT column. Conc., concentration of DNA in ng/μL. The original DNA was diluted 20-fold. The presented data are the average of 5–6 measurements taken from the same aliquot stored under indicated conditions.

more negative effect when the modified sequencing protocol was used.

The changes in signal strengths were much more complex and highly dependent on the duration time and storage solution. The higher signal strength value indicated that more of the "sequenceable" DNA form was accessible during the cycling step. However, a higher degree of DNA degradation resulted in overall shorter sequencing reads.

When DNAs were stored at other studied temperatures/solutions, there were no significant changes in read lengths, indicating that from the DNA sequencing point of view, samples can be stored for prolonged periods of time without any significant impact on the quality of the data. The changes in signal strength, though significant, have no effect on the quality of reads. In general, the data presented in this section indicate that it is "safer" to store DNA in TEsl (compared to storage in water) under all of the studied conditions.

| -20°C | | | -80°C | | |
|---|---|---|---|---|---|
| 260/280 | 260/230 | Conc. | 260/280 | 260/230 | Conc. |
| NA | NA | NA | NA | NA | NA |
| NA | NA | NA | NA | NA | NA |
| NA | NA | NA | NA | NA | NA |
| NA | NA | NA | NA | NA | NA |
| 1.90 ± 0.03 | 2.26 ± 0.01 | 57.95 ± 0.99 | 1.87 ± 0.02 | 2.26 ± 0.01 | 49.5 ± 3.50 |
| 1.92 ± 0.09 | 2.24 ± 0.08 | 58.45 ± 0.83 | 1.94 ± 0.05 | 2.25 ± 0.05 | 44.0 ± 2.81 |
| 1.92 ± 0.02 | 2.29 ± 0.04 | 74.57 ± 4.00 | 1.84 ± 0.05 | 2.26 ± 0.04 | 48.4 ± 1.91 |
| 1.90 ± 0.05 | 1.81 ± 0.04 | 62.82 ± 1.77 | 1.92 ± 0.06 | 1.87 ± 0.09 | 45.9 ± 2.86 |
| 1.82 ± 0.05 | 2.14 ± 0.03 | 46.67 ± 1.36 | 1.84 ± 0.01 | 2.21 ± 0.06 | 45.11 ± 0.66 |
| 1.88 ± 0.02 | 2.15 ± 0.02 | 46.18 ± 0.82 | 1.91 ± 0.03 | 2.31 ± 0.03 | 46.19 ± 3.18 |
| 1.85 ± 0.04 | 2.17 ± 0.02 | 42.40 ± 0.80 | 1.85 ± 0.05 | 2.21 ± 0.03 | 42.39 ± 0.15 |
| 1.87 ± 0.04 | 1.77 ± 0.01 | 49.03 ± 3.00 | 1.83 ± 0.03 | 1.91 ± 0.03 | 50.23 ± 2.57 |
| 1.76 ± 0.06 | 2.18 ± 0.07 | 47.33 ± 1.60 | 1.80 ± 0.02 | 2.22 ± 0.02 | 45.22 ± 1.84 |
| 1.99 ± 0.01 | 3.48 ± 0.69 | 49.01 ± 4.33 | 2.15 ± 0.08 | 7.39 ± 0.96 | 40.35 ± 1.55 |
| 1.76 ± 0.04 | 2.05 ± 0.17 | 46.34 ± 1.61 | 1.76 ± 0.03 | 2.08 ± 0.12 | 51.11 ± 4.26 |
| 2.07 ± 0.09 | 4.77 ± 0.81 | 44.56 ± 3.11 | 2.13 ± 0.03 | 3.59 ± 0.10 | 45.48 ± 1.31 |
| 1.82 ± 0.01 | 2.14 ± 0.03 | 46.43 ± 4.86 | 1.83 ± 0.02 | 2.23 ± 0.05 | 45.33 ± 0.80 |
| 1.94 ± 0.05 | 2.32 ± 0.05 | 43.45 ± 1.66 | 1.82 ± 0.04 | 2.40 ± 0.05 | 46.03 ± 1.80 |
| 1.81 ± 0.01 | 1.93 ± 0.05 | 46.55 ± 2.21 | 1.81 ± 0.04 | 2.26 ± 0.09 | 46.24 ± 0.68 |
| 1.86 ± 0.07 | 1.88 ± 0.05 | 46.73 ± 1.22 | 1.83 ± 0.03 | 1.83 ± 0.03 | 50.82 ± 1.44 |

The data presented in Figures 14-2 and 14-3 and Table 14-4 are important, at least when automation of the sequencing reaction is concerned. It is clear that there is no need to rigorously control the storage temperature of dye-terminator mix, as there were no significant changes in read lengths, signal strengths, or number of sequencing misreads even after eight days of storage at 37°C. This also has practical implications, as this enzyme can be safely shipped without taking any extreme temperature (like packing the shipment on dry ice) and time precautions. In another experiment, the dGTP v3.0 dye-terminator mix was stored for three days at room temperature without any effect on the quality of the DNA sequencing data (J.K., unpublished observations).

According to ABI's marketing literature, the BigDye terminator v3.1 works optimally with 5× 3.1 dilution buffer, which is quite expensive if purchased separately. We demonstrate that this buffer does not present any appreciable advantages, at least when the standard DNA sequencing template, pGem3zf, is used as shown in Figures 14.4 and 14.5. It remains to be determined if any advantages may be seen when using more challenging templates.

**Table 14-2.** Read lengths and signal strengths data for two DNAs stored at room temperature or at 4°C.

| | | | 23°C/RT | | | |
| | | | No HD | | +HD | |
| Time | DNA | Storage Solution | RL | SS | RL | SS |
|---|---|---|---|---|---|---|
| 0 Days | 1 | $H_2O$ | 917.7 ± 49.3 | 122.5 ± 22.2 | 955.3 ± 9.0 | 457.3 ± 56.9 |
| | | TEsl | 927.3 ± 15.3 | 153.4 ± 20.2 | 940.1 ± 15.3 | 705.7 ± 70.5 |
| | 2 | $H_2O$ | 929.7 ± 11.8 | 132.0 ± 15.3 | 938.0 ± 13.2 | 406.9 ± 65.6 |
| | | TEsl | 922.7 ± 19.0 | 123.1 ± 15.6 | 969.2 ± 12.9 | 475.1 ± 84.5 |
| 7 Days | 1 | $H_2O$ | 941.1 ± 18.9 | 1113.5 ± 214.4 | 940.4 ± 15.2 | 713.6 ± 152.2 |
| | | TEsl | 943.2 ± 20.6 | 292.7 ± 39.1 | 953.9 ± 23.4 | 1130.9 ± 137.8 |
| | 2 | $H_2O$ | 983.3 ± 9.3 | 866.1 ± 127.5 | 947.6 ± 15.7 | 602.8 ± 97.4 |
| | | TEsl | 951.2 ± 10.4 | 193.3 ± 29.9 | 973.3 ± 12.9 | 889.4 ± 216.7 |
| 14 Days | 1 | $H_2O$ | 918.2 ± 25.0 | 449.3 ± 103.4 | 905.3 ± 21.9 | 420.0 ± 61.6 |
| | | TEsl | 920.3 ± 26.7 | 219.6 ± 66.9 | 931.4 ± 14.9 | 1036.0 ± 279.6 |
| | 2 | $H_2O$ | 944.0 ± 15.0 | 569.8 ± 126.6 | 915.3 ± 46.3 | 498.1 ± 70.3 |
| | | TEsl | 934.2 ± 16.6 | 190.5 ± 27.7 | 937.8 ± 16.8 | 662.3 ± 208.8 |
| 28 Days | 1 | $H_2O$ | 849.0 ± 49.4 | 437.2 ± 109.5 | 734.6 ± 29.1 | 524.6 ± 84.2 |
| | | TEsl | 918.7 ± 14.8 | 419.0 ± 72.1 | 931.8 ± 20.6 | 1452.5 ± 361.8 |
| | 2 | $H_2O$ | 861.7 ± 38.3 | 611.3 ± 165.5 | 782.6 ± 46.5 | 579.6 ± 123.9 |
| | | TEsl | 900.8 ± 18.4 | 323.9 ± 81.3 | 954.2 ± 23.3 | 707.1 ± 159.9 |
| 56 Days | 1 | $H_2O$ | 464.9 ± 34.4 | 387.3 ± 60.7 | 366.7 ± 22.7 | 391.2 ± 48.4 |
| | | TEsl | 955.2 ± 14.1 | 362.6 ± 85.5 | 952.8 ± 26.3 | 1181.6 ± 293.5 |
| | 2 | $H_2O$ | 611.8 ± 59.9 | 470.0 ± 118.9 | 479.3 ± 32.7 | 482.3 ± 137.1 |
| | | TEsl | 970.8 ± 28.4 | 302.1 ± 64.2 | 960.6 ± 25.1 | 748.5 ± 261.5 |

Abbreviations are DNA #1, pGem3zf; DNA #2, p1467 plasmid; TEsl, Tris-Cl, 0.01 mM EDTA, pH 8.0; RL, read length; SS, signal strength; RT, room temperature; No HD, sequencing protocol without heat-denaturation; + HD, sequencing protocol with heat-denaturation; NA, not applicable.

| 4°C | | | |
| --- | --- | --- | --- |
| No HD | | +HD | |
| RL | SS | RL | SS |
| NA | NA | NA | NA |
| NA | NA | NA | NA |
| NA | NA | NA | NA |
| NA | NA | NA | NA |
| 937.1 ± 21.8 | 614.7 ± 117.0 | 942.8 ± 23.8 | 733.6 ± 146.1 |
| 929.2 ± 63.6 | 249.9 ± 62.1 | 956.7 ± 20.9 | 964.6 ± 328.8 |
| 977.7 ± 8.5 | 558.6 ± 144.1 | 976.1 ± 13.8 | 762.3 ± 104.7 |
| 963.9 ± 16.9 | 176.1 ± 25.8 | 978.5 ± 6.6 | 851.6 ± 183.0 |
| 889.7 ± 29.1 | 664.3 ± 241.2 | 903.2 ± 18.4 | 779.1 ± 89.7 |
| 903.6 ± 33.3 | 239.2 ± 29.9 | 921.4 ± 14.9 | 1035.1 ± 120.5 |
| 954.8 ± 16.3 | 709.1 ± 117.0 | 953.9 ± 23.5 | 793.3 ± 128.4 |
| 926.5 ± 22.3 | 176.1 ± 32.2 | 932.0 ± 17.9 | 770.3 ± 137.3 |
| 916.5 ± 29.8 | 1126.2 ± 274.5 | 934.4 ± 22.0 | 854.5 ± 92.3 |
| 914.9 ± 24.3 | 341.3 ± 101.3 | 935.3 ± 23.4 | 1263.1 ± 192.6 |
| 922.5 ± 25.4 | 1070.6 ± 264.1 | 932.3 ± 36.6 | 810.3 ± 327.9 |
| 919.2 ± 27.5 | 247.1 ± 76.8 | 935.7 ± 27.6 | 1045.8 ± 597.0 |
| 922.4 ± 53.0 | 1268.7 ± 285.9 | 906.4 ± 66.1 | 861.3 ± 165.4 |
| 917.5 ± 45.4 | 349.5 ± 96.7 | 929.7 ± 29.6 | 1353.7 ± 237.2 |
| 946.7 ± 17.7 | 1082.2 ± 263.4 | 931.3 ± 27.1 | 571.5 ± 176.8 |
| 928.5 ± 50.6 | 197.4 ± 27.5 | 944.1 ± 46.4 | 725.8 ± 223.9 |

**Table 14-3.** Read lengths and signal strengths data for two DNAs stored at −20°C and at −80°C.

| | | | −20°C | | | |
|---|---|---|---|---|---|---|
| | | | No HD | | +HD | |
| Time | DNA | Storage Solution | RL | SS | RL | SS |
| 0 Days | 1 | $H_2O$ | 917.7 ± 49.3 | 122.5 ± 22.2 | 955.3 ± 9.0 | 457.3 ± 56.9 |
| | | TEsl | 927.3 ± 15.3 | 153.4 ± 20.2 | 940.1 ± 15.3 | 705.7 ± 70.5 |
| | 2 | $H_2O$ | 929.7 ± 11.8 | 132.0 ± 15.3 | 938.0 ± 13.2 | 406.9 ± 65.6 |
| | | TEsl | 922.7 ± 19.0 | 123.1 ± 15.6 | 969.2 ± 12.9 | 475.1 ± 84.5 |
| 7 Days | 1 | $H_2O$ | 939.1 ± 22.5 | 249.8 ± 41.4 | 937.2 ± 18.5 | 803.2 ± 171.7 |
| | | TEsl | 937.1 ± 20.8 | 204.1 ± 39.6 | 953.6 ± 27.6 | 940.6 ± 106.6 |
| | 2 | $H_2O$ | 978.3 ± 9.9 | 228.3 ± 31.7 | 976.8 ± 15.1 | 526.7 ± 77.5 |
| | | TEsl | 962.6 ± 14.9 | 153.3 ± 21.0 | 980.1 ± 9.6 | 744.7 ± 130.8 |
| 14 Days | 1 | $H_2O$ | 918.8 ± 40.9 | 216.3 ± 38.4 | 922.2 ± 16.3 | 892.4 ± 135.6 |
| | | TEsl | 908.9 ± 25.4 | 226.3 ± 32.2 | 918.4 ± 16.3 | 1161.0 ± 199.6 |
| | 2 | $H_2O$ | 944.1 ± 15.3 | 345.4 ± 56.3 | 939.6 ± 22.8 | 855.8 ± 115.4 |
| | | TEsl | 915.0 ± 15.8 | 180.1 ± 40.2 | 928.1 ± 25.7 | 534.9 ± 156.2 |
| 28 Days | 1 | $H_2O$ | 902.4 ± 36.0 | 325.2 ± 88.3 | 923.8 ± 26.9 | 855.5 ± 189.4 |
| | | TEsl | 916.8 ± 32.7 | 316.8 ± 95.9 | 938.7 ± 19.3 | 1255.7 ± 284.5 |
| | 2 | $H_2O$ | 915.8 ± 32.0 | 382.7 ± 102.9 | 922.9 ± 33.8 | 833.1 ± 300.9 |
| | | TEsl | 905.5 ± 25.5 | 186.5 ± 40.6 | 937.5 ± 16.7 | 890.1 ± 328.6 |
| 56 Days | 1 | $H_2O$ | 900.3 ± 80.7 | 380.0 ± 71.9 | 916.4 ± 32.7 | 914.1 ± 174.6 |
| | | TEsl | 920.8 ± 26.9 | 278.7 ± 37.5 | 946.3 ± 23.5 | 1147.0 ± 188.1 |
| | 2 | $H_2O$ | 946.3 ± 23.5 | 372.2 ± 67.1 | 953.7 ± 20.7 | 728.3 ± 198.0 |
| | | TEsl | 934.5 ± 39.7 | 178.3 ± 28.7 | 958.8 ± 20.8 | 774.5 ± 158.2 |

Abbreviations are DNA #1, pGem3zf; DNA #2, p1467 plasmid; TEsl, Tris-Cl, 0.01 mM EDTA, pH 8.0; RL, read length; SS, signal strength; RT, room temperature; No HD, sequencing protocol without heat-denaturation; + HD, sequencing protocol with heat-denaturation; NA, not applicable.

| −80°C | | | |
| --- | --- | --- | --- |
| No HD | | +HD | |
| RL | SS | RL | SS |
| NA | NA | NA | NA |
| NA | NA | NA | NA |
| NA | NA | NA | NA |
| NA | NA | NA | NA |
| 863.3 ± 145.9 | 185.3 ± 52.9 | 906.4 ± 93.4 | 626.6 ± 187.9 |
| 929.3 ± 45.9 | 229.0 ± 45.2 | 945.8 ± 28.5 | 887.2 ± 264.4 |
| 961.7 ± 35.7 | 234.0 ± 50.0 | 966.6 ± 22.9 | 592.9 ± 158.7 |
| 926.4 ± 62.9 | 148.5 ± 42.3 | 955.0 ± 26.4 | 633.1 ± 189.2 |
| 869.2 ± 51.2 | 159.0 ± 72.4 | 915.7 ± 23.5 | 647.3 ± 190.2 |
| 910.6 ± 23.9 | 225.1 ± 23.8 | 922.1 ± 17.7 | 1152.3 ± 130.9 |
| 940.6 ± 38.0 | 261.7 ± 65.7 | 936.7 ± 24.0 | 843.9 ± 243.8 |
| 916.2 ± 20.4 | 163.0 ± 29.8 | 922.6 ± 16.7 | 568.8 ± 276.0 |
| 929.7 ± 21.7 | 358.4 ± 76.9 | 932.4 ± 28.4 | 1092.6 ± 141.8 |
| 927.8 ± 22.0 | 326.1 ± 61.8 | 938.4 ± 26.2 | 1275.5 ± 211.1 |
| 930.9 ± 23.1 | 379.9 ± 63.9 | 942.8 ± 22.1 | 1323.7 ± 388.5 |
| 933.2 ± 27.8 | 230.3 ± 65.8 | 940.0 ± 19.8 | 673.6 ± 252.2 |
| 918.7 ± 34.0 | 324.1 ± 68.4 | 922.8 ± 24.1 | 1014.1 ± 202.4 |
| 911.0 ± 26.3 | 242.1 ± 33.7 | 918.8 ± 33.4 | 974.6 ± 123.4 |
| 941.7 ± 29.3 | 296.3 ± 75.2 | 942.9 ± 27.9 | 702.1 ± 124.8 |
| 925.8 ± 53.4 | 198.2 ± 36.2 | 945.1 ± 27.7 | 703.7 ± 173.8 |

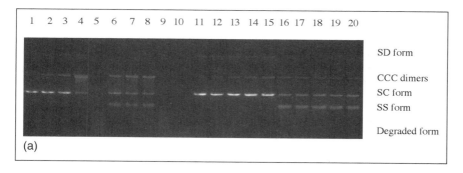

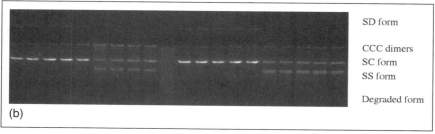

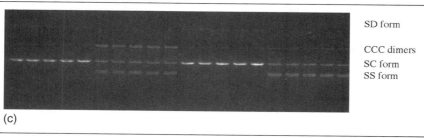

**Figure 14-1. Storage of plasmid DNAs under various conditions.** For all parts of this figure, lanes 1–10 indicate DNAs stored in H$_2$O. For lanes 11–20 DNAs were stored in TEsl. Samples 1–5 and 11–15 were untreated, samples 6–10 were denatured for 2.5 minutes at 98°C and samples 16–20 were denatured for 5 minutes at 98°C as described in the text. Lanes 1, 6, 11, and 16 were stored for 0 days*; lanes 2, 7, 12, and 17 for 1 week; lanes 3, 8, 13, and 18 for 2 weeks; lanes 4, 9, 14, and 19 for 1 month; and lanes 5, 10, 15, and 20 were stored for 2 months. *DNAs in lanes 1, 6, 11, and 16 were flash-frozen on dry ice on the day of the DNA preparation and stored at −80°C for two months. (a) Samples stored at room temperature; (b) samples stored at 4°C; (c) samples sotred at −20°C. Abbreviations are SD form, supercoiled dimer; oc form, nicked DNA; ccc, covalently closed circular form, that is, supercoiled DNA; ss form, single stranded form. This classification is according to Schmidt et al. (15) and Walther et al. (17).

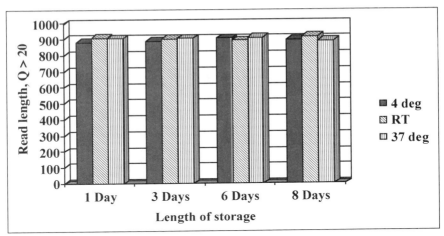

**Figure 14-2.** Effects of prolonged storage of Taq terminator mix on read length. Diluted aliquots of BigDye™-terminator mix v3.1 were stored at indicated temperatures for one to eight days. Each data point is the average of eight reads, and the standard deviation was in the range of 0.8% to 3.8% of averages. For the control, dye-terminator mix was stored at –20°C. The control Q ≥ 20 read length was 900.0 ± 24.0.

**Table 14-4.** Number of errors in different ranges after sequencing with BigDye™ terminator v3.1 stored under various temperatures and time conditions.

| Storage Time and Temperature | Number of Errors, Range | | | |
| --- | --- | --- | --- | --- |
| | 1–50 | 51–500 | 501–850 | 851–1000 |
| Control/–20°C | 1.0 ± 0.0 | 0.2 ± 0.4 | 0.0 ± 0.0 | 1.3 ± 1.0 |
| 1 Day/4°C | 0.3 ± 0.5 | 0.2 ± 0.4 | 0.0 ± 0.0 | 2.0 ± 1.1 |
| 3 Days/4°C | 0.8 ± 0.4 | 0.3 ± 0.8 | 0.2 ± 0.4 | 1.7 ± 0.5 |
| 6 Days/4°C | 0.5 ± 0.5 | 0.7 ± 0.8 | 0.2 ± 0.4 | 2.3 ± 2.0 |
| 8 Days/4°C | 0.8 ± 0.4 | 0.2 ± 0.4 | 0.0 ± 0.0 | 1.5 ± 0.5 |
| 1 Day/RT | 0.7 ± 0.5 | 0.3 ± 0.5 | 0.0 ± 0.0 | 2.0 ± 1.3 |
| 3 Days/RT | 1.0 ± 0.0 | 0.3 ± 0.8 | 0.0 ± 0.0 | 3.7 ± 3.1 |
| 6 Days/RT | 0.7 ± 0.5 | 0.2 ± 0.4 | 0.0 ± 0.0 | 1.2 ± 0.7 |
| 8 Days/RT | 1.0 ± 0.0 | 0.7 ± 0.5 | 0.0 ± 0.0 | 2.3 ± 0.8 |
| 1 Day/37°C | 0.8 ± 0.4 | 0.2 ± 0.4 | 0.0 ± 0.0 | 1.3 ± 0.5 |
| 3 Days/37°C | 0.8 ± 0.4 | 0.2 ± 0.4 | 0.0 ± 0.0 | 1.5 ± 0.8 |
| 6 Days/37°C | 0.8 ± 0.4 | 1.2 ± 1.2 | 0.0 ± 0.0 | 1.7 ± 0.5 |
| 8 Days/37°C | 0.7 ± 0.5 | 0.2 ± 0.4 | 0.0 ± 0.0 | 1.8 ± 1.2 |

RT = room temperature.

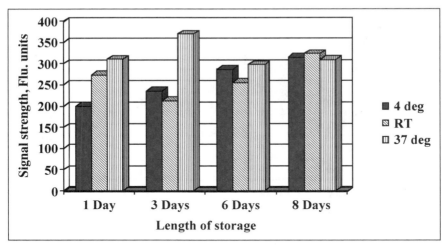

**Figure 14-3.** Effects of prolonged storage of Taq terminator mix on signal strength. Diluted aliquots of BigDye™ dye-terminator mix v3.1 were stored at indicated temperatures for one to eight days. Each data point is the average of eight reads and the standard deviation was in the range of 23% to 40% of averages. Signal strength is expressed in fluorescent units. The control value for signal strength is 235.4 ± 97.2 units.

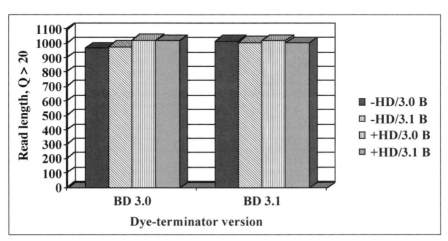

**Figure 14-4.** Various combinations of two versions of dye-terminator mixes with two different 5× dilution buffers; effect on read length. Each $Q \geq 20$ data point is the average of 12 reads, and the standard deviation is in the range of 0.6% to 6.0% of averages. The B in the key refers to the version of 5× dilution buffer.

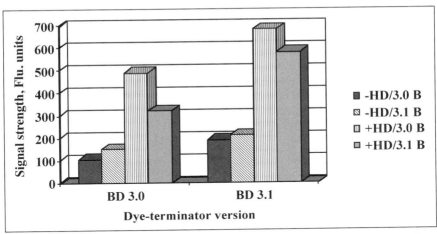

**Figure 14-5.** **Various combinations of two versions of dye-terminator mixes with two different 5× dilution buffers; effect on signal strength.** Each Q ≥ 20 data point is the average of 12 reads, and the standard deviation is in the range of 30% to 41% of averages.

## References

1. ABI PRISM® BigDye™ Terminator v3.1 Cycle Sequencing Kit. 2002. *Protocol.* Part number 4337035 Rev. A. Foster City, CA: Applied Biosystems.
2. Ausubel, F.M., Brent, R., Kingston, R.E., et al., eds. 1998. *Current Protocols in Molecular Biology.* New York, NY: John Wiley.
3. Demple, B., and Harrison, L. 1994. Repair of oxidative damage to DNA. *Annu Rev Biochem* 63: 915–948.
4. Evans, R.K., Xu, Z., Bohannon, K.E., Wang, B., et al. 2000. Evaluation of degradation pathways for plasmid DNA in pharmaceutical formulations via accelerated stability studies. *J Pharm Sci* 89: 76–87.
5. Ewing, B.G., Hillier, L., Wendl, M.C., and Green, P. 1998. Base-calling of automated sequencer traces using Phred. II. Error probabilities. *Genome Res.* 8: 186–194.
6. Friedberg, E.C., Walker, G.C., and Siede, W. 1995. *DNA Repair and Mutagenesis.* Washington, D.C.: ASM Press.
7. Imlay, J.A., and Linn, S. 1988. DNA damage and oxygen radical toxicity. *Science* 240: 1302–1308.
8. Kieleczawa, J. 2005. Controlled heat-denaturation of DNA plasmids. In: Kieleczawa, J. ed. *DNA Sequencing: Optimizing the Process and Analysis.* Sudbury, MA: Jones & Bartlett; 1–10.
9. Kieleczawa, J. 2005. Sequencing of difficult templates. In: Kieleczawa, J. ed. *DNA Sequencing: Optimizing the Process and Analysis.* Sudbury, MA: Jones & Bartlett; 27–34.

10. Klinman, D.M., Conover, J., Leiden, J.M., et al. 1999. Safe and effective regulation of hematocrit by gene gun administration of an erythropoietin-encoding DNA plasmid. *Hum Gene Ther* 10: 659–665.
11. Ledley, F.D. 1995. Nonviral gene therapy: the promise of genes as pharmaceutical products. *Hum Gene Ther* 6: 1129–1144.
12. Lindahl, T. 1993. Instability and decay of the primary structure of DNA. *Nature* 362: 709–715.
13. Sambrook, J., and Russell, D.W. 2001. *Molecular Cloning*, 3rd ed. Cold Spring Harbor, New York: Cold Spring Harbor Laboratory Press.
14. Schleef, M., Voss, C., and Schmidt, T. 2002. DNA drugs—production and quality assurance. *Engl Life Sci* 2: 157–160.
15. Schmidt, T., Friehs, K., Schleef, M., et al. 1999. Quantitative analysis of plasmid forms by agarose and capillary gel electrophoresis. *Anal Biochem* 274: 235–240.
16. Sun, N-E., Shen, B-H., Zhou, J-M., et al. 1994. An efficient method for large-scale isolation of plasmid DNAs by heat-alkali co-denaturation. *DNA Cell Biol* 13: 83–86.
17. Walther, W., Stein, U., Voss, C., et al. 2003. Stability analysis for long-term storage of naked DNA: impact on nonviral in vivo gene transfer. *Anal Biochem* 318: 230–235.

# 15    The DNABook™: A New Method to Store and Distribute Genome Resources

Midori Kobayashi[1] and
Yoshihide Hayashizaki[1,2,3]
[1]The DNABook Team, Preventure Program,
Japan Science and Technology Agency, Kanagawa,
Japan; [2]Genome Exploration Research Group and
the Core Group of the Genome Network Project,
RIKEN Genomic Sciences Center, RIKEN
Yokohama Institute, Kanagawa, Japan;
[3]Genome Science Laboratory, Discovery and
Research Institute, RIKEN Wako Main Campus,
Saitama, Japan

Biomolecules are usually stored in a buffer solution at low temperature to avoid the loss of their physiological activities. Compared to other biomolecules, DNA is chemically and physically stable; however, its solution is also stored in a freezer to avoid degradation by potentially present nucleases. DNA samples are usually delivered at low temperature, typically as *E. coli* transformants in tubes or titer plates in a freezer box filled with dry ice. To keep the DNA samples at low temperature, refrigerators, freezers, or cold rooms are essential in any bioscience laboratory. If the number of samples is large, then a substantial number of freezers taking considerable space is required to store the samples in addition to high maintenance costs. Furthermore, finding a particular sample may be a time-consuming and tedious task. Thus, conventional freezing systems are not ideal for storage and distribution of large numbers of samples.

The number of genome and transcriptome experiments has increased exponentially in recent years, producing an enormous quantity of genome resources including physical clones, PCR fragments, RNAi libraries, etc. Almost 1500 genome-sequencing projects of about 500 eukaryotic organisms have either been completed or are in progress (http://www.genomesonline.org; accessed July 2005). In addition, a number of cDNA sequencing projects is under way for organisms of scientific or practical importance (21). Furthermore, a large number of RNAi constructs that are effective in knock-down analysis of *Drosophila melanogaster* and *Caenorhabditis elegans* have been reported (2, 10), and siRNA expression libraries targeting tens of thousands of genes in human and mouse have been constructed (14). In this post-genomic era, genome network analyses of protein-protein interactions, protein-DNA interactions, RNA-DNA interactions, RNA-protein interactions are becoming the mainstream of biomedical research. For such analyses, a large number of clones, rather than a single clone, are used. To effectively carry out the genome network analysis, the scientific community must be able to quickly share high-quality genome resources and information. Following the innovative progress in computer technology, sequence information is digitally stored in databases and the information can be rapidly distributed to the scientific community via the Internet.

However, the system of distributing physical genome resources has not progressed accordingly. Clones are stored in a large number of freezers and the delivery is still achieved using tubes or plates stuffed into a freezer box with dry ice. There has been a need for new technology for the storage and distribution of genome resources. In our view, there are two keys to an improved distribution of physical biological resources:

1. An ability to dry samples for their convenient handling at room temperature.
2. An ability to "print" multiple copies of the same resource for simultaneous worldwide distribution.

This chapter describes a novel approach to solve these issues.

## DNABook™ Technology

### Concept

The idea for the DNABook came from the Genome Exploration Research Group (GERG) at the RIKEN Genomic Science Center (GSC) while working on the RIKEN Mouse Encyclopedia project, which began in 1995. The aim of the project is to systematically collect the physical full-length cDNA clones covering almost all of the mouse transcriptome by developing a series of innovative technologies and analysis systems. Among the many achievements of the project, one notable example is the cap-trapper technology for the construction of high-quality full length cDNA libraries (3–8). The specific goals of the projects are to supply: (a) a clone bank of all mouse full length cDNAs; (b) a cDNA sequence database with functional annotation; (c) genome mapping of the cDNAs; (d) an expression profile database of these cDNA clones; (e) a protein-protein interaction database; and, finally, (f) an encyclopedia in which physical clone resources and their information are combined. To generate such an encyclopedia, the problem of storage and distribution system of the numerous clones needed to be solved. The DNABook prints both information and clones together on special paper, and may solve the problem of creating a genome encyclopedia (9).

In the DNABook, DNA samples are spotted onto a paper where the corresponding information is also printed. One book of about 200 pages can carry several tens of thousands of samples with associated information. Currently, spotted DNA is predicted to be stable at least for two or three years at room temperature (9, 12), so a huge number of clones potentially can be stored in a laboratory. It is also very convenient for users, because they only cut or punch out a part of the spot from the page and then recover the DNA from it using PCR or *E. coli* transformation.

Table 15-1 shows different mediums of disseminating information, and the significance of the DNABook for the distribution of genome resources combined with corresponding information. Information in early civilizations was carved in stone, pressed in clay, or hand-written on paper. This hand-crafted information was limited by its distribution to only a small number of people and often contained many errors. In the middle of the fifteenth century, Johannes Gutenberg invented and developed printing technology, and subsequently many people could share identical information. They could read and examine the first printed document, the Bible, and discuss issues with a common reference. Printing technology is still important for supporting modern journalism and mass communication; however, it is rapidly being replaced by information technology (IT) using electronic media, particularly in the scientific community, which is able to distribute more information faster.

The distribution of genome resources by the conventional method using tubes and plates with dry ice is analogous to the handwritten information technology of distribution in the past; it can handle a limited amount of information and is prone to mistakes. In the bioscience community, many researchers have found it difficult to quickly obtain identical samples. The advent of the DNABook potentially may change this situation. This technology enables the publication of many copies of genetic resources with their associated information, which then can be distributed to many researchers in a short time. Users can carry out various bioscience experiments using identical samples and then discuss their results obtained from a common reference. We expect that the DNABook technology will further progress this exciting era of bioscience.

**Table 15-1.** Methods of information and genome resources distribution.

| Methods | Handwriting | Printing | IT (digital) | The DNABook™ | Conventional Method for Genome Resources |
|---|---|---|---|---|---|
| Media or vehicle | Paper | Paper | Electron | Paper | Tubes and plates |
| Distributed object | Information | Information | Information | Information and genome resources | Genome resources |
| Amount (in types, sizes, and copies) | Small | Large | Large | Large | Small |

## Materials and Methods

### Paper for DNA Printing

Currently, the medium used for DNA printing is preferably a water-soluble, 60MDP paper (Mishima Paper Co., Ltd., Tokyo, Japan) that is comprised of two kinds of fibers: wood pulp and water-soluble carboxyl-methyl cellulose (CMC). This paper dissolves in water at room temperature within 35 seconds (12).

### Plasmid Preparation

Mouse cDNA clones obtained from GERG, RIKEN (RIKEN mouse cDNA) were used for the tests. Plasmid DNA of RIKEN mouse cDNA was purified using a Qiagen Spin Miniprep Kit (Qiagen, Inc., Valencia, CA, USA). DNA was dissolved in TE (10 mM Tris-HCl, 1 mM EDTA, pH 8.0) and the concentration was adjusted to 0.1 µg/µL. About 0.1 µL of the solution was spotted five times for each spot on the 60MDP sheet (the total amount of DNA in each spot is about 0.05 µg).

### Polymerase Chain Reaction

After air-drying the spotted paper for about 30 minutes, the DNA was extracted as follows: a piece of the paper (4 mm × 4 mm) containing the DNA spot was cut out and placed into a 0.2 mL PCR tube. Fifty microliters of PCR solution containing 1.5 units of KOD Plus DNA Polymerase (TOYOBO Co., Ltd., Osaka, Japan), 1× KOD Plus PCR Buffer, 0.2 µM of PCR primers, 0.2 mM each of dNTPs, and various concentrations (1 mM–7.5 mM) of $MgCl_2$ was added, and the reaction started. The following cycling conditions were used: 2 minutes at 94°C followed by 29 cycles [(94°C, 1 minute) (60°C, 1 minute) (68°C, 75 seconds)]. The final extension was for 15 minutes at 74°C. Aliquots of PCR solutions were analyzed on 1% agarose gel electrophoresis using standard molecular biology techniques (12).

### Recovery of Spotted DNA

Three RIKEN mouse cDNA with various insert sizes (722, 2418, and 5438 base pairs) spotted onto 60MDP paper were extracted and recovered using PCR. All cDNA clones tested were amplified successfully at higher $Mg^{2+}$ concentrations of around 5.3 mM. Higher $Mg^{2+}$ is needed because 60MDP binds cations (12). Furthermore, an aliquot of solution containing dissolved DNA sheet could transform *E. coli* without any additional purification (12).

Next, 93 randomly selected RIKEN mouse cDNA clones of various insert sizes (732–4896 bp) were spotted on 60MDP paper and then extracted. The resulting samples were amplified using PCR and almost all of the cDNA inserts (98%) were successfully amplified. These results suggest that almost all cDNA clones can be applied to DNA sheets (12).

## The Spotted DNA: Preservation of DNABooks

DNABooks must survive the various conditions present throughout the bookbinding process at the publisher, shipment to users, and their subsequent storage. Temperature, pressure, humidity, light, and physical rubbing could affect the quality of the DNABook. So far, no problems have been encountered when they have been stored at room temperature. DNA sheets treated at 140°C for several seconds, 100°C for 10 minutes, or –40°C for 14 hours all had the cDNA inserts successfully recovered using PCR. High-pressure conditions from 8.8 to 17 MPa did not disturb the recovery of DNA. These results suggest the DNA sheets remain undamaged under adverse environmental conditions (such as high temperatures and pressures) during bookbinding and the low temperatures during an air transport.

DNA sheets were tested to evaluate if rubbing stress on the sheets could cause problems such as inhibiting PCR amplification, or cross contamination with neighboring samples. DNA sheets spotted with three kinds of cDNA plasmids were stored in a humidified incubator or inserted into a book with other DNA spotted sheets, and then shaken strongly using a rotating shaker (at 180 rpm) at 37°C with 70% humidity for 14 hours. The spotted DNA samples were extracted and amplified using PCR. All DNA inserts (722, 2418, and 5438 bp) were recovered. No contamination of DNA spots with neighboring spots was observed. One component of 60MDP paper, the water-soluble CMC, allows DNA to be incorporated and trapped by the CMC gel when spotted. In addition, CMC might be able to keep DNA bound tightly within the DNA sheet after drying, thus preventing the release of spotted DNA (12).

DNA recovery via PCR was not possible after exposure to ultraviolet light (30 min irradiation). So, light affects DNA integrity, indicating that DNABooks should not be exposed to sunlight for prolonged periods of time (9).

As described above, exposure to humidity of 70% for a short time did not damage the spotted DNA; however, it is preferable to keep the DNABook in dry conditions, for example, by storing in a sealed plastic bag with a drying agent such as silica gel. Because the paper used in DNABooks is water soluble, users must be careful not to spill water over the pages.

From these tests results, it is estimated that DNABooks may be stable for two or three years under normal storage conditions. However, the exact durability of DNABooks cannot be precisely predicted at this time, but will be revised by extrapolation of ongoing stability tests.

## Manufacturing the DNABook

The publishing process of the DNABook can be divided into two parts: the "wet process," which consists of preparing the plasmid DNA, and the "dry process," which is the editing. The wet process includes re-array of *E. coli* clones, purification of plasmid DNA, adjustment of DNA concentration, quality check, transfer of the samples from a 96-well plate to a 384-well plate, and spotting. The editing process includes laying out the spotting page; writing, editing, and layout of the text page; preparation of the cover page; proofreading; printing; and binding.

## Printing the DNA

The special automated printer for DNA samples was originally developed by the GERG, RIKEN, and The Japan Science and Technology Agency (JST) Preventure Group. The printer is more similar to a DNA chip arrayer than a usual printer (Figure 15-1a). An aspirator was adopted as the paper feeder, because the physical characteristics of 60 MDP paper used for the DNABook made paper feeding difficult (9). The printer has a robotic arm that dips an array of tiny pinheads (Figure 15-1b) into the sample solutions in a standard format of 96- or 384-well plate. Then it stamps the samples onto a sheet of the paper that is marked with a preprinted spotting grid, at a rate of three sheets per minute. Programming can control the address of each clone in the grid. After spotting, the sheets are dried and bound. The major problem of printing a large number of samples is contamination of each sample. To avoid contamination of the samples, the disposable printing pins are changed at the same time as when the set of DNA samples on the master sample plate are changed. The JST Preventure Group is developing a higher throughput type of printer.

## Trial Production

### *First DNA Sheet Published in* Genome Research

The DNABook concept was first published in the *Genome Research FANTOM* special edition (12). In this trial, twelve RIKEN mouse cDNAs

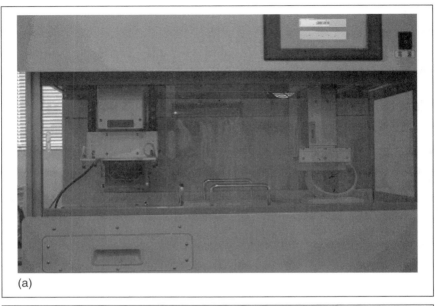

(a)

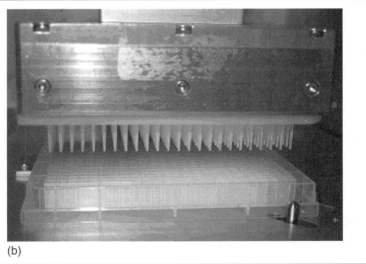

(b)

**Figure 15-1.** Special printer designed to automatically print DNA samples. (a) It appears more like a DNA chip arrayer than a normal printer. (b) A robotic arm dips pinheads into the sample solutions in a standard 96- or 384-well plate, and then stamps the sample solution in a dot onto a sheet of special paper that has been preprinted with a grid.

encoding enzymes of TCA cycle were printed on the 60MDP paper with their related information, that is, their length and accession numbers, and was distributed to many readers.

This trial was also intended as a field test to learn about the tolerance of the DNA sheet to temperature, humidity, mechanical sliding, rubbing stress, etc., which the sheet might encounter during practical use. Readers were asked to extract the DNA samples spotted on the sheet and amplify them using PCR or transformation of *E. coli*, and to report the results to the authors.

### Examples of DNABooks

*RIKEN Mouse Genome Encyclopedia™*

The first hardcover DNABook was produced in April 2003 (Figure 15-2a). Mouse is one of the best animal models of human. By choosing mouse as the subject organism, many kinds of full-length cDNA libraries could be constructed from a wide variety of tissues, including those that might not be available for human, for example, very early embryonic stages and pre-implantation embryos. These full-length cDNA clones would be useful in developing and analyzing various disease-model mice. In this

**Figure 15-2.** Different DNABooks that have been published to date.

book, 60,770 mouse full-length cDNA clones (20) were printed. A single page contains two identical sets of clones (one set contains 384 clones). Each spot contains 5 ng of DNA (sequence and annotation information can be obtained by accessing the RIKEN Web site: http://fantom2. gsc.riken.jp/). The second edition was published in September 2003. This time a loose-leaf type of DNABook, where individual pages can be easily replaced, was produced.

## RIKEN Human cDNA Encyclopedia: The Metablome DNABook™

The second DNABook (Figure 15-2b), published in June 2003, focused on a particular field: human full-length cDNA clones related to the metabolome. These cDNAs encode basic proteins important for sustaining life. A set of cDNA clones involved in metabolic functions could allow the in vitro reconstruction of the process. Such a set could be useful for the study of drug effects on the metabolic process. The cDNAs were prepared from 1211 selected clones in the KEGG (Kyoto Encyclopedia of Gene and Genomes developed by Professor M. Kanehisa) database (http://www.geome.ad.jp/kegg/) (11) using RT-PCR method. The 5'-untranslated readings (5'-UTRs) of these clones were removed for easy transcription. The characteristic point of this DNABook is that metabolic pathways are printed with physical clones encoding enzymes involved in the pathway.

## AQUA DNABook™: Microsatellite Marker and Disease Diagnostic Marker

The Kanagawa Prefectural Fisheries Research Institute, Tokyo University of Marine Science and Technology, and GERG, RIKEN produced this DNABook in March 2004 (Figure 15-2c). In contrast to other DNABooks, where full-length cDNAs were printed, this DNABook contains 217 pairs of primers (20–30 bases oligomers) of microsatellite markers of a flatfish. The markers that detect highly polymorphic alleles located on the linkage map are of great use in the linkage analysis of loci related to characters such as taste or resistance to disease. The book also contains 24 pairs of primers for diagnostics of 17 infectious diseases of several fish caused by viruses, bacteria, or parasites including koi herpes virus, which was prevalent at the time in Japan.

## RIKEN Arabidopsis cDNA Encyclopedia DNABook

This book (Figure 15-2d) focuses on the transcription factors of *Arabidopsis thaliana*, and contains 1069 genes that encode transcription factors in the RIKEN *Arabidopsis* Full Length (RAFL) cDNA collection (3–8, 15, 16,

22). RAFL cDNA collection, which is the largest collection of full-length cDNAs in the *Arabidopsis* community, was made by GERG in collaboration with the Plant Functional Genomics Research Group of RIKEN, GSC, in December 2004 and was deposited in the RIKEN Bio Resources Center (http://rarge.gsc.riken.jp/). Some transcription factors of higher plants regulate their target genes responding to environmental stress, such as high/low temperature or drying, and to stimulation by hormones. Transcription factors also relate to organ development in plant flowers, leaves, or roots. Thus, analysis of transcription factors of a model plant such as *Arabidopsis* is important for breeding applications for agricultural or horticultural crops.

## Rice Full-Length cDNA Encyclopedia™

Rice, one of the world's major staple foods, is grown in a wide range of areas including some with cultivation difficulties. Efforts to develop resistance to biotic or abiotic stress and to increase yield are currently in progress in various breeding programs; however, biological mechanisms to control yield have not been elucidated sufficiently. To obtain a more complete knowledge of the genetic constituents of rice, cDNA projects were initiated by a collaborative team of the National Institute of Agrobiological Science, GERG of RIKEN, and the Foundation for Advancement of International Sciences for cDNA analysis. In the rice full-length cDNA project, 32,127 clones selected from over 170,000 clones were completely sequenced and mapped to the genome sequence (13). This DNABook (Figure 15-2e) was published in April 2005.

## The Pyro DNABook™

To produce novel and functional biomaterial, the program of the NanoLEGO initiative, focused on "self-organized" proteins, has begun. NanoLEGOs are artificial proteins based on protein-protein interaction data selected by large-scale screening. Proteins derived from the highly thermophilic bacterium, *Pyrococcus horikoshii*, are suitable for NanoLEGO because of their high stability. In addition, they are easily purified to almost a single band after heat treatment in an *E. coli* expression system, while *E. coli* proteins are denatured and aggregated. From the 1390 genes cloned from *P. horikoshii*, 960 clones encoding cellular proteins were selected. Protein-protein interaction screening of these genes was performed using a mammalian cell two-hybrid system (18, 19). One hundred seventy types of interactions, including 71 kinds of self-interactions were detected (Usui et al., unpublished data). The DNABook of these genes (Figure 15-2f) has been published in June 2005.

## Progress and Problems of DNABooks

### Progress and Potential Applications of the DNABook

Genome and transcriptome analyses of many organisms are being carried out as described above and certainly more organisms will be subjects of genome-wide analysis in the future. Various kinds of the DNABooks are planned for publication, including: (a) a genome encyclopedia of multiple organisms, which covers whole genome or transcriptome; (b) clones related to some special functions or organs (for example, a "kinase book" or "brain book," and so on); (c) primers for genetic analysis or diagnostics; (d) cDNAs cloned under promoters of expression vectors, because these clones are readily available to express and more convenient to users.

In addition, DNA printing technology in the not too distant future could be utilized in life science journals as the DNA could be printed directly onto the relevant pages of scientific papers, so researchers would not be limited by inaccessibility of DNA clones and authors are not troubled about distributing the clones to other researchers (9).

Moreover, this DNA technology might become essential to public biological resource centers, which have a vital role in archiving and distributing biological resources. Indeed, some biological resource centers are using the DNABooks as a medium to distribute genome resources (21).

### Problems

With the projected future increase in the number of copies and increasing subjects of new DNABooks, some problems will arise and of concern are issues surrounding copyright. Conventional interpretations of copyright laws do not extend to DNA-type of materials when conveyed on paper. Ensuring adequate regulation is likely to prove difficult because the DNA-Books combine a conventional use of printed material with a completely new one. However, we believe that this problem will be resolved as the DNABook gains approval and popularity in the worldwide scientific community.

The potential biohazard problem of printing recombinant bacteria or genetic materials encoding pathogenic or toxic proteins also must be examined. Current DNABook handling is in compliance with Japanese laws on the control of recombinant DNA.

## FTA® Technology

The alternative representative of other technologies for storage and distribution of nucleic acids is FTA® paper developed by Whatman, Inc.

**Table 15-2.** Comparison of FTA® paper and the DNABook™.

|  | FTA® paper | The DNABook™ |
|---|---|---|
| Medium | Paper | Paper |
| Number of genome resources | Large | Large |
| Number of copies of each item | 0 | Large |
| Number of pages | A single sheet | A single sheet or multiple sheets (can be bound as a booklet or a book) |

(Middlesex, UK). The development of this technology started as a method of storage and delivery of blood samples. DNA in blood samples is used for diagnosis of genetic diseases and for the determination of paternity or human identification. Blood samples are spotted and dried onto a piece of filter paper at a point of collection, and then sent on to a central facility where the DNA in the samples is purified and analyzed.

In FTA paper, a composition of chemicals is absorbed onto a solid media such as a cellulose-based matrix or nylon membrane. The chemicals are: (a) a monovalent weak base such as Tris; (b) a chelating agent such as EDTA; (c) an anionic detergent such as SDS; (d) either uric acid or urate. When a sample (e.g., blood) is applied to the paper, the absorbed chemicals induce lysis of blood cells or cells of potentially contaminated pathogens, denature proteins to expose DNA, and protect the exposed DNA from degradation (1). The exposed DNA is immobilized to filters of the matrix and stored at room temperature. This paper is suitable for storing genetic materials in crude samples, for example, cell suspensions such as blood, or buccal cells, body liquids, homogenates of tissues, etc. Recently, the GenVault Corporation (17) developed a novel 384 plate for genetic samples that has a small disc of FTA paper inserted into each well. A large number of such plates are stored and retrieved by robotic devices in an automated archiving system.

Table 15-2 compares the FTA paper with the DNABook as a media of genome resource distribution. Both can distribute a large number of different kinds of samples. The major difference between them is in the copy numbers of samples. The FTA paper can distribute many kinds of samples to one user, but cannot distribute multiple copies of the same sample to many users, which in our fast-paced bioscientific world is equivalent to handwritten information of the past (Table 15-1). On the other hand, the DNA printing technology utilized in the DNABook is able to make many copies of each sample. Also, FTA paper exists as individually separated pieces of paper, whereas the DNABook may be set in single sheet or multiple sheets as a booklet or a large book.

## Future Considerations

With the ongoing explosion of the genome-wide analysis of life, the numbers of genome resources will continue to increase, and the demands for their mass storage and distribution will grow. Thus, a room-temperature type of storage and distribution system of genome resources will be indispensable in the post-genomic era. The technology of storage and distribution of genome resources on a solid support, which is being established for DNA, as described above, will also be applied to other biomolecules such as RNA or proteins. Furthermore, functional complexes of biomolecules, such as in the ribosome or proteasome, organelles such as mitochondria or chloroplast, and a cell itself, might be dried on paper and stored or distributed at room temperature. By combining the technology of drying biological resources and nanotechnology, various Lab-on-Chips might be generated. For example, by adding a drop of buffer to a chip containing dried cDNA clones related to a special biochemical process and ribosomes and protein synthesizing enzymes, a physiological reaction would be reproduced on the chip. In the future it is possible researchers in smaller laboratories will popularize the "pot-noodle" approach for the distribution of biological resources.

*Acknowledgments*

This work was supported by Research Grants for Preventure Program C of Japan Science and Technology Agency (JST), Core Research for Evolutional Science and Technology of JST, the RIKEN Genome Exploration Research Project, and the Genome Network Project; from the Ministry of Education, Culture, Sports, Science, and Technology of the Japanese Government to Y.H. We thank Mr. Julian Gough for proofreading the manuscript.

## References

1. Belgrader, P., Del Rio, S.A., Turner, M., et al. 1995. Automated DNA purification and amplification from blood-stained cards using a robotic workstation. *BioTechniques* 19: 426–432.
2. Boutros, M., Kiger, A.A., Armknecht, S., et al. 2004. Genome-wide RNAi analysis of growth and viability in *Drosophila* cells. *Science* 303: 832–835.
3. Carninci, P., and Hayashizaki, Y. 1999. High-efficiency full length cDNA cloning. *Methods Enzymol* 303: 19–44.

4. Carninci, P., Kvam, C., Kitamura, A., et al. 1996. High-efficiency full-length cDNA cloning by biotinylated CAP trapper. *Genomics* 37: 327–336.

5. Carninci, P., Nishiyama, Y., Westover, A., et al. 1998. Thermostabilization and thermoactivation of thermolabile enzymes by trehalose and its application for the synthesis of full-length cDNA. *Proc Natl Acad Sci USA* 95: 520–524.

6. Carninci, P., Shibata, Y., Hayatsu, N., et al. 2000. Normalization and subtraction of cap-trapper-selected cDNAs to prepare full-length cDNA libraries for rapid discovery of new genes. *Genome Res* 10: 1617–1630.

7. Carninci, P., Shibata, Y., Hayatsu, N., et al. 2001. Balanced-size and long-size cloning of full-length, cap-trapped cDNAs into vectors of the novel λ-FLC family allows enhanced gene discovery rate and functional analysis. *Genomics* 77: 79–90.

8. Carninci, P., Westover, A., Nishiyama, Y., et al. 1997. High efficiency selection of full-length cDNA by improved biotinylated cap trapper. *DNA Res* 4: 61–66.

9. Hayashizaki, Y., and Kawai, J. 2004. A new approach to the distribution and storage of genetic resources. *Nature Rev Genet* 5: 223–228.

10. Kamath, R.S., Fraser, A.G., Dong, Y., et al. 2003. Systematic functional analysis of the *Caenorhabditis elegans* genome using RNAi. *Nature* 421: 231–237.

11. Kanehisa, M., Goto, S., Kawashima, S., and Nakaya, A. 2002. The KEGG databases at GenomeNet. *Nucleic Acids Res* 30: 42–46.

12. Kawai, J., and Hayashizaki, Y. 2003. DNABook. *Genome Res* 13: 1488–1495.

13. Kikuchi, S., Satoh, K., Nagata, T., et al. 2003. Collection, mapping, and annotation of over 28,000 cDNA clones from *japonica* rice. *Science* 301: 376–379.

14. Paddison, P.J., Silva, J.M., Conklin, D.S., et al. 2004. A resource for large-scale RNA-interference-based screens in mammals. *Nature* 428: 427–431.

15. Seki, M., Carninci, P., Nishiyama, Y., et al. 1998. High-efficiency cloning of *Arabidopsis* full-length cDNA by biotinylated CAP trapper. *Plant J* 15: 707–720.

16. Seki, M., Narusawa, M., Kamiya, A., et al. 2002. Functional annotation of a full-length *Arabidopsis* cDNA collection. *Science* 296: 141–145.

17. Shon, J. 2005. New best practices for biosample management: moving beyond freezers. *Am Biotech Lab* 23: 10–15.

18. Suzuki, H., Fukunishi, Y., Kagawa, I., et al. 2001. Protein-protein interaction panel using mouse full-length cDNAs. *Genome Res* 11: 1758–1765.

19. Suzuki, H., Saito, R., Kanamori, M., et al. 2003. The mammalian protein-protein interaction database and its viewing system that is linked to the main FANTOM2 viewer. *Genome Res* 13: 1534–1541.

20. The FANTOM Consortium and the RIKEN Genome Exploration Research Group Phase I and II Team. 2002. Analysis of the mouse transcriptome based on functional annotation of 60,770 full-length cDNAs. *Nature* 420: 563–573.

21. Weaver, T., Maurer, J., and Hayashizaki, Y. 2004. Sharing genomes: an integrated approach to funding, managing and distributing genomic clone resources. *Nature Rev Genet* 5: 861–866.
22. Yamada, K., Lim, J., Dale, J.M., et al. 2003. Empirical analysis of transcription activity in the *Arabidopsis* genome. *Science* 302: 842–846.

# 16

# Assessing the Quantity, Purity and Integrity of RNA and DNA Following Nucleic Acid Purification

William W. Wilfinger, Karol Mackey, and Piotr Chomczynski
*Molecular Research Center, Inc., Cincinnati, Ohio, USA*

The technology that is currently employed for the extraction and purification of nucleic acids has expanded since the original studies of Chirgwin et al. (4) and Chomczynski and Sacchi (5). In these original reports, strong chaotropic agents such as guanidinium thiocyanate and phenol were employed to denature proteins and inactivate nucleases. The solubilized protein fraction was subsequently separated from the nucleic acids via basic biochemical strategies that included differential solubility, solvent partition, centrifugation, and selective precipitation methodologies. Although contemporary extraction methodologies such as the modified-single step method (6, 25) still employ chaotropic agents, alternative nucleic acid extraction technologies now include a variety of different biochemical approaches that include adsorption to silica- or glass-based fiber spin columns or affinity binding to insoluble cellulose matrix, latex beads, and paramagnetic particles (1, 39). Over 30 companies currently market a plethora of nucleic acid extraction kits based on these diverse technologies (8). Detailed technical information and a comprehensive list of publications are routinely available at the manufacturer's Web sites.

The selection of a nucleic acid extraction methodology is generally based on pragmatic considerations that include extraction sample cost, time commitment, technical complexity of the methodology, quantity of the starting biological sample, and suitability and adaptability of the methodology for high-throughput automation. Advancements in nucleic acid extraction technology over the last ten years enable individuals with

*DNA Sequencing II: Optimizing Preparation and Cleanup*
Edited by Jan Kieleczawa
©2006 Jones and Bartlett Publishers

relatively limited technical experience to successfully extract nucleic acids from most biological tissues. Historically, whenever RNA or DNA extracts could be used to "get a result," additional information was rarely collected that would provide an accurate assessment of the purified nucleic acids or helpful insights into appropriate corrective measures that could be employed when troubleshooting aberrant results.

As molecular biology protocols have become more standardized, it is easier to document the importance of nucleic acid quality in obtaining reproducible results. For example, it has been established that transient transfection studies require "endotoxin-free ultrapure" DNA for optimal results (17, 33), while microarray-based differential gene expression studies have been shown to be extremely sensitive to changes in RNA integrity (2, 22). It is increasingly clear that a more comprehensive awareness and understanding of nucleic acid quality is critical to the interpretation and evaluation of the results obtained from sensitive molecular biology applications. The purpose of this chapter is to provide a brief overview of the instrumentation and methods that are currently available for evaluating purified nucleic acids, and to discuss the assumptions and limitations of the data that are obtained with this technology. Second, we present useful recommendations for documenting and standardizing the quality of isolated nucleic acid preparations.

## Overview of the Instruments and Criteria Employed to Assess the Quality of Nucleic Acid Extracts

In 1942, Warburg and Christian (43) utilized a standard laboratory spectrophotometer to estimate nucleic acid contamination in a purified enzyme preparation. In our current "molecular biology" applications, this "two-wavelength" method is conversely applied to assess protein contamination in nucleic acid extracts. For example, the absorbance at 260 nm is employed for nucleic acid quantitation and the $A_{260}/A_{280}$ ratio is computed to provide information relating to extract purity (e.g., protein contamination). Contemporary laboratory molecular biology protocols recommend using an aliquot of the purified nucleic acid extract for obtaining absorbance measurements at 260 and 280 nm because it is generally understood that these measurements provide accurate quantitative and qualitative assessments of nucleic acid extracts. RNA extracts that display an $A_{260}/A_{280}$ ratio of 1.8 to 2.0 and a ribosomal RNA 28S/18S ratio of ~2:1 (e.g., RNA integrity) are considered to be suitable for molecular biology applications such as Northern blotting, RNase protection assay, RT-PCR, etc. In contrast, the recommended $A_{260}/A_{280}$ ratio for DNA extracts is ~1.6 to 1.8 with an expected electrophoretic mobility >20 kb. Although more sophisticated laboratory instruments have been developed for obtaining nucleic acid absorbance measurements (Table 16-1), many investigators have a limited understanding of the various factors that influence these measurements and the extent to which this information may be meaningful in evaluating nucleic acid extracts. Major disadvantages of employing absorbance spectroscopy for the quantification of nucleic acids are the amount of sample that is required for quantitation and the inability of this methodology to distinguish between RNA and DNA in the same sample (i.e., low sensitivity and specificity).

To better address sensitivity and specificity issues, highly sensitive fluorometric assays were developed for nucleic acid quantitation. Although the fluorescent dyes that are used in these assays can discriminate between RNA and DNA as well as other extract contaminants, the assays require purified nucleic acid standards, dedicated instruments (Table 16-1), and they do not provide any information relating to extract purity.

The suitability of an RNA sample for first strand cDNA synthesis depends on the purity of the extract as well as the integrity of the mRNA molecule. Whenever large quantities of total RNA are available (e.g., 1–10 μg), nucleic acids can be efficiently separated by agarose gel electrophoresis and RNA integrity can be evaluated by staining the separated nucleic acids with a suitable detection dye such as ethidium bromide (34). However, in order to avoid smeared bands and anomalous band artefacts, the nucleic acids must be thoroughly solubilized, relatively free of protein contamination, and appropriate electrophoresis conditions must be maintained during the separations. The subjective evaluation of the staining intensity of the ribosomal 28S/18S band ratio provides a cursory view of the RNA integrity. When available, gel-imaging instruments can be employed to provide a quantitative assessment of the rRNA band profiles (Table 16-1). Because agarose gel electrophoresis requires the commitment of laboratory time and resources as well as microgram quantities of RNA for visual inspection and reliable quantitation, this procedure is unsuitable for many clinical applications. In applications where nanogram quantities of nucleic acid are available, chip-based instruments such as the Agilent Bioanalyzer (Table 16-1), or the BD Clontech™ RNA/cDNA Quality Assay (3) are employed to evaluate 3′, 5′ mRNA integrity.

## Assessment of Nucleic Acid Purity and Content by Spectrophotometric Methods: Theory and Assumptions

The spectrophotometric measurement of nucleic acid absorbance is widely employed as a rapid and inexpensive method for the quantitative and qualitative assessment of purified nucleic acid samples. Because absorbance measurements are routinely accepted as the "gold standard"

**Table 16-1.** Laboratory instruments currently used to evaluate nucleic acid extracts.

| Instrument | Cost | Purity | Integrity | Quantitation | Comments | Ref. |
|---|---|---|---|---|---|---|
| **UV Absorbance Spectrophotometer** | | | | | | |
| Lab Spectrometer | 4–10 K | G | — | G | Usefulness of the data is dependent upon defined measurement conditions. | 19, 38 |
| GeneQuant II | 3–5 K | G | — | G | Instrument dedicated for nucleic acid quantification. | 41 |
| Nano Drop | 7–8 K | S | — | G | Small sample volume, data capture and analysis ability. | 31 |
| Microplate Spectrometer | 25–30 K | G | — | G | Suitable for automation. | 18, 19 |
| **Fluorometer and Fluorescence Spectrometry** | | | | | | |
| DyNA Quant | 3–5 K | — | — | G | RNA & DNA dyes provide specificity during quantitation. | 41 |
| Picofluor | 2 K | — | — | G | RNA & DNA dyes provide specificity during quantitation. | 29 |
| Microplate Reader | 30 K | — | — | G | RNA & DNA dyes provide specificity during quantitation. | 18, 19, 36 |
| **Dual Application Instruments** | | | | | | |
| Gel Electrophoresis | 1–2 K | — | L | L | Subjective approximation of nucleic acid integrity and quantitation. | 15, 35, 36 |
| Agilent Bioanalyzer | 25–30 K | — | S | G | Currently, state of the art instrumentation for assessing nucleic acid integrity. | 26, 32 |
| Gel Imaging Scanner | 5–70 K | — | G | G | Nucleic acid quantification following agarose gel electrophoresis. | 11, 28, 35, 36 |

L, limited; G, good; S, superior.

Reference citations were selected to provide the reader with a broad overview of equipment manufacturers in specific areas. Many of these references are available in electronic format and they provide a general survey of the various instrument options as well as the manufacturer's Internet addresses. These links can be used to obtain more detailed information directly from the manufacturer. Selected instruments that are specifically tailored for nucleic acid applications are also included.

for evaluating nucleic acids, molecular biologists should understand the theoretical basis for these measurements; an accurate assessment of extract quality is pivotal to establishing their "suitability" for subsequent molecular biology applications.

The theory associated with absorbance spectroscopy is based on the Beer-Lambert law. Simply stated, when a beam of monochromatic UV light is passed through a solution containing light-absorbing molecules such as RNA or DNA, there is a linear relationship between light absorbance and nucleic acid concentration as illustrated by the following equation:

$$A = \varepsilon * b * c$$

where A is absorbance, $\varepsilon$ is the wavelength-dependent molar absorptivity coefficient, b is the light path length, and c is the analyte concentration. To accurately quantify the light absorbing molecules at their respective absorbance maxima (260 and 280 nm, respective $\gamma_{max}$ for nucleic acid and protein), there is an assumption that the solution does not contain any other molecules that will absorb light at the same wavelengths. Unknown contaminants or factors that affect absorbance readings at the designated wavelengths will lead to inaccurate concentration estimates and erroneous results.

Commercial extraction reagents routinely employ strong chaotropic agents such as phenol, guanidine, and a variety of salts; therefore, it is also important to verify that purified nucleic acids are free of these contaminants. After recording the absorbance at the designated $\gamma_{max}$ wavelength for nucleic acid (260 nm) and protein (280 nm), the "$A_{260}/A_{280}$" ratio is computed. Classically, DNA and RNA extracts that display respective $A_{260}/A_{280}$ ratios of 1.8 and 2.0 are considered to be "free" of protein contamination. This assumption, however, can be misleading. Based on theoretical calculations, Glasel (12) reported that nucleic acid extracts routinely employed for molecular biology applications might still contain significant quantities of protein contamination. Glasel further concluded that the "venerable" $A_{260}/A_{280}$ ratio was a very poor predictor of protein contamination in nucleic acid preparations, because the absorptivity coefficient of the nucleic acid is so much greater than that of the contaminating protein; the protein absorbance contributes minimally to the computed ratio. For example, the absorptivity coefficient for protein and DNA at 280 nm is ~1 and 10, respectively, for 1 mg/mL solutions (12). Furthermore, the absorbance of a pure nucleic acid preparation at 280 nm is >50% of the recorded 260 nm absorbance. So protein absorbance in nucleic acid extracts can be obscured by the significantly larger nucleic acid absorbance that is routinely measured at 280 nm. Therefore, in accordance with Warburg and Christian's (43) original application, it should be understood that the $A_{260}/A_{280}$ ratio is a much better predictor of nucleic acid contamination in protein extracts than protein contamination in nucleic acid samples (10, 12, 20, 21).

Manchester (20, 21), Stulnig and Amberger (40), and Wilfinger, Mackey, and Chomczynski (44) have discussed other problems associated with the use of UV absorbance measurements for the estimation of protein and nucleic acid concentrations. For example, in older spectrophotometers that utilize scanning monochromators and a single photomultiplier tube, a one-nm calibration shift in the spectrometer absorbance wavelength to either 259/279 or 261/281 nm will produce a respective 5% to 6% decrease or increase in the $A_{260}/A_{280}$ ratio (20, 44). Although this problem has been virtually eliminated in newer spectrophotometers that employ dual beam optics and either charge-coupled or charge-transfer device array detectors (38), many investigators that routinely use older spectrophotometers are generally unaware of the necessity for maintaining proper instrument calibration.

In addition to instrument calibration issues, the absorption spectra of nucleic acids are strongly affected by temperature, pH, and salt concentration. The characteristic absorption of UV light in the region of 260 nm by the conjugated double bonds of the heterocyclic DNA and RNA bases serves as the basis for the detection and quantification of nucleic acids by absorption spectroscopy (38). Therefore, the combined effects of the diluent pH and salt concentration that is used for spectrophotometric measurement has a significant influence on the observed $A_{260}/A_{280}$ ratio (27, 44). Simply changing the pH of the solvent used to dilute nucleic acids for spectrophotometric measurements can modulate base ionization (tautomerization) and affect $\gamma_{max}$. For example, acidic solvents such as DEPC-treated water, shift both the nucleic acid and protein absorbance maxima to a higher wavelength, thereby decreasing the ability to detect protein contamination in nucleic acid extracts (27, 44).

The consequence of a shift in the diluent pH and salt concentration on the 260 and 280 nm absorbance and the $A_{260}/A_{280}$ ratio is illustrated in Figure 16-1. The dilution of purified RNA or DNA in DEPC-treated water at pH 5.5 for spectrophotometric measurement resulted in an $A_{260}/A_{280}$ ratio of ~1.5; however, if an aliquot of the same nucleic acid was diluted in as little as 0.2 mM $Na_2HPO_4$ at pH 7.5, the $A_{260}/A_{280}$ ratios increased to 2.1 and 1.75, respectively. The second consequence that is clearly demonstrable from Figure 16-1 is the pronounced effect that increasing salt concentration has on the recorded absorbance at 260 and 280 nm. Apparently, the pH/salt-induced tautomer state of the purine and pyrimidine bases significantly affects their ability to absorb UV light. Although RNA absorbance at 260 nm is only slightly diminished in the presence of 1 to 10 mM $Na_2HPO_4$, salt concentrations above 0.01 mM $Na_2HPO_4$ significantly suppress both DNA and protein absorbance. The dilution of DNA in TE buffer (10 mM Tris, 1 mM EDTA, pH 8.0) for spectrophotometric

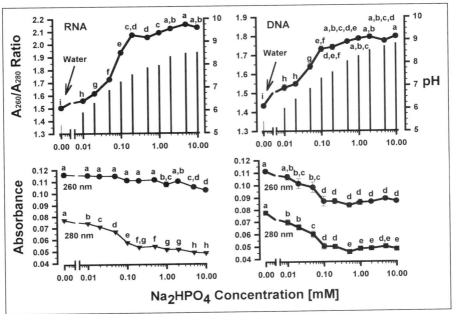

**Figure 16-1.** Diluent pH and $Na_2HPO_3$ concentration affect the spectrophotometric quantitation of RNA, DNA and protein absorbance, and the calculation of an $A_{260}/A_{280}$ ratio. A 10 mM $Na_2HPO_4$ solution was prepared without pH adjustment and diluted as indicated. Nucleic acids (5 μL) were diluted to a final volume of 1 mL for spectrophotometric nucleic acid quantitation (HACH Company, Loveland, CO, USA). The pH of the diluted $Na_2HPO_4$ solution is depicted on the right ordinate of the top panels (solid bars) and the corresponding 260 and 280 nm absorbance of rat liver RNA (left panel) and calf thymus DNA (right panel) is presented in the bottom panels in relation to the designated $Na_2HPO_4$ concentration. Rat liver RNA was extracted with TRI Reagent in accordance with the manufacturer's protocol (25). RNA and calf thymus DNA (Sigma-Aldrich, St. Louis, MO, USA) were initially solubilized with RNase-free water. Treatment means identified with different letters are significantly different ($P < .05$) and all data are reported as mean ± 1 standard error of the mean (SEM); $N = 3$. (RNA data presented previously appeared in Wilfinger, W.W., Mackey, K., and Chomczynski, P. 1997. Effect of pH and Ionic Strength on the Spectrophotometric Assessment of Nucleic Acid Purity. *BioTechniques* 22: 474–481 and are reproduced by permission of Informa Life Science Publishing.)

quantitation suppresses absorbance at 260 and 280 nm by 20% to 30% thereby underestimating both DNA and protein content (27, 44).

In addition to being a relatively insensitive indicator of protein contamination (10, 12, 20, 21, 44), other contaminants such as phenol can influence the $A_{260}/A_{280}$ ratio (40). Phenol, a strong chaotropic agent that is found in many nucleic acid extraction solutions, has a $\gamma_{max}$ at 270 nm (Figure 16-2). Although the $A_{260}/A_{280}$ ratio of a pure phenol solution is 2.0, increasing

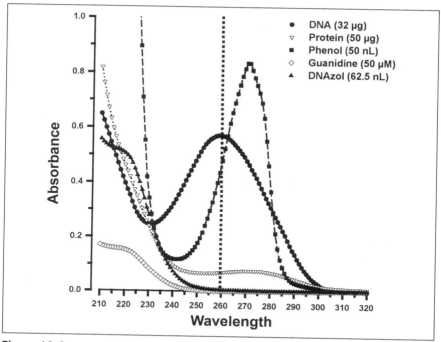

**Figure 16-2.** Absorbance spectra of purified calf thymus DNA, rat liver powder protein, guanidine thiocyanate, phenol, and DNAzol®. The test materials were solubilized and subsequently diluted in 1 mM $Na_2HPO_4$ at pH 8.5. Nucleic acid absorbance scans were performed in a total volume of 1 mL of the $Na_2HPO_4$ buffer in the presence of the designated amount of the specified test material. The absorbance profiles represent the mean absorbance of two separate scan determinations. DNAzol is a patented DNA isolation reagent (7).

amounts of phenol in the presence of a nucleic acid lowers the $A_{260}/A_{280}$ ratio (40). Stulnig and Amberger (40) claim that the calculation of an $A_{260}/A_{270}$ ratio would provide a more accurate and sensitive indicator of phenol contamination in RNA samples that have been purified by acid guanidinium thiocyanate-phenol-chloroform extraction methodology.

Inorganic salts generally do not display significant absorbance at wavelengths above 250 nm. As noted in Figure 16-2, nucleic acids show a $\gamma_{min}$ at ~230 nm, while protein, phenol, and salts tend to display greater absorbance at this wavelength. Because of the enhanced absorbance of various contaminants relative to the $\gamma_{min}$ of the nucleic acid, a number of investigators have proposed evaluating absorbance at lower wavelengths (10, 13, 14), and newer instruments that are designed for nucleic acid absorbance measurements display either the $A_{260}/A_{230}$ ratio or an $A_{260}/A_{240}$ ratio as a standard data measurement (Table 16-1) (13, 19, 23, 31, 41).

Because nucleic acid absorbance is minimal at 230 nm, absorbance from protein, phenol, and salt contamination would be more easily detected than at 280 nm, and thereby provide additional information for the assessment of nucleic acid purity (Figure 16-2) (10). Calculating an $A_{260}/A_{230}$ ratio at wavelengths that represent the $\gamma_{max}$ (260 nm) and $\gamma_{min}$ (230 nm) of pure nucleic acids produces a ratio of ~2.3 to 2.5. The presence of contaminants that absorb at lower wavelengths significantly elevates total absorbance at 230 nm relative to the absorbance at 260 nm, thereby lowering the computed ratio. Therefore, the evaluation of nucleic acid absorbance at 260 nm in relation to absorbance at 230 nm provides additional sample purity information that would not be obtainable by simply utilizing the $A_{260}/A_{280}$ ratio.

## Nucleic Acid Quantitation

Ultraviolet absorption does not distinguish between RNA and DNA, so accurate quantitation is dependent on the separation of the two nucleic acids. Although minuscule amounts of DNA contamination may be detectable in purified RNA samples following RT-PCR, this contamination is not adequate to contribute significantly to RNA absorbance. In contrast, degraded RNA is not detectable in PCR but DNA extracts containing degraded RNA will display a significantly greater A260 absorbance. For example, a sample containing 1000 RNA fragments, 6 bp or smaller would display a greater A260 absorbance than six undegraded mRNA molecules containing 1000 bp. The increased UV absorption of this hydrolyzed RNA (i.e., hyperchromic effect) can artificially elevate both the A260 absorbance and the $A_{260}/A_{280}$ ratio (44).

Although spectrophotometric absorbance measurements are used routinely for DNA quantitation, the $A_{260}$ values may be inflated if the samples have not been treated with RNase. Commercial extraction solutions routinely employed for DNA isolation are designed to solubilize and stabilize DNA while RNA is simultaneously degraded. If any degraded RNA co-isolates with the DNA, it can contribute significantly to the DNA absorbance, although it may be visually undetectable following agarose gel electrophoresis. This situation is clearly illustrated in the plant DNA data presented in Table 16-2. Irrespective of the tissue source or the solvent employed for spectrophotometric analysis, DNA content is overestimated by 49% to 118% as a result of RNA contamination.

In addition to these inflated DNA concentration estimates, sample purity is also influenced by the hyperchromic absorbance contribution from the degraded RNA in the sample. Note that the DNA extracts without RNase-treatment that were subsequently diluted in 1 mM $Na_2HPO_4$ display $A_{260}/A_{280}$ ratios ranging from 1.76 to 1.82. Based solely

**Table 16-2. Effects of solvents and degraded RNA on the spectrophotometric quantification of genomic DNA.**

| Tissue | RNase Trt. | Water | | | | | 1.0 mM Na$_2$HPO$_4$ | | | | | % Change DNA H$_2$O vs Na$_2$HPO$_4$ |
|---|---|---|---|---|---|---|---|---|---|---|---|---|
| | | DNA (µg/µL) | A$_{260/280}$ | A$_{260/230}$ | A$_{230/280}$ | PQ | DNA (µg/µL) | A$_{260/280}$ | A$_{260/230}$ | A$_{230/280}$ | PQ | |
| Lettuce | − | 1.638 39%↓ | 1.538 | 2.012 | 0.756 | 2.661 | 1.439 47%↓ | 1.766 | 1.807 | 0.978 | 1.848 | 12%↓ |
| | + | 0.999 | 1.502 | 2.107 | 0.714 | 2.951 | 0.756 | 1.685 | 1.777 | 0.950 | 1.871 | 24%↓ |
| Tomato | − | 1.919 43%↓ | 1.634 | 2.153 | 0.758 | 2.840 | 1.688 54%↓ | 1.822 | 1.953 | 0.937 | 2.084 | 12%↓ |
| | + | 1.091 | 1.452 | 2.008 | 0.723 | 2.777 | 0.774 | 1.724 | 1.728 | 0.990 | 1.745 | 29%↓ |
| Potato | − | 1.453 33%↓ | 1.633 | 2.075 | 0.787 | 2.637 | 1.227 43%↓ | 1.771 | 1.856 | 0.955 | 1.943 | 16%↓ |
| | + | 0.977 | 1.492 | 1.846 | 0.807 | 2.287 | 0.702 | 1.675 | 1.583 | 1.065 | 1.486 | 28%↓ |
| Mean % Change | | 38%↓ | | | | | 48%↓ | | | | | 27%↓ |

Plant genomic DNA was extracted with DNAzol-ES (7) and solubilized in 10 mM Tris, 1 mM EDTA buffer at pH 8.0. Aliquots of the isolated DNA were diluted for quantitation to a final volume of 1 mL with distilled laboratory water or 1 mM Na$_2$HPO$_4$ buffer, pH 8.5. The arrows represent the percent change in DNA content before and after RNase treatment within the six treatment groups. The arrows in the far right column represent the percent change in the DNA content among the water and Na$_2$HPO$_4$ diluents.

on the $A_{260}/A_{280}$ ratio, one would conclude that these samples are relatively pure. However, after the residual amounts of degraded RNA are removed, the $A_{260}/A_{280}$ ratios decrease to between 1.69 and 1.72. In the presence of RNA contamination, the sample $A_{260}/A_{280}$ ratios are artificially increased. Although these RNase-treated DNA extracts are still suitable for molecular biology applications, they may still contain some protein and or polysaccharide contamination that would otherwise remain undetectable prior to RNase treatment.

Because of the problems associated with obtaining accurate quantitative estimates of DNA content by absorbance spectrophotometry, an assortment of fluorochromes have been developed for the quantitation of RNA and DNA (35, 36). Dyes such as Hoechst 33258®, POPO™, BOBO™, TOTO®, YOYO®, and PicoGreen® have been employed to quantitate double-stranded DNA (29, 35, 36, 41) and RiboGreen® (35, 36) has been used for RNA and DNA quantitation. Typically, dyes bind to DNA by either inserting "flat" chemical moieties between adjacent bases (ethidium bromide, propidium iodide), or they display elongated structures that wrap around the DNA by inserting into the minor groove of the double helix (Hoechst 33258, DAPI) (35). At a neutral pH and in the presence of salt, the fluorescence intensity of the Hoechst dye is directly proportional to the DNA content of the sample if the average adenine:thymine (A:T) bp ratio of the DNA standard is similar to the DNA sample being quantified. Moreover, the fluorescence signal is not altered by protein or residual RNA contamination. Therefore, accurate estimates of DNA content can be obtained regardless of the sample purity.

PicoGreen Reagent can also be used for DNA quantitation (35, 36). This dye is reported to detect as little a 1 ng of double-stranded (ds) DNA and display a linear dynamic fluorescence range with DNA concentrations up to 1000 ng/mL. Although the fluorescence signal is increased in the presence of phenol, ethanol, chloroform, and proteins, and decreased in the presence of NaCl, the sensitivity of the dye enables sufficient dilution, so that common contaminants do not significantly influence DNA quantitation. Moreover, the specificity of the PicoGreen dye enables investigators to quantitate dsDNA in the presence of equimolar concentrations of either single-stranded (ss) DNA or RNA with minimal effect on total fluorescence. Although a DNA standard curve must be prepared, the sensitivity and specificity of the dyes makes them ideally suited for high-throughput automated quantitation in microplate formats (24, 36).

RiboGreen, although less specific, shows similar enhanced fluorescence sensitivity to RNA and DNA in a dynamic range of 15 to 1000 ng/mL (35, 36). Like PicoGreen, this dye is also well suited for high-throughput automation. However, because all of these dyes bind strongly with their respective nucleic acid, they are potential carcinogens and/or mutagens and must be handled with care. Newer DNA and RNA binding dyes such as SYBR Green I and II appear to exhibit lower levels of mutagenicity in the Ames test and, therefore, may not be as toxic as the dyes that bind more strongly to the nucleic acids (36). Regardless of the dye used for quantitation, an appropriate nucleic acid standard curve must be employed to convert the observed fluorescence signal into concentration units. Although this is not an issue when large numbers of samples are processed, it does become a nuisance when a multiple point standard curve must be prepared in order to quantitate the nucleic acid content of one or two samples. Although fluorescent dyes can provide sensitive and accurate quantitative determinations, these fluorochromes are generally sensitive to photobleaching and they must be protected from ambient light. Finally, fluorometric nucleic acid quantitation does not provide any information relating to extract purity.

## Nucleic Acid Integrity

Because absorbance measurements do not evaluate RNA quality, most laboratories employ agarose slab gel electrophoresis in conjunction with ethidium bromide staining to evaluate the integrity of RNA (34–36). In view of the fact that ribosomal RNA constitutes more than 95% of the total extractable RNA, the ribosomal 28S and 18S bands are routinely scrutinized to provide an indirect assessment of messenger RNA integrity. Based on visual inspection, extracts that displayed an ~2:1 ratio of 28S to 18S rRNA bands are considered to contain "intact" mRNA. However, in practice, rRNA ratios ranging anywhere from 1:1 to 2:1 are suitable for most molecular biology applications. Although the subjective visual inspection of RNA integrity is generally adequate for Northern analysis and RT-PCR applications, agarose gel electrophoresis is not practical for assessing the integrity of small quantities of total RNA (i.e., <5 µg total RNA) and unsuitable for more demanding applications such as differential gene expression analysis.

To separate smaller quantities of nucleic acid and thereby improve the sensitivity of the detection methodology, capillary electrophoresis was initially employed (8, 16, 30, 37, 42). More recently, microfabricated chip electrophoretic separation technology coupled with laser-induced fluorescence nucleic acid detection has made it possible to evaluate picogram quantities of nucleic acid on the Agilent 2100 Bioanalyzer (32). Because of this improvement in sensitivity, it is now possible to critically evaluate the integrity of a minuscule amount of nucleic acid and provide an objective assessment of RNA integrity. For example, as a nucleic acid sample undergoes electrophoretic separation on a Bioanalyzer chip, output from the fluorescence detector is stored in memory and the fluorescence signal is subsequently displayed as an electropherogram, which is an electronic representation of the captured nucleic acid fluorescent signal during the time period in which the electrophoretic separation was performed.

Newly developed Agilent software utilizes a proprietary algorithm to analyze defined regions of the electropherogram (26). As an RNA sample is degraded, the fluorescence associated with the 28S and 18S ribosomal bands decrease in proportion to an increase in the fluorescent signal of smaller and faster eluting RNA fragments in other regions of the electropherogram. Based on the mathematical analysis of the RNA elution profile, the Bioanalyzer software assigns an RNA integrity number (RIN) to the sample based on a numbering system from 1 to 10, where 1 is the most degraded profile. The Agilent software as well as the "Degradometer" software developed by Auer et al. (2) provides a more thorough and objective mathematical assessment of RNA integrity than has been previously possible from image analysis agarose gel scanning instruments (11). Because the number of "hits" that are observed following differential gene expression studies was linked directly to RNA integrity, the assignment of an RIN value or a "degradation factor" number will enable investigators to objectively select high quality RNA for their studies (2, 22, 26), and thereby decrease the number of false positive responses.

## An Additional Absorbance Measurement Improves Nucleic Acid Purity Assessment

Even though the collection of absorbance measurements at the $\gamma_{max}$ for nucleic acids and protein (e.g., 260 and 280 nm, respectively) provide some useful information relating to nucleic acid content and purity, investigators and instrument manufacturers recognize that evaluating absorbance at the nucleic acid $\gamma_{min}$ provides additional useful information (10, 13, 14, 23, 31, 41). To clarify how various chemicals that are routinely employed in nucleic acid extractions might influence absorbance measurements, we have done spectral scans of purified DNA in 1 mM $Na_2HPO_4$ and in the presence of increasing quantities of four different contaminants. In Figure 16-3, $A_{260}$ absorbance data and the $A_{260}/A_{280}$, $A_{230}/A_{280}$, $A_{260}/A_{270}$, and $A_{260}/A_{230}$ ratio estimates are given for DNA in the presence of various contaminants. Because most commercial spectrophotometers that are specifically designed for nucleic acid quantitation routinely provide absorbance measurements at 230, 260, and 280 nm wavelengths, we have selected several different absorbance ratios within a 50 nm region of the UV spectrum to determine if they provide additional data that would be useful in evaluating the purity of nucleic acids.

As expected, increasing quantities of DNA increase the $A_{260}$ absorbance as depicted in Figure 16-3a. Increasing quantities of either protein or phenol also increase the 260 nm absorbance (Figure 16.3b, c). In

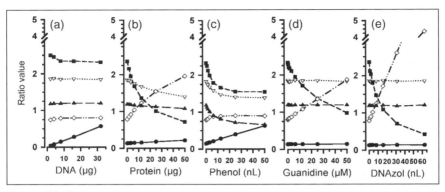

**Figure 16-3.** Effects of increasing quantities of DNA (a), or a constant amount of DNA and increasing quantities of either protein (b), phenol (c), guanidine thiocyanate (d), or DNAzol (e) on 260 nm absorbance (●) and the $A_{260}/A_{280}$ (▽), $A_{260}/A_{230}$ (■), $A_{230}/A_{280}$ (◇) and $A_{260}/A_{270}$ (▲) ratios. Aliquots of the specified test solutions were prepared and diluted with 1 mM $Na_2HPO_4$ at pH 8.5 to a volume of 1.0 mL for spectrophotometric assessment. Each data point represents the absorbance value or the computed ratio that was obtained from two separate spectral scans. DNAzol is a patented DNA isolation reagent (7).

the presence of either protein or phenol contamination both the $A_{260}/A_{280}$ and the $A_{260}/A_{230}$ ratios decrease. However, the $A_{260}/A_{230}$ ratio displays a more precipitous rate of decline than the $A_{260}/A_{280}$ ratio, thereby demonstrating that it is "more sensitive" to either protein or phenol contamination than the $A_{260}/A_{280}$ ratio. Only in the presence of phenol does the $A_{230}/A_{280}$ ratio remain virtually unchanged, but it increases significantly as a result of protein contamination. The decrease in the $A_{260}/A_{230}$ ratio is due to the increased absorbance of protein and phenol relative to nucleic acid at wavelengths below 240 nm (Figure 16-2). In contrast, the $A_{230/280}$ ratio was unchanged because the 230 and 280 nm absorbance both increased proportionately in the presence of phenol contamination (Figure 16-2). Therefore, the simultaneous decrease in the $A_{260}/A_{280}$ and the $A_{260}/A_{230}$ ratio without a concomitant increase in the $A_{230/280}$ ratio provides circumstantial evidence for phenol contamination. Confirmation of phenol contamination can be further documented by a decrease in the $A_{260}/A_{270}$ ratio (40), but the relative rate of decline of this ratio is modest and similar to that observed for both the $A_{260}/A_{280}$ and $A_{260}/A_{230}$ ratios (Figure 16-3c). Although protein contamination also lowers the $A_{260}/A_{270}$ ratio the relative rate of decline is not as precipitous as that noted in the presence of phenol (Figure 16-3, c vs. b).

In contrast to phenol and protein, which show significant absorbance in the 260 to 280 nm range, guanidine contamination does not increase the $A_{260}$ absorbance (Figure 16-3d). Furthermore, the $A_{260}/A_{280}$ ratio is unaffected by these contaminants, thus giving the false impression the DNA extracts are "pure." However, the $A_{230}/A_{280}$ ratio increases significantly and the $A_{260}/A_{230}$ ratio decreases. Because the $A_{260}/A_{230}$ ratio decreases more precipitously in the presence of all of the contaminants evaluated in this study, it provides a superior overall estimate of general contamination than the $A_{260}/A_{280}$ ratio. It is also apparent that these various contaminants influence $A_{260}/A_{230}$, $A_{260}/A_{280}$, and $A_{230}/A_{280}$ differently, thus providing some indication as to the nature of the contaminant.

The inverse relationship between the $A_{260}/A_{230}$ and the $A_{230}/A_{280}$ ratios that are obtained from nucleic acid extracts containing various contaminants is depicted graphically in Figure 16-4. In a defined solvent of known pH and ionic strength, pure DNA standards with an absorbance 0.1 to 1.0 display a linear $A_{260}/A_{230}$ and $A_{230}/A_{280}$ ratio of ~2.389 and 0.779, respectively (Figure 16-3). The quotient of these two ratios is ~3 (e.g., 2.389/0.779 = 3.07). In every case in which known amounts of protein, phenol, guanidine, or DNAzol®were added to DNA as depicted in Figure 16-3, the quotient of the $[A_{260}/A_{230}]/[A_{230}/A_{280}]$ ratio fell below 2.8 (Figure 16-5). Based on this observation, we propose that the quotient of the $[A_{260/230}]/[A_{230/280}]$ ratios be considered as a "purity quotient" (PQ) for evaluating contamination in nucleic acid extracts:

$$PQ = [A_{260}/A_{230}]/[A_{230}/A_{280}]$$

Irrespective of the $A_{260}/A_{280}$ ratio, extracts that display a PQ value >2.8 are essentially free of protein, guanidine, and phenol contamination. For computational purposes, PQ can be mathematically simplified and expressed as:

$$PQ = \{[A_{260}]*[A_{280}]\}/[A_{230}]^2$$

However, because the solvent that is employed for spectrophotometric measurement can affect nucleic acid absorbance, a comparison of absorbance ratios or PQ values will only be valid for samples that are diluted in the same spectroscopy solution. Furthermore, if this computation can be proved to be a reliable prognosticator of nucleic acid contamination in a broad array of biological extracts, in the future it may be possible to use PQ data to establish a "purity" baseline that may predict whether or not an extract will be suitable for a particular molecular biology application such as RT-PCR, real time PCR, gene chip analysis, etc.

## Suitability of Spectrophotometric Absorbance Measurements for Evaluating Nucleic Acid Purity

When applied appropriately, absorbance measurements can be used to collect reliable information relating to the purity of nucleic acid prepara-

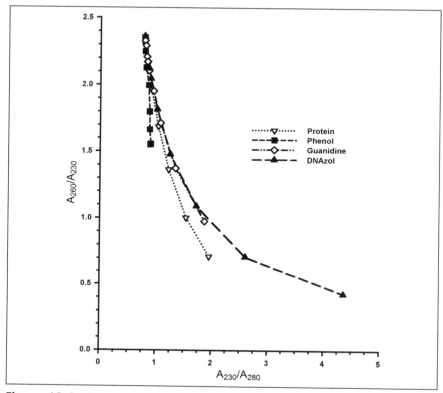

**Figure 16-4.** Reciprocal changes in the relationship of the computed calf thymus DNA $A_{260}/A_{230}$ and $A_{230}/A_{280}$ ratios in the presence of increasing quantities of designated contaminants. The computed $A_{260}/A_{230}$ and $A_{230}/A_{280}$ ratio data presented in Figure 16-3 is plotted as a function of increasing protein (0–50 μg), phenol (0–50 nL), guanidine thiocyanate (0–50 μM), or DNAzol (0–62.5 nL) contamination. As noted in Figure 16-3, in the presence of increasing quantities of the various contaminants, the $A_{260}/A_{230}$ ratio decreases and $A_{230}/A_{280}$ ratio increases in numeric value. A demonstrable curvilinear relationship exists between the $A_{260}/A_{230}$ and $A_{230}/A_{280}$ ratios for each contaminant except phenol, which displays a more linear association. In contrast, however, the $A_{260}/A_{230}$ and $A_{230}/A_{280}$ ratios that are computed with increasing quantities of purified control calf thymus DNA standard remain constant at ~2.39 and 0.78, respectively. All of the ratios were recorded in the presence of a 1 mM $Na_2HPO_4$ at pH 8.5. DNAzol is a patented DNA isolation reagent (7).

tions. Spectrophotometers are inexpensive, absorbance measurements can be rapidly recorded, and the resulting data can be easily archived with most modern instruments (Table 16-1). However, if these measurements are to have any meaningful value in the laboratory as well as in the published literature, it is essential that laboratory investigators solubilize their nucleic acids in an application appropriate solvent such as DEPC-treated

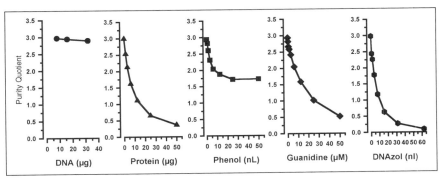

**Figure 16-5.** **Protein, phenol, guanidine thiocyanate, and DNAzol contaminants lower the nucleic acid purity quotients.** Increasing quantities of the designated contaminants were added to a constant amount of calf thymus DNA standard (8 μg) and the absorbance readings were recorded in duplicate in 1 mL of 1 mM $Na_2HPO_4$ at pH 8.5. The $A_{260}/A_{230}$ and $A_{230}/A_{280}$ ratios were utilized to calculate a purity quotient for DNA in the absence of the contaminant and in the presence of increasing quantities of the designated contaminants. In all cases, the computed purity quotient declined in response to increasing quantities of contamination.

water, formamide, TE-buffer, etc., and then select a standardized solvent for dilution and spectrophotometric assessment of nucleic acid purity.

Based on the data provided in this report, we recommend for consideration a 1.0 mM $Na_2HPO_4$ solution as the diluent for spectrophotometric measurements. The suggested ionic strength is known to have a minimal influence on RNA absorbance (Figure 16-1) while providing a pH that gives a stable and reproducible estimate of the $A_{260}/A_{280}$, $A_{260}/A_{230}$, and $A_{230}/A_{280}$ ratios. In addition, although the presence of a limited amount of salt affects the DNA $\gamma_{max}$ absorbance (Figure 16-1), this diluent provides greater stability for estimations of PQ than would otherwise be possible in water or salt solutions while still providing sufficient sensitivity to detect contaminants such as protein, guanidine, and phenol (Figure 16-5). Furthermore, $Na_2HPO_4$ is an inexpensive salt that is commonly available in most laboratories. The stock solution is easily prepared; it does not require pH adjustment and it is stable for long periods of time.

After selecting an appropriate diluent, such as 1 mM $Na_2HPO_4$, nucleic acid quality control data at 230, 260, and 280 nm is collected and archived for all laboratory extractions. We recommend using these absorbance measurements to calculate nucleic acid yield, the $A_{260}/A_{280}$, $A_{260}/A_{230}$, and $A_{230}/A_{280}$ ratios and the purity quotient for each extraction performed in the laboratory. The data that we have presented demonstrate that the individual computed ratios provide some clues relating to the general nature of a nucleic acid contaminant while the PQ provides a sensitive broad-spectrum assessment of the sample purity. These ratio calculations can be easily computed from absorbance data that have been imported into a computer spreadsheet or an automated laboratory information management system (LIMS). Moreover, if it can be demonstrated that these ratio calculations and the PQ value provide a reliable assessment of nucleic acid purity, instrument manufacturers could easily provide these additional data as an automatic calculation with their standard nucleic acid absorbance readings.

Nucleic acid quality control data that are collected and archived can be further correlated with the results that are obtained from various molecular biology applications. After an adequate number of extractions have been performed to establish meaningful statistics (i.e., 10–15 extractions), confidence envelopes could be computed, and a quality control "threshold" could be established for various laboratory applications. Moreover, this information could be used to optimize extractions and improve experimental results by identifying nucleic acid extracts that are significantly different from previous results (e.g., outliers) and that may be inappropriate for some sensitive molecular biology applications. The application of a more rigorous assessment of nucleic acid extracts will improve laboratory efficiency, decrease costs, and provide a more meaningful approach to troubleshooting extraction problems.

## Assessment of Nucleic Acid Extracts: Summary and Conclusions

Ultimately, the selection of methodology and instrumentation is always a delicate balance between addressing the immediate research requirements of a laboratory with the finite financial resources that are available to the investigator. Because of the increase in knowledge and development of new technology in the area of molecular biology, it is difficult to predict how research may evolve over the next decade. In this regard, however, it is important to recognize the enormous potential for differential gene expression analysis to provide meaningful diagnostic and therapeutic direction for the treatment of human and animal diseases. Therefore, any technology that has the potential to provide a more accurate assessment of nucleic acid extracts will be extremely important because it can improve the reliability and cost effectiveness of these molecular biology techniques.

As stated in the introduction of this chapter, it would be extremely beneficial to routinely perform a more thorough quality control assessment of nucleic acid extracts. Moreover, absorbance spectroscopy, when employed under defined experimental conditions in conjunction with microchip electrophoretic separation technology, currently provides the most sensitive analytical and diagnostic appraisal of the quantity, quality, and integrity of nucleic acid extracts. Instruments such as the NanoDrop®

ND-1000 (31) spectrophotometer have the capacity to perform absorbance measurements on 1 or 2 μL of undiluted extract and provide concentration estimates ranging from 2 to 3700 ng/μL. Absorbance scans ranging from 220 to 750 nm can be captured and archived, and these data can be subsequently utilized for appropriate absorbance ratio calculations. Because of the extremely small sample requirements, this instrument is ideally suited for assessing the purity and quantity of extracts that otherwise would be prohibitive with a standard spectrophotometer. Moreover, the measurements do not involve the destruction of the sample, so it could theoretically be recovered and subsequently used for integrity analysis in the Agilent 2100 Bioanalyzer. Although the current NanoDrop® spectrophotometer is not suited for high-throughput automation, more expensive microplate spectrophotometers could be utilized to address this specific requirement (18, 19, 36).

The routine assessment of nucleic acid integrity with microchip-based electrophoretic separation technology, although potentially useful, is probably unnecessary for most routine molecular biology applications. The initial expense of the Bioanalyzer and the relatively high costs of consumables, such as the RNA 6000 LabChip Kit, limit the immediate day-to-day use of this instrument to microarray laboratories. Classical agarose gel electrophoresis will continue to provide a subjective "snapshot" of the integrity of nucleic acid extracts until the more sensitive microcapillary-based analytical instruments and PCR-based methodology (3) becomes more affordable or until this technology becomes more useful in a broader array of separation applications.

Finally, a more critical appraisal of nucleic acid extracts will provide more consistent and reproducible extract results, and thereby augment the development and troubleshooting of new and more sensitive molecular biology applications. We hope that this review will stimulate discussion and result in the general adoption of universal criterion upon which nucleic acid purity, integrity, and quantitation will be based.

## References

1. Athas, G. 1999. Pure and simple: Genomic and plasmid DNA purification techniques. *Scientist* 13: 24–29.
2. Auer, H., Lyianarachchi, S., Newsom, D., et al. 2003. Chipping away at the chip bias: RNA degradation in microarray analysis. *Nat Genet* 35: 292–293.
3. BD Bioscience. 2003. *A Simple PCR-Based Alternative for Assessing Human RNA and cDNA Quality.* Technical Bulletin, RNA/cDNA Quality Assay. Camarillo, CA: BD Science.
4. Chirgwin, J., Przybyla, A., MacDonald, R., et al. 1979. Isolation of biologically active ribonucleic acid from sources enriched in ribonuclease. *Biochemistry* 18: 5294–5299.

5. Chomczynski, P., and Sacchi, N. 1987. Single-step method of RNA isolation by acid guanidinium thiocyanate-phenol-chloroform extraction. *Anal Biochem* 162: 156–159.

6. Chomczynski, P. 1993. A reagent for the single-step simultaneous isolation of RNA, DNA and protein from cell and tissue samples. *BioTechniques* 15: 532–537.

7. Chomczynski, P., Wilfinger, W., and Mackey, K. 1998. Isolation of genomic DNA from human, animal, and plant samples with DNAzol reagents. In: Fox, C.F., and Connor, T.H., eds. *Biotechnology International*. San Francisco: Universal Medical Press; 185–188.

8. Chua, J. 2004. A buyer's guide to DNA and RNA prep kits. *Scientist* 18: 38–39.

9. Cohen, A.S., Najarian, D.R., Paulus, A., et al. 1988. Rapid separation and purification of oligonucleotides by high-performance capillary gel electrophoresis. *Proc Natl Acad Sci U S A* 85: 9660–9663.

10. Fleck, A., and Begg, D. 1965. The estimation of ribonucleic acid using ultraviolet absorption measurements. *Biochim Biophys Acta* 108: 333–339.

11. Fletcher, T.S., Weiss, T.L., Wang, Y.E, et al. 1999. Quantitative analysis of DNA, RNA and protein by digital imaging. In: Connor, T.H., Weier, H., and Fox, F., eds. *Biotechnology International II*. San Francisco: Universal Medical Press; 281–285.

12. Glasel, J.A. 1995. Validity of nucleic acid purities monitored by 260nm/280nm absorbance ratios. *BioTechniques* 18: 62–63.

13. Held, P. 2003. *The Importance of the 240nm Absorbance Measurement: The A260/A280 Ratio Just Isn't Enough Anymore*. Winooski, VT: Bio-Tek Instruments.

14. Huberman, J.A. 1995. Importance of measuring nucleic acid absorbance at 240nm as well as at 260 and 280nm. *BioTechniques* 18: 636.

15. Jaffe, S. 2004. Lab matters. Electrophoresis: the state of the gel. *Scientist* 18: 45–48.

16. Krzysztof, S., and Wlodzimierz, K.J. 2002. RNA structure analysis assisted by capillary electrophoresis. *Nucleic Acids Res* 30: e124, 1–8.

17. Le Borgne, S., Mancini, M., Le Grand, R., et al. 1998. In vivo induction of specific cytotoxic T lymphocytes in mice and Rhesus macaques immunized with a DNA vector encoding an HIV epitope fused with hepatitis B surface antigen. *Virology* 240: 304–315.

18. Malik, T. 2004. Buyer's guide to microplate readers: absorbance readers still reign, as fluorescence and luminescence options gain in popularity. *Scientist* 18: 34–35.

19. Malik, T.J. 2004. UV spectrophotometers buyers' guide: consider optical configuration, spectral range, sample format, and analytical tools. *Scientist* 18: 39–40.

20. Manchester, K.L. 1995. Value of A260/A280 ratios for measurement of purity of nucleic acids. *BioTechniques* 19: 208–210.

21. Manchester, K.L. 1996. Use of UV methods for measurement of protein and nucleic acid concentrations. *BioTechniques* 20: 968–970.

22. Marx, V. 2004. RNA quality: defining the good, the bad, and the ugly. *Genomics Proteomics* 4: 14–21.

23. McGown, E.L. 2000. UV absorbance measurements of DNA in microplates. *BioTechniques* 28: 60–64.
24. McGown, E., and Su, M. 2000. DNA quantification in the LMax™ microplate luminometer. Molecular Devices Application Note 37. Available at http://www.moleculardevices.com. Accessed July 2005.
25. Molecular Research Center, Inc. 2004. TRI Reagent®-RNA/DNA/protein isolation reagent. Product Technical Information, TR-118. Cincinnati: Molecular Research Center.
26. Mueller, O., Lightfoot, S., and Schroeder, A. 2004. RNA Integrity Number (RIN)-Standardization of RNA Quality Control. Agilent Technologies Application Note. Palo Alto, CA: Agilent Technologies.
27. Okamoto, T., and Okabe, S. 2000. Ultraviolet absorbance at 260 and 280 nm in RNA measurement is dependent on measurement solution. *Int J Mol Med* 5: 657–659.
28. Peirce, J. 2003. Advancing with gel documentation systems: digital acquisition of information for under $20,000. *Scientist* 17: 39–41.
29. Quast, B. 2001. A compact, handheld laboratory fluorometer. *Am Biotech Lab* 8: 68.
30. Righetti, P.G., Gelfi, C., and D'Acunto, M.R. 2002. Recent progress in DNA analysis by capillary electrophoresis. *Electrophoresis* 23: 1361–1374.
31. Robertson, C. 2003. Spectrophotometry's next-generation technology. *Am Biotech Lab* 21: 23–24.
32. Salowsky, R., and Hengler, A. 2002. *High Sensitivity Quality Control of RNA Samples Using the RNA 6000 Pico LabChip Kit.* Agilent Technologies Application Note. Palo Alto, CA: Agilent Technologies.
33. Schorr, J., Moritz, P., Seddon, T., and Schleef, M. 1995. Plasmid DNA for human gene therapy and DNA vaccines. Production and quality assurance. *Ann N Y Acad Sci* 772: 271–273.
34. Sharp, P.A., Sugden, B., and Sambrook, J. 1973. Detection of two restriction endonuclease activities in *Haemophilus parainfluenzae* using analytical agarose-ethidium bromide electrophoresis. *Biochemistry* 12: 3055–3063.
35. Sinclair, B. 2000. An eye for a dye: safe and sensitive new stains replace ethidium bromide for routine nucleic acid detection. *Scientist* 14: 31–35.
36. Singer, V.L., Jin, X., Jones, L., et al. 1998. Sensitive fluorescent stains for the detecting nucleic acids in gels and solutions. In: Fox, C.F., and Connor, T.H., eds. *Biotechnology International.* San Francisco: Universal Medical Press; 267–276.
37. Skeidsvoll, J., and Ueland, P.M. 1996. Analysis of RNA by capillary electrophoresis. *Electrophoresis* 17: 1512–1517.
38. Smutzer, G. 2001. Spectrophotometers: an absorbing tale. *Scientist* 15: 27–28.
39. Stull, D., and Pisano, J.M. 2001. Purely RNA: New innovations enhance the quality, speed and efficiency of RNA isolation techniques. *Scientist* 15: 29–31.
40. Stulnig, T.M., and Amberger, A. 1994. Exposing contaminating phenol in nucleic acid preparations. *BioTechniques* 16: 403–404.
41. Teare, J.M., Islam, R., Flanagan, R., et al. 1997. Measurement of nucleic acid concentrations using the DyNA Quant™ and the GeneQuant™. *BioTechniques* 22: 1170–1174.

42. Ulfelder, K.J., Schwartz, H.E., Hall, J.M., and Sunzeri, F.J. 1992. Restriction fragment length polymorphism analysis of ERBB2 oncogene by capillary electrophoresis. *Anal Biochem* 200: 260–267.
43. Warburg, O., and Christian, W. 1942. Isolation and crystallization of enolase. *Biochem Z* 310: 384–421.
44. Wilfinger, W.W., Mackey, K., and Chomczynski, P. 1997. Effect of pH and ionic strength on the spectrophotometric assessment of nucleic acid purity. *BioTechniques* 22: 474–481.

# 17 Essential Software and Other Tools Used in Modern Biology Laboratories

Jan Kieleczawa,[1] Deven Atnoor,[1]
Russ Carmical,[2] Andrew Hill,[1] Fei Lu,[2]
Ken Paynter,[2] and Paul Wu[1]
[1]Wyeth Research, Cambridge, Massachusetts,
and [2]Seqwright, Inc., Houston, Texas

Almost any major biochemical advance in the last three decades was made possible by the development of new methods, scientific instrumentation, computers, bioinformatics, and other tools. This enormous progress can be easily illustrated, for example, by comparing the first edition of a famous textbook by Lehninger (11) to its third edition (17). The first edition contained only one detailed molecular drawing of a protein (myoglobin) in contrast to the over 200 detailed structures illustrated in the third edition. In the mid 1970s, the first molecular biology techniques, for example, the joining of two foreign fragments of DNA, and the first DNA sequencing projects (13, 23) were just getting off the ground. By the year 2000, we had witnessed full-blown activities that led to the completion of the majority (as of now there are still about 300 small sequence gaps) of the sequence for the entire human genome (8, 20, 26).

This chapter describes a number of methods, instruments, computer programs, and other tools that are used in today's modern biology laboratories to help yield ever more accurate and sophisticated data. Most modern molecular biology laboratories are interdisciplinary (i.e., simultaneously working on DNA, proteins, cells, etc.), but this chapter concentrates on describing several tools that can be found in a typical laboratory with the main focus on the DNA research. These tools are described in as logical an order as possible, starting with the quantitation of DNA, storing the data in databases, and analysis. We are fully aware

*DNA Sequencing II: Optimizing Preparation and Cleanup*
Edited by Jan Kieleczawa
©2006 Jones and Bartlett Publishers

that this is only a small fraction of methods and tools that are commonly used, and a broader description can be found in many textbooks (2, 12, 17) and other specialized monographs (3, 5–7, 14, 19, 21, 22, 24), to select just a few. The choice of material presented in this chapter was influenced by the experience and preferences of the contributing authors and, thus, does not necessarily imply that they are the only types available or are the best option on the market. Any omission of other tools, methods, programs, etc., is purely unintentional.

## Determination of Quantity and Quality of DNA

Typically, the first assay performed by a scientist after the isolation of a DNA (be it plasmid, cosmid, PAC, BAC, PCR fragment, phage, or any other type) is to assess its quantity, quality, and integrity. The most widespread techniques used are the agarose gel electrophoresis of the DNA samples followed by imaging with either classical Polaroid or digital systems. The advantage of using Polaroid systems is their simplicity. They are still routinely used in many laboratories as the images are easy to acquire and are convenient to place in laboratory notebooks. However, the process can be laborious and complicated when, for example, a publication or presentation needs an agarose gel image. Therefore, digital imaging, although technologically more demanding, is a much better option. A number of systems are on the market that can be used for digital imaging of agarose gels. Here we describe one such system, Kodak's EDAS 290* package (Eastman Kodak Company, Rochester, NY, USA).

### EDAS 290 Imaging System

This system utilizes digital DC290 camera (1792 × 1200 pixels image resolution) and the Electrophoresis Documentation and Analysis Software (EDAS) to capture and analyze a variety of images. It is available both for Macintosh (Mac; OS 8 and higher) and personal computer (PC; Microsoft Windows 2000 and higher) platforms and for 110 V and 220 V power systems. A more detailed description of system requirements is on the Kodak Web site (http://www.kodak.com/go/scientific).

Once an image is acquired using the Kodak or any other TWAIN capture device (assuming that the images can be saved in tiff or jpeg formats), the EDAS software can perform a number of analyses, including:

---

*Effective May 1, 2004, the Kodak EDAS 290 system was replaced by a more sophisticated Gel Logic imaging system featuring charge coupled device (CCD) instead of a digital camera.

- Finding lanes and DNA bands on an agarose gels (lanes and bands can be manually added)
- Calculating DNA sizes and concentrations with all numerical values (providing that the appropriate mass and size markers were used)
- Lane labeling and in general annotating an image
- Profile viewing
- Viewing Southern, Northern, and Western blots
- Microplate assays
- Quantitative PCR
- Thin layer chromatography
- Image adjusting (horizontal, vertical, and custom flipping and rotating)
- Image inverting
- Colony and plaque counting
- Resizing an image
- Exporting an image in a variety of formats (tiff, gif, jpeg, bmp)
- Archiving and creating databases of images

An example of an image taken with Kodak's DC290/EDAS 1D system is shown in Figure 17-1. A variety of options allow for precise manipulation and display of an agarose gel image that can be easily prepared for documentation or publication.

## Other Imaging Systems

There are a number of different imaging systems that can accommodate just about any need and application. Here we list a few of these systems and you are encouraged to visit these Web sites to learn about the specific features of each system:

- Versadec 4000/5000 from BioRad (http://www.biorad.com)
- Fla 3000 from FujiFilm (http://www.fujifilm.com)
- GelDoc-it from UVP (http://www.uvp.com)

The other common method to determine the amount of DNA (and RNA) is by UV spectrophotometry (see Chapter 16). This section describes two very useful devices that are commonly found in many laboratories.

## NanoDrop® ND-1000 Spectrophotometer

Template quality and quantity are of paramount importance to the success of a DNA sequencing reaction. The quality of the template not only relies on the integrity of the DNA, but also on the presence or absence of contaminants. In addition to the adverse effects that contaminants might have

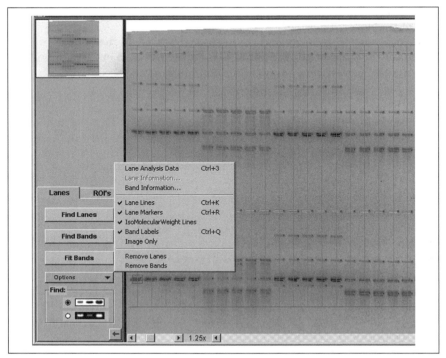

**Figure 17-1.** Image of an agarose gel taken with Kodak's DC290 camera and analyzed with EDAS 1D software. The inverted image displays found lanes and bands. Many other features to enhance an image are readily available from an assortment of menus.

on enzyme efficiency, overestimation of the template concentration may occur if the contaminants add to a sample's optical density at 260 nm ($OD_{260}$). Incorrect estimates of template concentration can result in two types of failures. Too much template can cause substrate saturation and result in an uneven signal distribution and shorter reads ("top-heavy," that is, a lot of signal at the beginning then fading quickly), whereas too little template can result in complete failure of the reaction. Two methods are typically employed to assess both quality and quantity of DNA templates: ultraviolet/visable (UV/Vis) spectrophotometry and agarose gel electrophoresis. If the sample is relatively pure (i.e., without significant amounts of contaminants such as proteins, phenol, agarose, or other nucleic acids), then UV/Vis spectrophotometry is easy and accurate. If contaminants are suspected or the integrity of the template is in question, then gel electrophoresis is the preferred method to assess quality and quantity, despite being less accurate and more labor-intensive. Subse-

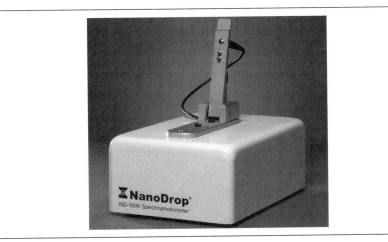

**Figure 17-2.** Image of a NanoDrop® ND-1000 Spectrophotometer.

quent paragraphs discuss the state-of-the-art instrumentation capable of performing the required measurements for DNA sequencing.

Since its inception during World War II, UV/Vis spectrophotometry has proven to be a standard measurement and quality control method for a large variety assays and sample types. As other technologies have evolved to enable greater sample manipulation and detection, this fundamental tool has not kept pace, remaining basically unchanged until now.

The relatively recent advent of microspectrophotometry may help enhance the usefulness of this quantitation method. A sample retention system developed by NanoDrop Technologies (Wilmington, DE, USA; http://www.nanodrop.com) has improved upon the basic design of the UV/Vis spectrophotometer. The NanoDrop ND-1000 Spectrophotometer (Figure 17-2) uses the surface tension properties of the sample to hold the sample in place during the measurement cycle, eliminating the need for cuvettes or capillaries. During each measurement cycle, the sample is assessed at two different path lengths (1.0 mm and 0.2 mm), resulting in a broad dynamic range (2 ng/µL to 3700 ng/µL for dsDNA). The 0.2 mm path allows absorbance measurements 50 times more concentrated than can be obtained using traditional 1-cm path cuvette systems, virtually eliminating the need to perform dilutions. Its ability to assess at-strength samples while consuming only 1 µL of sample enables investigators to explore possibilities previously prohibited by cuvette-based spectro-photometry.

To take a measurement, 1 µL of a sample is pipetted directly onto the lower optical (measurement) surface (Figure 17-3a). An upper optical fiber

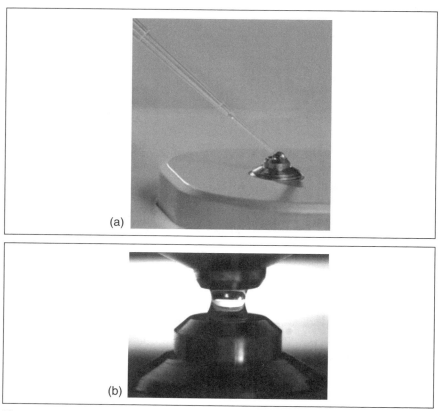

**Figure 17-3. Using a NanoDrop ND-1000 Spectrophotometer.** (a) Loading 1 µL of sample. (b) A 1 µL sample "bridging" the optical fibers.

automatically engages the sample, forming a liquid column of mechanically controlled path length (1 mm) that is held in place by the sample's surface tension (Figure 17-3b).

Direct contact of the sample with the optical surfaces reduces light diffraction. Preparing the instrument for another measurement simply involves cleaning the sample off with a lab wipe before loading the next sample.

The NanoDrop ND-1000 Spectrophotometer uses a xenon flash lamp as the light source. The most commonly used light sources for measurements in the short UV (~260 nm) are arc sources, including xenon continuous arc, xenon flashlamp, and deuterium arc lamps. Deuterium and xenon continuous arcs have relatively short lives (~800 to 1000 h) and must be warmed up to be stable enough to make good measurements. In

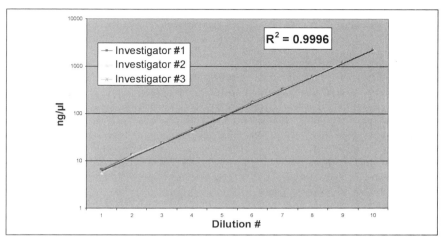

**Figure 17-4.** A stock solution of pGEM (2300 ng/µL) was serially diluted by three independent investigators and the concentration was measured with the Nanodrop spectrophotometer. The resulting concentration was then plotted against the dilution number and a Pearson's correlation coefficient was calculated.

contrast, xenon flashlamps require no stabilization time and fire only during the measurement cycle, thus significantly increasing the life of the lamp (~50,000 measurements).

The internal spectrometer of the ND-1000 is a 2048-element linear silicon CCD array with a wavelength range of 220 to 750 nm, a wavelength accuracy of 1 nm, and a wavelength resolution of 3 nm (FWHM; at Hg 546 nm) or full width half maximum as compared to mercury as a reference. The instrument's precision is 0.003 absorbance units and the absorbance accuracy is 2% (at 0.76, absorbance at 257 nm). Unlike scanning spectrophotometer systems, the fixed CCD array allows for a rapid assessment of the full UV/Vis spectrum of the sample (~10 to 12 seconds).

Plotting independent measurements of a serial dilution of plasmid DNA from 2300 ng/µL to 6 ng/µL results in Pearson's correlation coefficient ($R^2$) of 0.999 (Figure 17-4), indicating good precision and reproducibility at both the upper and lower limits of detection.

Nucleic acid absorbance spectrum measurements are made from 220 nm to 350 nm. The absorbance spectrum, relevant ratios, and calculated concentrations are automatically displayed (Figure 17-5). Each measurement is archived into a graphing spreadsheet such as MSExcel.

Microspectrophotometry is rapidly becoming the quantitation standard in service-based DNA sequencing laboratories because of its preci-

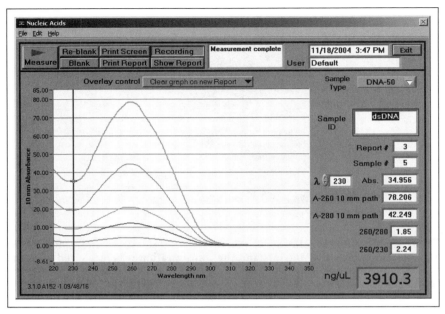

**Figure 17-5.** NanoDrop ND-1000 Spectrophotometer software display of nucleic acid absorbance spectra with various ratios and calculated sample concentrations. The data is automatically stored in graphics spreadsheets. The researcher may also "Record" the current set of measurements for immediate viewing.

sion, accuracy, and ease-of-use. Biomolecular core facilities are subject to samples varying greatly in concentration and quality. A rapid and reliable means of sample quality control (QC) is imperative to the success of downstream reactions.

### Biophotometer®

The Biophotometer (Eppendorf AG, Hamburg, Germany) is a very compact spectrophotometer that is specifically designed for various nucleic acids and protein determinations. Each determination method (ssDNA, dsDNA, RNA, protein) has its own key, and the parameters and conversion factors are automatically set when a method is selected. The standard 1-cm path cuvette can accommodate as little as 10 µL of a solution, but more reliable data are obtained when 60 µL of a solution is used. The other product features include:

- Fast measurements (~2 sec/sample)
- Background correction at 320 nm for UV methods

- Automatic calculation of sample dilution (if dilution factor was entered)
- Storage of the last 100 measurements
- Storage of all calibrations
- Checking for sample consistency
- Data transfer to a PC

For more information on this product, the reader is encouraged to visit the manufacturer's Web site (http://ww.eppendorf.com).

## Other Spectrophotometric Systems

As in the case of imaging systems, there are even more different spectrophotometers to choose from. Below is a list of just a few additional spectrophotometers that are commonly used in many laboratories. Again, for the newest model of a particular device, the reader is encouraged to contact the specific manufacturer.

- DU® series 500 UV/Vis from Beckman (http://www.beckman.com)
- 2100 series from Geneq (http://www.geneq.com)
- U-3010 from Hitachi (http://www.hitachi-hta.com)
- Lambda 650/850/950 series from Perkin-Elmer (http://www.perkinelmer.com)
- UV-2501/3101 from Shimadzu (http://www.shimadzu.com)

In addition to the agarose gel electrophoresis and spectrophotometric methods, scientists can use other approaches to determine the concentration of nucleic acids. The following is a partial list of these methods:

- Assays with a variety of fluorometric dyes like PicoGreen®, Hoechst®, etc.
- DNA Dipstick™ from Invitrogen (http://www.invitrogen.com)
- DyNA Quant 200 from Amersham (http://www.amershambiosciences.com)
- Picofluor™ from Turner Biosystems (http://www.turnerbiosystems.com)
- Spectra Max® microplate readers from Molecular Devices (http://www.moleculardevices.com)

Each method offers distinct advantages (speed, simplicity, accuracy, etc.), but they may also lack some other important features, for example, an ability to distinguish between DNA and RNA, sample integrity, possible contaminations, etc. Therefore, it is critical for the scientist to choose the best method(s) to suit a project's specific needs.

## Databases to Store and Analyze DNA Information

This section easily could be expanded into a full chapter or even an entire book, but here we concentrate on two aspects of information storing: basic information about a DNA of interest (size, concentration, resistance marker, cloning vector, etc.) and its analysis (i.e., blasting against all known related data sets in various databases).

To our best knowledge, there is no commercial database that is designed just to store a basic set of information about prepared DNA. Here we show that a very limited knowledge of, for example, the Access database available as a part of the Microsoft Office suite, can lead to a creation of a useful tool. Figure 17-6 shows an example of a screen with a pertinent data about newly prepared plasmid DNA. If needed, other fields can be added to accommodate any other site-specific requirements.

However, there are number of commercial and in-house developed LIMS systems that are specifically geared toward DNA sequencing applications. Two such LIMS have been described in great detail in earlier papers (10, 18). Since their publication (10, 18), many new features have been added to both LIMS systems. The most important new function for Geospiza's Finch-Server (http://www.geospiza.com) is the addition of a project-based data organization module that targets the scalability and automation of high-throughput DNA sequencing decision support systems. Specifically, the new module and add-on applications organize chromatogram files with reference sequence data and analysis pipelines (workflows) to automate the assembly of reads into contig maps (contigs), comparison of contigs (and/or reads) to reference sequences, and detailed reporting of results through high level summaries, graphics, and drill down tables. Geospiza's new Finch software addresses numerous applications such as high-throughput confirmatory sequencing, de novo sequencing, site directed mutagenesis, and DNA sequencing-based genotyping.

For a 4D-based system (4th Dimension, Inc.), the detailed calculation of GC content and its linkage to an appropriate chemistry as well as a priori scanning of a provided reference sequence for a potential difficult region (e.g., direct and inverted repeats, hairpins, homoplymers, di- and tri-nucleotide repeats, etc.) decreases the number of reactions needed to complete a specific sequencing projects. The detailed description of the 4D LIMS system will be presented in a separate publication in the near future. Another commercially available system is the dnatools LIMS database. The dnaLIMS™ by dnaTools™ (http://www.dnatools.com) is a Web-based LIMS used to capture DNA sequencing requests, create sample sheets, store and manage sequencing results, distribute the results to the users, and create usage reports and accounting summaries. It is designed so that all of the input forms, reports, and process flow can be

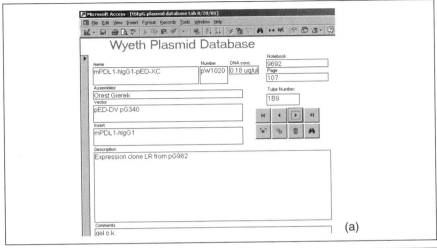

**Figure 17-6. Example of a simple database for storing information about DNAs.** (a) Detailed information about a particular clone. (b) An overview for a set of clones.

customized to the specific requirements of a DNA sequencing core facility. Requests are entered using a customized sequencing request form that dynamically creates a table sized to the number of reactions being entered. Requests also can be entered using plate-based forms or via upload and import of electronic spreadsheets. Plate-based and import forms can directly create sample sheets or deposit the requests into a queue. The dnaLIMS has a viewer so users can display chromatograms with quality scores (phred and KB-caller) as well as built in interfaces to automatically run Phred, Phrap, Consed, Blast, and GCG applications. The dnaLIMS has a series of add-on modules to extend the LIMS into other laboratories within the same organization. These modules include Oligo Ordering, Peptide Synthesis, Protein Sequencing, Mass Spec, Real Time PCR, Histology, Cytometry, Proteomics, Library Services, and Media Prep. All of these modules have a consistent look and provide an easy tool for institution-wide tracking and accounting.

## Sequence Analysis Packages

A variety of tools are on the market that can help to analyze a molecule of interest in great detail. Here we focus on a small number of programs that are useful in analyzing DNAs. Some of these programs are geared toward a limited number of superbly performed tasks (e.g., Sequencher) for sequence editing and assembly, and others are more comprehensive and encompass many different analytical functions (e.g., DNASTAR, or Vector NTI) to perform protein structure prediction, primer design, multiple sequence alignments, etc.

A few of the features of selected programs are listed below. Further details can be found on the manufacturer's Web sites or literature.

The advent of commercially available sequence analysis tools that can operate on a desktop computer has been a great facilitator of research for many laboratories and institutions, because not everyone has access to nor experience with the Unix/Linux-based applications that dominate the larger genomics communities. Here we look at several popular desktop applications for analyzing sequence data.

### Sequencher™: Software Overview

Sequencher (current version 4.5) is a sequence analysis and assembly program with multipurpose functionality, simple graphical user interfaces (GUIs), and fast processing speeds that operates on a standard PC or Mac desktop computer. The program was introduced in 1991 by Gene Codes Corporation (Ann Arbor, MI, USA; http://www.genecodes.com), and has followed a *best practices* approach to software development, adding feature and functionality enhancements as its large and growing user base has continued to demand more out of the product. Sequencher works well as both a de novo and resequencing analysis tool. A quick overview of its key features includes:

- Sequence trace viewing
- Contig assembly (multiple assembly options and strategies)
- Comparative sequence analysis
- Sequence trimming and editing by quality score (Phred/KB)

- Restriction maps, motif definitions, OFR maps, and protein translations
- Second peak heterozygote calling
- Numerous and flexible file importing and exporting abilities
- Report generating
- Multiple visualization tools

Over the course of its nearly 15-year history, Sequencher had been cited in hundreds of publications ranging in disciplines from forensic science to cancer research. Perhaps contributing most to the program's success is the relative simplicity in getting acquainted with and using Sequencher. Out of the box, Sequencher's core assembly function can be used with little or no instruction. Importing chromatogram sequences can be done via drag-and-drop into the main window editor or through "import" functionality. The program accepts multiple formats including: .ab1, .scf, .phd, GenBank, FastA, EMBL, ASCII, GCG, and most other common sequence text file types. A window asking about sequence trimming can be set to open after each file import. Options for trimming include trim by quality (Phred and KB), the number of ambiguous (N) base calls, and length. Sequence specific (vector) trimming is also available. Initially, the default trim criteria are set and are suitable for most common projects.

Once the sequences have been imported and trimmed, selecting multiple traces can be done using standard PC/Mac commands such as Ctrl-A or dragging over files with the mouse pointer and Sequencher's assembly algorithm takes a few seconds for most projects and up to 15 minutes for all-at-once assemblies of something as large as a full-length BAC. Several adjustments to the assembly algorithm and its parameters can be made depending on the user's application. Contig viewing and editing is performed through the text assembly view, and edits (with the exception of deletions) can be set to distinctly contrast (bold color or case change) with the unedited sequence. Similar to importing, exporting can be done in a number of formats including contigs and text consensus.

Sequencher would not be the program it is without its assembly functions and graphical visualizations (Figure 17-7; color Plates 12 and 13). Users can choose several assembly algorithms including *Dirty* (default), *Clean*, and *Large Gap*. Parameters include aligned sequence *match percentage* and *minimum overlap* between fragments. There are also several options for optimizing assembly times for users working with very large constructs and sample sets. Furthermore, Sequencher's *Assembly by Name* function (version 4.5) allows the user to define sample naming conventions via common and custom delimiters, and to perform auto-assembly into groups. For instance, a user can import forward and reverse reads from 400 samples, and within seconds assemble them into 400 separate contigs.

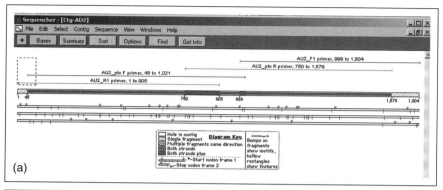

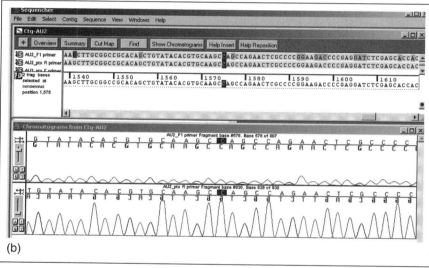

**Figure 17-7.** **Various displays of data by Sequencer program.** (a) Overview of a finished contig, where green indicates forward and red indicates reverse sequenced strands. Numbers indicate positions of individual fragments in a contig. Completeness of coverage is indicated by colors and texture of the bar below the overview. (See Plate 12.) (b) Light-blue coloring of a sequence at the top indicates good quality of data; the darker the blue color, the lower the quality of data. (See Plate 13.) The lower part shows chromatograms corresponding to a sequence fragment and allows a visual inspection of data. Sequencher allows for a number of view options that can be turned on or off to suit particular needs.

The Large Gap algorithm, as the name suggests, is especially useful when aligning sequences containing large gap deletions against a reference sequence. Similarly, the function is excellent for aligning cDNA against genomic sequence to get a quick look at splice sites. Combined with the software's *Motif Definitions* and *Mark Feature* functionality, users can generate coding region maps.

For mutation and resequencing studies, Sequencer has two features: *Secondary Peak Search* (mixed base/heterozygote identification) and *Reference Sequence*. The *Secondary Peak Search* option allows the user to define the threshold for calling the second peak under the major form and assigning that base position an IUPAC ambiguity code. Upon executing the command, Sequencer recalls all the base assignments within the contig or sequence that fall inside the threshold range. These base changes are displayed as standard edits.

The *Reference Sequence* command allows the user to define a reference sequence, make created contig edits against the reference without altering it, and generate an automatic variation report for the consensus versus the reference. The report is fairly basic, detailing nucleotide positions and base changes. The report is not hyperlinked back to the sequence assembly, so it is more easily used to generate a report following creation instead of navigating to basecalls for review.

Sequencer has a number of other visualization tools such as *Restriction Map*, amino acid *Translation*, *Mark* (highlight) *Feature*, and *Motif Definitions*. The names of these functions are pretty self-explanatory. The motif definition feature is somewhat limited compared to other programs in that it is restricted to a small number of motifs; however, Gene Codes has slated this for enhancement in the 2006 Sequencer upgrade. In the interim, simply maintaining a concatenated FastA file of motifs, such as a primer sequence library, and assembling it against a construct will accomplish nearly the same effect, unless duplications within the construct being assembled against are involved.

One of the Sequencer's highlights is its reporting function. Users can print sequence graphical contig overviews showing open reading frames (ORFs), directional sequence coverage, and construct size as well as sequence text alignments in a number of fonts, font colors, and font sizes. Users can also print and export consensus sequence, restriction maps, amino acid translations, and a table detailing sequence variation against a reference. For those working with Phrap, Sequencer can import .ace files and generate any of these reports.

Sequencer in many regards is also a data management tool. Large amounts of sequence data can be pulled into a single Sequencer Project File (.spf). Sequence data are replicated in the program, so the original trace files are never compromised when editing. Also, Sequencer maintains an original impression of all unedited sequence, so users may also *Revert To* (original) *Experimental Data* in the event of some major editing glitch. Data can be grouped using the assembly functions or archived into data sets called *Refrigerators*. Sequences and contigs may also be assigned *Labels* and be grouped and sorted with these tags. Standard selecting and sorting by name, creation, and modification dates and sequence length are also included. The overall ability of Sequencer to maintain large data sets in a single project file without compromising the software's speed and performance also makes it a great conduit for sharing data among research groups.

Sequencher has done a formidable job of establishing itself as the "all things to everybody" sequence analysis tool. Few other software suites are used as universally in such a breadth of laboratory environments as this program. Still, the Sequencher is not perfect, and Gene Codes continues to build functionality into its software with every upgrade and release. A good example would be the *Assembly by Name* feature, which was released in 2005 with v.4.5, yet many of Sequencher's competitors had already included this function in their software years earlier.

Furthermore, Sequencher has not yet addressed primer design functionality, a clear gap in its niche of being a multifunctional tool. Such algorithms are fairly commonplace and could be extremely beneficial, for example, when combined with Sequencher's *Large Gap* algorithm for aligning cDNA to genomic DNA.

Despite some minor shortcomings, one cannot argue with the staying power and the overall appeal and success that Sequencher has enjoyed within the genomics community. New users should not be too concerned with having to abandon an obsolete platform any time soon because Sequencher has a 15-year history of continual functional enhancements and additions, while other "popular" programs that in the past had fulfilled a similar need are no longer supported by their creators. For being a platform that can adeptly perform a broad range of tasks within a single laboratory or be used as a standardized conduit for sharing sequence data among different laboratories, Sequencher definitely excels (25).

## Mutation Surveyor™: Software Overview

Mutation Surveyor (current version is 2.52) is a DNA sequence analysis program from SoftGenetics (State College, PA, USA; http://www.softgenetics.com) focused on identifying mutations and polymorphisms. What makes this software stand out from other sequence comparison tools is that it uses a point-to-point physical trace comparison for detecting variation instead of a standard base call. The result is an algorithm that more closely mimics the process that researchers use during a manual creation of base calls and results in automated mutation analyses with fewer false positives and negatives than most other programs. Furthermore, a real gem of the software is its patented heterozygous insertiondeletion (hetero-indel) peak deconvolution function, which can pull apart a mixed peak profile into separate traces, allowing the user to clearly view the sequences independently as (nearly) clean data (Figure 17-8).

Overall, the Mutation Surveyor is user-friendly and the amount of orientation time required to start performing analyses is minimal or, thanks

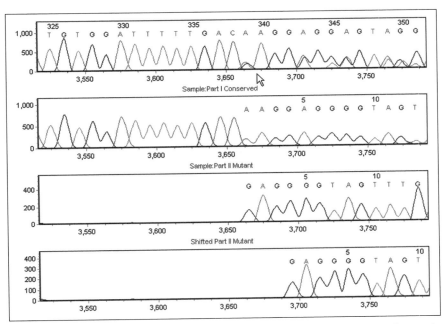

**Figure 17-8. Hetero-indel deconvolution.** The top panel is the original sample trace containing a 3 base-pair heterozygous insertion-deletion (AAG). The next panel down is the conserved view. The lower panels show the mutant and shifted mutant forms.

to the software's defaults, nearly nonexistent. Setting up a project is as easy as importing sequence trace files (.ab1, .scf, and .gz formats) and executing the run command. The program does require a reference sequence for comparative analysis, but the default parameters are set to automatically download GenBank sequences from a database URL maintained by SoftGenetics, which seems to contain all the human sequence. If security is a concern or if a sequence is not present in the public domain, the database URL from which Mutation Surveyor automatically downloads the reference can be changed or deactivated. A reference can be manually imported as well.

The program can analyze up to 400 traces at a time requiring approximately 30 minutes per analysis. Smaller sample sets are analyzed much faster. Samples are assembled into contigs and a variation report is automatically generated. The analysis defaults recognize most standard sequence file naming delimiters, so forward reads from patient 1 are automatically matched to reverse reads from the same patient. Custom naming conventions may also be defined.

Another great feature is Mutation Surveyor's automated batching capability with its Auto Run function. The user may define a time and network destination folder, such as the DNA sequencer run destination folder, for the software to import all the data and begin analyzing without any human involvement.

Following analysis, multiple graphical display options allows the user to easily navigate through large amounts of sequence data and quickly find areas of variation. These displays include a branching-directory navigation window, sequence text display, sequence trace display, Basic Graphic Display (mutation frequency overview), and mutation electropherogram and data table.

Reports contain the typical parameters and options: sample name, nucleotide change, position, quality score, amino acid change, etc. Variations are color coded to denote high probability versus low probability, and all fields are hyperlinked back to the appropriate position in the sequence trace for fast confirmation/rejection of auto-calls. Reporting formats can be customized and exported in various text file formats. SoftGenetics even offers to create custom reports if customer's requirements are not met by one of the default outputs.

Mutation Surveyor's point-to-point physical comparison is accomplished using five independent parameters:

- *Mutation Heights*: the height of the mutation peak above the point at which a mutation is registered by the software.
- *Overlapping Factor*: the amount of overlap between two peaks before a mutation is registered.
- *Dropping Factor*: the relative intensity drop of a sample's peak relative to the reference.
- *SnRatio*: signal-to-noise ratio.
- *Mutation Score*: a measure of the probability of error (confidence value).

Defaults are predefined, but users have the option of customizing any of these parameters. SoftGenetics claims their patented algorithm can achieve automation mutation calling accuracy of 99.5% with bidirectional sequencing. Overall, the calling accuracy is greater compared to programs that rely strictly on base call confidence scores.

Mutation Surveyor is particularly adept at detecting hetero-indels and, therefore, is an excellent tool for tumor sample resequencing. SoftGenetics states that the software can detect this type of mutation down to 5% of the major form. In addition, the deconvolute function allows the user to clearly see the indel sequence, which is typically much harder to decipher using other programs.

This function is also useful in a more universal sense for pulling apart peak profiles in a regular mixed sequence. For example, Figure 17-9 shows data from a clone containing sequence duplication. Before analysis, it was

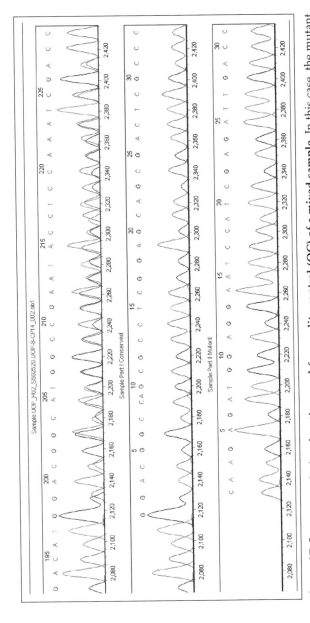

**Figure 17-9. Deconvolution function used for quality control (QC) of a mixed sample.** In this case, the mutant sequence is not shifted left or right because the mutant is not a true hetero-indel. Having a reference sequence for the expected sequence, Mutation Surveyor can pull the expected sequence from the unknown (mutant) sequence. BLASTing the unknown sequence reveals its identity.

unknown exactly how and where this duplication occurred. Sequencing from the duplication into the vector junction revealed a mixed sequence that had been pulled apart into two separate clean traces using the deconvolute function. This differs from the hetero-indel application in that the second trace profile is not matched to the reference and a gapped alignment is not created. The result is a very quick and easy analysis of the duplication and identification of its exact location and flanking sequence.

In terms of overall utility, accuracy, and ease of use, Mutation Surveyor is undoubtedly among the premier comparative sequence analysis tools for the desktop PC (16).

### SeqScape® Software: Overview

SeqScape (current version is 2.5) is a sequence analysis program from Applied Biosystems (Foster City, CA, USA; http://www.appliedbiosystems.com) designed to address variant identification and SNP discovery. The basic functionality of the software is that it analyzes sequence trace data (base calls and quality scores), compares it against a reference sequence, and generates a report of all variants in the sequence(s) versus the reference. The hardware specifications are not excessively demanding, allowing SeqScape to operate on most PCs, and with modest size sample sets, the JAVA-based program can perform analysis within few minutes. SeqScape is the tool of choice of Applied Biosystems (ABI) and has been positioned as an integrated component of ABI's suite of products, equipment and software for DNA sequencing and analysis.

For those familiar with ABI's software, the overall look and feel of SeqScape is similar to Sequencing Analysis Software and MicroSeq® ID Software, and it shares many of the same options and graphical user interfaces (GUIs). On the surface, it looks like ABI just took their Sequencing Analysis Software and added an assembly function to create another niche product offering. This is because the backbone of SeqScape is the KB™ Basecaller (KB), which was introduced in 2002, alongside the 3730/3730xl model DNA sequencer and has since been incorporated into ABI's full sequence analysis pipeline (including Sequencing Analysis Software, SeqScape and MicroSeq ID). Similar to other programs described in this chapter, KB/SeqScape allows the user to define threshold limits for mixed base identification (e.g., 62% of the major form). Low quality bases can be trimmed (*Clear Range*) from sequences and whole traces can be rejected (*Filter*) based on overall sequence quality values (*QVs*) assigned by KB.

SeqScape can be optionally integrated into ABI 3130/3130xl and 3730/3730xl Data Collection Software v3.0 for auto-analysis directly from the sequencer, which can be especially useful for processing in 96-well batches rather than larger data sets, which tend to slow down SeqScape's processing capabilities. The software seems to work optimally in project sizes of 96 samples or less without sacrificing central processing unit (CPU) speed.

It should also be mentioned that SeqScape is optionally bundled to ABI's VariantSEQr™ Resequencing System. Purchasers of the VariantSEQr probe sets receive a CD-ROM with a preconfigured SeqScape analysis template specific for the purchased probe sets. This is a nice enhancement because setting up a new analysis template in SeqScape is not as straightforward as in other programs. In addition, SeqScape will recognize and automatically trim M13 tails from the sequence introduced by the M13-tailed primer sets. (At the time of writing this section, we are uncertain about the future of the VariantSEQr system as the ABI's listing of this product is "currently unavailable.")

Upon first opening SeqScape, the user may notice that the program requires an access I.D. and password. While this may seem trivial and at times a nuisance, ABI has been one of the few companies to address the U.S. Food and Drug Administration's (FDA's) guidance for 21 CFR 11 in relation to DNA sequence analysis software. For individuals working in a regulatory environment, this is a major concern. In addition to access control, SeqScape has an optional audit trail function as a part of its custom reporting function (*Report Manager*), which records all created base edits and the reasons for change.

From the point of installing SeqScape to actually using the program adeptly is not as intuitive and straightforward as with the other desktop analysis programs mentioned above. ABI supplies an arsenal of product literature and a one-inch thick user guide, rather than the few release notes supplied for the Mutation Surveyor and Sequencher. Nonetheless, after digesting the initial orientation, the software becomes much easier to use and is especially convenient for having to reanalyze traces for peak spacing, base calls, etc., because the KB Basecaller is integrated.

Creating the initial Reference Data Group (RDG) template is perhaps one of the more tedious aspects of using SeqScape. The user must import a reference sequence and define layers (units of analysis such as exons only or introns only) and non-overlapping regions of interest (exon 1, exon 2, etc.), which are grouped together into active layers. SeqScape has tried to make this process easier by defining the RDG information automatically using the feature table from a GenBank (.gb, .gbk) entry. All introns, exons, promoters, and other features are clearly displayed by the software as layers and attractive visual overviews automatically upon importing the GenBank file. This is a very nice feature of SeqScape and one that would be nice to see replicated in other, similar analysis programs. However, without a GenBank reference file or a well-defined feature table, the RDG creation process takes much longer. In general, working with large sample sets and few RDGs is not a problem, but vice versa (small sample sets and many RDGs) can become extremely time consuming because of the involved RDG creation process. Once a RDG is created, it can be reused and applied to any project.

Once the initial project template is created, importing and analyzing data are fairly mundane. Sample naming conventions and delimiters can be defined in SeqScape for automatic grouping of data sets, as in the aforementioned programs. Multiple visualization tools allow the user to view and pilot through analyzed data in the project navigator using a branching tree format as layer overviews displaying *known* (previously confirmed) and *all* (confirmed and unconfirmed) variants, as graphical contig alignments, as text alignments with QV bars, and as sequence traces (Figure 17-10). From here, the user can easily plot to variant base calls, make base edits, delete/remove selections, and adjust clear ranges (CRs), if necessary. Again, it is best to keep the project sizes limited to 96 since SeqScape reanalyzes the data after every base edit. For instance, a project with 100 specimens and 400 sequences has an analysis time of 10 seconds per base edit on a 2.4 GHz Pentium 4 with 1 GB of RAM.

SeqScape's Report Manager is fairly comprehensive and generates ten reports with each project analysis: AA Variants, Analysis QC, Audit Trail, Base Frequency, Genotyping, Library Search, Mutations, RDG, Sequence Confirmation, and Specimen Statistics.

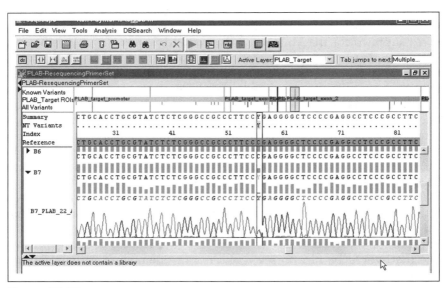

**Figure 17-10.** Graphical visualization of layer overviews and sequence alignments with quality values (QV) bars.

Reports are hyperlinked to the Project Navigator, so clicking on a linked report entry takes the user directly to the sequence for immediate visual confirmation or rejection.

The most important criterion for evaluating SeqScape is the degree of accuracy of the calls mutations and, specifically, heterozygotes. For 50:50 mixtures, the program calls mixed bases pretty accurately, although there is still typically the need to manually edit or analyze out a fair amount of false positives. SeqScape also has the ability to detect hetero-indels, identify the indel sequence, and assign a quality score as well. Sequence trace deconvolution is currently unavailable due to SoftGenetic's patent.

Dropping the threshold for calling the secondary peak height to ≤25% in the Analysis Protocol Manager will substantially increase the false positive rate. This is expected, although in general the software is not as forgiving as other programs in the amount of baseline noise it can tolerate before miscalling heterozygotes. Nonetheless, with a little practice and very clean sequence data (which is optimal and preferred for heterozygote identification), SeqScape is more than a capable tool for comparative sequence analysis. Furthermore, ABI supports SeqScape by providing a wealth of data and literature validating the accuracy and many re-sequencing applications of this software (1).

These three programs—Sequencher, Mutation Surveyor, and Seq-Scape—represent some of the most popular commercially available desktop sequence analysis tools. No one towers over the rest as each has its own share of strengths and weaknesses (Table 17.1). These software tools have enjoyed success for a reason and the manufacturers continue to improve their function on behalf of their users. It is worthwhile to point out that the information provided here is a high-level overview of the basic functionality and features of each program, and users are encouraged to evaluate each the these programs to determine which best suits their needs.

Demo copies of the programs mentioned in this chapter can be obtained by visiting the manufacturers' Web sites.

## Vector NTI Overview

Vector NTI Advance (current version is 9.0) is a DNA and protein sequence analysis and data management suite of software programs from Invitrogen Bioinformatics software solutions (Invitrogen, Carlsbad, CA, USA; formerly Informax, Inc.; http://www.invitrogen.com). Vector NTI is available for both Mac and Windows operating systems.

The core functionality of Vector NTI is the database management software for keeping track of DNAs, proteins, oligos, enzymes, and even BLAST and PubMed literature searches. As such, keeping track of sequences becomes a tractable task, allowing for sorting and searching by

Exploring - Local Vector NTI Database

Table  Edit  View  Analyses  Align  Database  Assemble  Tools  Help

M-CSF

All Subsets | Subset M-CSF of DNA/RNA Molecules

DNA/RNA Molecules (MAI...
- Invitrogen vectors
- Annotation Jamoree
- B7-H3 retree
- Butyrophilin like
- Costim Collection Ligands
- Costim Collection Receptor
- export
- Follistatin related
- Gateway & Topo tools
- Gateway Vectors
- General Vectors
- GPI
- Group 2
- Ig constructs
- IIGSF3 project
- IgV project
- m.hBart1 project
- m/hB7-H3 project
- CTLA4 mutants
- Manufacturing Vectors
- GFP, YFP, CFP, DsRed
- pG construct sequences
- PD-1 project
- Costim Ligand Pep
- RAGE
- OMHsIgG1 Project
- Miscellaneous
- Search Results
- Sequenced Vectors
- VNTI class
- Acc2
- 2q33 genomic
- Tags

| Name | Length | Form | Storage Type | Author | Original Author | Modified |
|---|---|---|---|---|---|---|
| hCSF1 BC021117 | 2851 | Linear | Basic | Paul Wu | UNKNOWN | 01/18/05 05:44PM |
| hCSF1 Celear Genomic GA_x5YUV32VTKB_015 | 110000 | Linear | Basic | Paul Wu | Paul Wu | 02/08/05 07:55AM |
| hCSF1 genomic 51458073_066 | 110000 | Linear | Basic | Paul Wu | Paul Wu | 01/26/05 04:32PM |
| hCSF1 NM_000757 exons | 2749 | Linear | Basic | Paul Wu | UNKNOWN | 01/26/05 12:04PM |
| hCSF1 NM_000757 exons+hM-CSF NM_000757 | 2749 | Linear | Basic | Paul Wu | Paul Wu | 01/26/05 03:02PM |
| hCSF1 var 2 NM_172210 | 1519 | Linear | Basic | Paul Wu | UNKNOWN | 01/21/05 06:07PM |
| hCSF1 var 3 NM_172211 | 1855 | Linear | Basic | Paul Wu | UNKNOWN | 01/21/05 06:07PM |
| hCSF1 var 4 NM_172212 | 2545 | Linear | Basic | Paul Wu | UNKNOWN | 01/21/05 06:07PM |
| hCSF1R public genomic 3' 37550092_107 | 110000 | Linear | Basic | Paul Wu | Paul Wu | 02/09/05 11:08PM |
| hCSF1R public genomic 5' 37550092_106 | 110000 | Linear | Basic | Paul Wu | Paul Wu | 02/09/05 11:07PM |
| hM-CSF Celera Exons | 2749 | Linear | Basic | Paul Wu | UNKNOWN | 02/08/05 08:00AM |
| hM-CSF NM_000757 | 2749 | Linear | Basic | Paul Wu | UNKNOWN | 12/01/04 10:19AM |
| hM-CSF R Origene | 2919 | Linear | Basic | Paul Wu | UNKNOWN | 02/09/05 10:48PM |
| hM-CSF receptor NM_005211 | 3985 | Linear | Basic | Paul Wu | UNKNOWN | 12/07/04 09:26AM |
| hMC51R public exons | 3985 | Linear | Basic | Paul Wu | UNKNOWN | 02/09/05 11:17PM |
| mCSF (ERD) | 2185 | Linear | Basic | Paul Wu | UNKNOWN | 02/03/05 05:02PM |
| mCSF (ERD) public mouse exons | 2185 | Linear | Basic | Paul Wu | UNKNOWN | 02/03/05 05:06PM |
| mCSF1 BC066200 | 3252 | Linear | Basic | Paul Wu | UNKNOWN | 01/18/05 05:47PM |
| mCSF1 BC066205 | 3252 | Linear | Basic | Paul Wu | UNKNOWN | 01/18/05 05:46PM |
| mCSF1R BC036343 | 3699 | Linear | Basic | Paul Wu | UNKNOWN | 01/18/05 05:24PM |
| mCSF1R BC050024 | 3822 | Linear | Basic | Paul Wu | UNKNOWN | 01/18/05 05:36PM |
| mCSF1R BC066863 | 3793 | Linear | Basic | Paul Wu | UNKNOWN | 01/18/05 05:27PM |
| mCSF1R IMAGE30436119 | 3989 | Linear | Basic | Paul Wu | Paul Wu | 02/10/05 05:51PM |
| mCSF1R IMAGE30436119 public exons | 3989 | Linear | Basic | Paul Wu | UNKNOWN | 02/10/05 06:30PM |
| mM-CSF NM_007778 | 3922 | Linear | Basic | Paul Wu | UNKNOWN | 12/01/04 10:19AM |
| mM-CSF NM_007778 public exons | 3922 | Linear | Basic | Paul Wu | UNKNOWN | 02/10/05 01:33AM |
| mM-CSF R NM_007779 | 3665 | Linear | Basic | Paul Wu | UNKNOWN | 12/15/04 01:09PM |
| pSP65+mMCSF (ERD) | 2185 | Linear | Basic | Paul Wu | UNKNOWN | 02/10/05 01:42AM |
| pSP65+mMCSF (ERD) public exons | 2185 | Linear | Basic | Paul Wu | UNKNOWN | 02/10/05 12:37AM |
| pW1653 pOTB7+hCSF1 | 4661 | Circular | Constructed | Paul Wu | Paul Wu | 02/09/05 10:44PM |
| pW1654 pCMV6-XL4+hCSF1R | 8487 | Circular | Constructed | Paul Wu | Paul Wu | 02/09/05 11:43PM |
| pW1654 pCMV6-XL4+hCSF1R insert | 4314 | Linear | Basic | Paul Wu | Paul Wu | 02/09/05 11:33PM |

**Figure 17-11.** **Vector NTI Database Explorer window showing files within the M-CSF subbase.** An overview-type of information (length, form, author, etc.) is easily accessible in this view.

name, date created, or other field specified by the user (Figure 17-11). Files can be stored in folder-like subbases and dividers within the integrated personal database of Vector NTI. The sequences themselves can be imported from a variety of formats including GenBank, GenPept, EMBL/SWISS_PROT, and FASTA, and even directly from any text ASCII file. Likewise, sequences can be exported into the same variety of formats.

The main way of working with the DNA/RNA or protein sequences is using the molecule viewer (Figure 17-12). The power of Vector NTI lies in the dynamic interface of the molecule view that allows the user to view the sequence, annotation, and a graphical representation of the annotation in three panes within one window. Clicking on a feature in the graphics window will automatically highlight the sequence text below. Users are able to add their own annotation to their sequence with relative ease.

Included in Vector NTI are sets of powerful automated tools to aid in standard molecular biological analyses and manipulations such as open reading frame identification searches, DNA to protein translations, PCR primer design, and restriction enzyme maps. More advanced features include construct design strategies using restriction sites to "cut and paste" compatible DNA ends to generate complete plasmid maps. Invitrogen's popular Gateway cloning system is now complemented in Vector

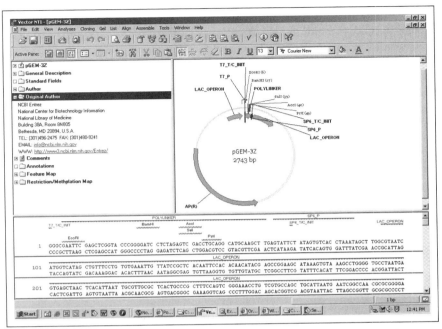

**Figure 17-12. Vector NTI Molecule Display window for GenBank entry AF414120.** This view shows three panes representing different details for this molecule. The top left panel provides information useful for referencing. The top right panel is a graphic representation with restriction sites and other features. The bottom panel shows a comparison to other molecules in GenBank with additional annotations.

NTI as a functionality that allows for quick parallel cloning of your gene of interest into multiple vectors with virtual LR or BP cloning reactions.

Many other programs exist in the Vector NTI suite of programs. Below is a list of just a few of the more useful ones.

- AlignX performs multiple sequence alignment for either protein or DNA sequences. It features sequence comparison dot plots, guide trees, and multicolor alignment charts, and has dynamic links back to Vector NTI main window.
- BioAnnotator hosts a large number of protein and nucleic acid analysis tools including ProSite, PFAM, BLOCKS, and Proteolytic cleavage.
- ContigExpress is used for assembling DNA sequencing project data as chromatograms from automated sequencing machines. Contigs generated in ContigExpress can be viewed in Vector NTI's main window by a simple drag and drop.

- GenomeBench, as the name suggests, is used for genomic DNA analysis. Data from UCSC, Ensembl or other DAS-compliant servers can be downloaded and viewed in GenomeBench. Mapping cDNA sequences to genomic DNA sequences can be done in GenomeBench using Sim4 or Spidey.
- 3D-Mol is a program for viewing, manipulating, and calculating three-dimensional molecular structures based on files found in the Brookhaven Protein Data Bank database.

The Vector NTI's comprehensive functions come with a cost, as it is generally more expensive than other sequence analysis software packages. However, considering the needs of a highly productive, modern molecular biology laboratory, Vector NTI fills an important need in organizing, archiving, and analyzing a whole host of molecular biology data.

### DNAStar Overview

This is a PC and Mac desktop package consisting of seven separate units (current version is 6.0):

1. EditSeq. This is a comprehensive unit used for editing, importing/ exporting of sequences and adding annotations. It is capable of: importing/exporting data in a variety of popular formats including fasta and GenBank, and translating a DNA molecule using any of 14 different provided genetic codes (if desired, a user can add its own code), identify ORFs, create, and modify annotations using GenBank's pre-selected keys.
2. GeneQuest. This helps to find genes (uses Borodovsky's Markov method) and their regulatory motifs as well as other patterns and structures in a primary sequence. It can predict intron/exon boundaries using species-specific patterns for splice variants. In addition, it has typical features for sequence analysis displaying: restriction sites (and simulated graphical image of an agarose gel separation), dyad, direct and inverted repeats, codon usage, and RNA folding.
3. MegAlign. Primarily used for pairwise (Dotplot, Lipman-Pearson, Martinez Needleman-Wunsch, and Wilbur-Lipman algorithms) and multiple (Clustal V, Clustal W, and Jotun Hein algorithms) sequence alignments. It can reconstruct phylogeny and calculate sequence similarity and distance.
4. PrimerSelect. It contains a sophisticated (although quite comparable to other such programs) set of tools for designing and analyzing primers used for PCR, as hybridization probes or sequencing. Many parameters (Tm, lengths, GC-content, etc.) can be easily adjusted to

avoid potential pitfalls such as hairpins or formation of primer-dimers.

5. Protean. Enables a display of four predicted secondary stru...res (alpha helix, beta sheet, coil structures, and turns) and other physical characteristics of protein sequences with an impressive suite of analysis tools, including amphiphilicity, antigenicity, charge density, hydropathy, and surface probability (each of these tools has many different methods).

6. SeqBuilder. Allows for the creation (and construction) of plasmid maps and sequence editing and analysis. The partial list of features in this unit include restriction sites finding, frequency of cuts, type/class of enzyme, overhang compatibility, ORFs viewing, translation of DNA with variety of genetic codes, and finding related sequences with Blast.

7. SeqMan II. Used for sequence assembly and management of contigs. This unit can read variety of formats and download sequences directly from NCBI. It also can be used to read and edit assemblies created by Phrap and Sequencher (this is a particularly useful new feature in this release) and perform many routine-editing tasks like removing vector sequences and trim poor quality bases.

Each of the above modules contains many more features that are nicely integrated in an office-like suite: the files generated in one unit can be relatively easily imported to other units with preserved annotations and other added characteristics. Files can be imported and exported in a variety of popular formats. All displays are quite flexible, can be customized for a particular purpose, and are publication-ready. For the full range of DNASTAR capabilities and features and a fully functional demo version of this program, visit the manufacturer's Web site (http://www.dnastar.com).

No description of sequence analysis tools would be complete without mentioning a venerable GCG/Seqweb package (also know as Wisconsin package, currently in the care of Accelerys; http://www.accelerys.com). This is a powerful and comprehensive Unix-based software with more than 140 individual tools of which over half are accessible through Web-based interface. An excellent description of this package is presented elsewhere (13).

## Sequence Repository Databases

A centralized sequence repository is necessary for scientists to exchange, interpret, and compare their results. It allows for an efficient, effective, long-term storage and provides a vehicle for generating workflows for post-processing that enable a unique common reference for communication. The analysis results generated by downstream processing of the sequence(s) can be effectively communicated with other analyses, technologies, and groups. The analysis as well as annotations generated during post-processing also can be stored with the sequences, hence futher improving the value of the sequence.

One of the most widely used public domain sequence repositories is the set of databases (Entrez) produced by the National Center for Biotechnology Information (NCBI). GenBank (part of Entrez) contains publicly available DNA sequences for more than 140,000 named organisms (4). The database is produced in an international collaboration between NCBI-GenBank (http://www.ncbi.nlm.nih.gov/), EMBL-EBI (European Bioinformatics Institute; http://www.ebi.ac.uk/embl/) (9), and the DDBJ (DNA Database of Japan; http://www.ddbj.nig.ac.jp/) (9) named the International DNA Database Consortium. The sequences are submitted by individual sequencing laboratories or by large-scale sequencing projects. Each of the three groups collects a portion of the total sequence data reported worldwide, and all new and updated database entries are exchanged between the groups daily. The unique identifiers generated by these groups are being used universally for identifying the sequence.

In the early days of GenBank, two things caused people to want to submit their data to a centralized data repository. They realized that having it all in one place would benefit the entire scientific community. That was the carrot. However, there was also a stick. All of the publishers of major bioscience journals made it a requirement for publication that sequenced data, contained within or referred to an article, could not be published until it had been submitted to GenBank.

When scientists submit data to GenBank/EMBL/DDBJ, their data are confidential until their work is published. This helps to dispel concerns that the availability of their data in GenBank before publication may compromise their work. The sequence record is released when the article referring to the sequence is published.

All three sequence repositories also have a wide array of resources for querying the sequence and associated information. A diagrammatic representation of the available resources at NCBI is shown in Figure 17-13 (color Plate 14). Each of these nodes is richly connected to the others, making unique relationships among its members. The system hence serves as a great discovery tool.

These sites also provide a plethora of tools, not only for database searches and sequence retrieval, but also for a variety of other purposes such as sequence analysis, similarity and homology, structure analysis, and protein function analysis.

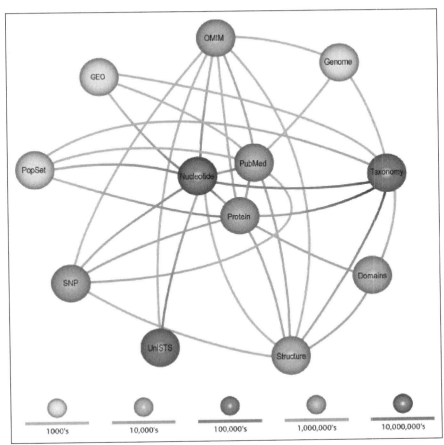

**Figure 17-13. Entrez Search and Retrieval System Nodes,** © **NCBI.** Different colors represent the number of entities in each category. (See Plate 14.) Abbreviations are: Nucleotide, number of sequenced bases deposited in GenBank; PubMed, number of citations; Protein, number of known proteins; OMIM, Online Mendelian Inheritance in Men, a database of human genes and genetic disorders; Genome, number of genomes sequenced; Taxonomy, number of known species; Domains, number of identified protein domains; Structures, number of solved protein structures; UniSTS, number of unique short tagged sequences; SNP, number of identified single nucleotide polymorphisms; PopSet, a set of DNA sequences collected to analyze the evolutionary relatedness of a population; GEO, Gene Expression Omnibus, a data repository of high-throughput gene expressions and hybridization arrays. This figure was downloaded from the NCBI Web site. Used with permission.

The full release database(s) or daily updates can be downloaded either directly from the above-mentioned Web sites or from http://www.ncbi.nlm.nih.gov/Ftp/index.html. A wide range of software tools can be downloaded for building internal analytical capabilities (ftp://ftp.ncbi.nih.gov/toolbox/). The majority of commercial entities have opted for local copies of the publicly available sequences either in the form of flat files or in a more structured relational database. The rationales for local copies are manyfold: confidentiality, proprietary annotation and analysis, ability to query and analyze the data from the confines of the firewall, ability to utilize proprietary toolset, creating workflows around the data, and the potential to integrate with the internal dataset.

From the beginning, Genbank has stored the sequences using a flat-file data structure. Each entry or record in GenBank may contain identifying, descriptive, and genetic information in ASCII-format files. This flat-file format has seen signs of strain with the growth of genetic data. With advancing technologies and an ever-improving knowledge base, especially one growing as quickly as genetic data, it's difficult for the design of a data store such as Genbank to keep abreast. Despite several attempts to change the format of stored data, the flat file structure remains a default option.

One other way of storing the data is in the structured relational database. The majority of other users have adapted some form of relational database storage. These databases typically have a relational structure imposed upon the data, including associated indices, links, and a query language. These databases can be used for storage and retrieval of the sequence and related data.

A sequence stored in a central repository may need to be pre-processed to meet the storage requirements and optionally post-processed for generating analysis and other annotation associated with it. These steps can be performed manually. However, it is beneficial to have some of the processing steps automated. The workflows can automate some of the processing steps, including validation checks, vector contamination, automated sequence assemblies, sequence analysis (including BLAST), translations, and annotation, to name a few. A variety of proprietary workflows exist at all levels, including workflows post-submission to NCBI for analysis and annotation. These workflows can be used for additional analysis of the sequences and sequence sets for characterization and classification.

In the current age of the life sciences, investigators have to interpret many types of information from a variety of sources: lab instruments, public databases, gene expression profiles, raw sequence traces, single nucleotide polymorphisms, chemical screening data, proteomic data, putative metabolic pathway models, and many others. The sequence data storage is just part of the solution to the Life Science data integration problem. Solutions may include a federated database approach, memory-mapped data structures, indexing flat files, and data warehousing.

In order for successful and fruitful life science data integration, a standard vocabulary and a standard set of tools for data access must to be present on top of any data storage solution offered. Descriptions of possible configurations are presented on the Internet (http://www.iscb.org/ismb2000/tutorial_prg.html, and http://www.iscb.org/ismb2000/tutorials/griffiths.html).

A number of tools are required to effectively make use of the sequence data present in the sequence repositories. These include:

- *Query Tools*: Required to help users ask meaningful biological questions across multiple domains and transfer the integrated data to a visualization tool for complex analysis.
- *Data Browsers*: Required to help users understand what is contained in the integrated data source. The Browser should lead users to an intuitive query interface.
- *Visualization Tools*: Required to help users sort through large volumes of integrated data, finding patterns, and trends that would otherwise go unnoticed.
- *Data Mining Tools*: Required by advanced users to automatically and intelligently search the integrated database to find ways to understand the data, predict future outcomes from it, and extract knowledge leading to new discoveries.

### Spotfire

Spotfire (http://www.spotfire.com; current version 8.1.1) is a data visualization and analysis system with wide usage in the life sciences industry. Founded in 1996, the Spotfire software has evolved from a stand-alone visualization application to a more powerful client-server system. The client software runs under the Microsoft Windows operating system and is capable of data retrieval, dynamic visualization, and analysis. In biology and the life sciences, potential applications for the software include retrieval, visualization, and analysis of data from high-throughput screens, plate-based biological assays (e.g., siRNA screens, ELISA), gene expression data from microarray experiments, and single nucleotide polymorphism (SNP) studies.

#### Data Retrieval

Spotfire can read data stored in formatted text files (e.g., CSV, tab-delimited text) and Microsoft Excel spreadsheets (XLS files) as well as querying data from relational databases (Microsoft Access or ODBC data sources). Through server-side data access functions, Oracle relational-type of databases can also be queried. The ability to query all of these data sources from a single software package makes Spotfire a potential point of data integration in any organization that maintains multiple data stores.

## Dynamic Visualization

The single most powerful aspect of Spotfire is its dynamic visualization capability. Once data are retrieved by the client, users can dynamically interact with visualizations of the data. Available visualizations include scatter plots, 3-D scatter plots, bar charts, histograms, pie charts, line charts, and heat maps. Figure 17-14 shows an example of a 3-D scatter plot. In all visualizations, data points can be colored, shaped, sized, or labeled based on continuous or discrete record values in the data. In addi-

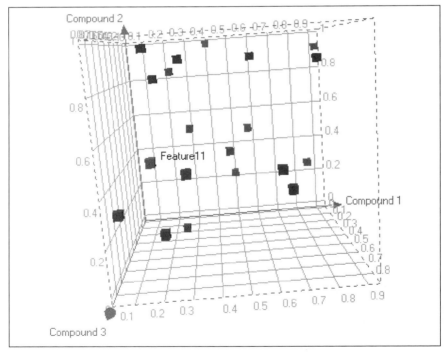

**Figure 17-14.** Example of a Spotfire three-dimensional scatter plot of hypothetical data from a mulitparametric cellular assay measuring 20 cellular features in cultured cells after treatment with three compounds (corresponding to the axes Compound 1, 2, and 3). The program can dynamically show labels for a user-selected datapoint, in this case the datapoint corresponding to "Feature 11."

tion, all the visualizations allow dynamic filtering and drill-down capability. Dynamic filtering is implemented by query devices. These devices are sliders, checkboxes, and others that are associated with each column in a dataset. Through the query devices, users can select any data subset or range, and the corresponding visualization is immediately updated to show only the selected data. If users click or select data points in any visualization using a pointing device, all information for the associated data records is shown in a *Details-On-Demand* window. This allows rapid drill-down to specific observations of interest. The combination of these features makes Spotfire extremely powerful for rapid data exploration in datasets containing tens to hundreds of thousands of records.

## Analysis

Early versions of the software focused on data visualization alone, but in subsequent releases, the analysis capabilities of Spotfire have been expanded through the concepts of *Tools and Guides*. These can be thought of as add-ons to the client software that are provided from the server and perform specific analysis tasks. Spotfire has organized Tools and Guides into groups based on typical customer applications; these groups are termed "*DecisionSites.*" For example, *DecisionSite for Functional Genomics* provides Tools and Guides that perform data formatting and analysis tasks that are commonly encountered in genomics applications such as data pivoting and clustering. On the client side, users select a DecisionSite that is available on a server and the corresponding Tools and Guides are then available for use on any dataset that is loaded into the client's system. Importantly, users can develop and deploy their own Tools and Guides to carry out custom analyses, using either programing methods or macro-like recording options. To provide more sophisticated analytical tasks, it is possible to create Tools and Guides that integrate with popular analysis tools such as Splus (http://www.insightful.com), R (http://www.r-project.org), or MATLAB (http://www.mathworks.com).

## Ingenuity Pathways Analysis

One of the central challenges in bioinformatics is turning complex genomic datasets into actionable knowledge or hypotheses for more focused experimentation. For example, many genomic methods such as comparative sequence analysis, gene expression microarrays, and proteomics yield large sets of genes of interest that contain hundreds or thousands of genes. These lists require biological annotation to make them scientifically useful, but the lists are too large for scientists to explore using one-gene-at-a time literature searches. Ingenuity Pathways Analysis (IPA; http://www.ingenuity.com; current version is 3.0) is a

Web-delivered application that enables biologists to discover, visualize, and explore therapeutically relevant networks significant to their experimental results, such as gene expression array data sets. The program is offered by Ingenuity on a subscription-access basis.

Users of IPA enter lists of gene identifiers (including Affymetrix, GenBank, Entrez Gene, and UniGene identifiers) into the system. IPA maps these gene identifiers to its knowledge base, and then provides several analyses. The Ingenuity knowledge base contains a large amount of biological information about gene function and gene interactions derived from a human review of the literature as well as automated analysis of journal abstracts. Available analyses in IPA include:

- *Network Explorer.* The application generates hypothetical networks containing user genes based on biological interactions between those genes at the transcriptional or protein levels. Networks are generated by a graph-building algorithm that links genes according to known biological relationships. An example is shown in Figure 17-15.
- *Functional Analysis.* The biological functions of genes are cataloged using IPA's gene ontologies, and the most prevalent or statistically enriched biological functions are identified and highlighted for the user.
- *Canonical Pathways.* User genes are overlaid onto canonical pathways from the literature to provide biological context for the user's gene list.

In many of its views, IPA provides extensive linkages to specific literature references and public literature sources (i.e., PubMed, etc.). This allows users to rapidly explore prior knowledge about their genes of interest.

In summary, IPA is a powerful annotation tool for large-scale genomic datasets with the ability to provide non-obvious insights into the biology of complex gene sets.

### Other Tools

A variety of publicly available as well as proprietary tools exists that can carry out complex data analysis or more of these requirements. This list is growing rapidly, and better and faster tools are being developed continuously.

Among hundreds of other commercial tools that can be used in a modern biology laboratory are:

- Ariadne Genomics (http://www.ariadnegenomics.com). Another pathway analysis software package integrating biological networks between proteins, genes, and tissues.

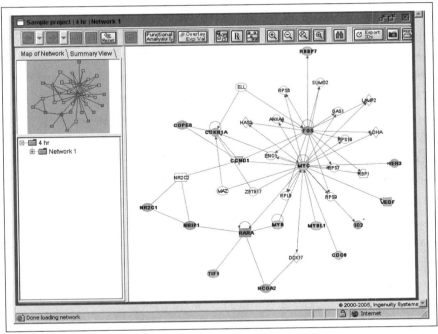

**Figure 17-15.** **Example of an Ingenuity Network analysis.** A dataset containing 145 gene identifiers and their corresponding expression values after stimulation of a breast cancer cell line was uploaded as an Excel spreadsheet using the template provided in the application. Each gene identifier was mapped to its corresponding gene object in the Ingenuity Pathways Knowledge Base. A cutoff was set to identify genes whose expression was significantly differentially regulated. These genes were then used as the starting point for generating biological networks. The diagram shows one of nine biological networks that were generated by the application based on literature associations among the genes. Shaded nodes indicate genes from the input gene list, and clear nodes are other genes that are functionally linked to user genes by biological or biochemical relationships. Data are taken from an IPA user tutorial dataset, which is available from the application's online help function.

- DNAsis Max from Hitachi (http://www.hitachisoft-bio.com). A comprehensive desktop package for sequence analysis.
- Oligo 6 (http://www.oligo.net) and Primer Designer (http://www.scied.com) packages. Comprehensive tools for designing and analyzing probes used for PCR, DNA sequencing, and hybridization.
- SRS platform from Lion Biosciences (http://www.lionbioscience.com). A comprehensive set of tools to analyze complex genome data. Integrates data from Celera, Incyte, and Derwent.

In addition, many bioinformatics departments of major universities post numerous software tools (typically free) that are as good as any commercial package, and these quite often are accompanied by extensive documentation and even source code. Of these, we wish to mention just a handful of sites: Indiana University (http://bioportal.cgb.indiana.edu); University of California at Santa Cruz (http://genome.ucsc.edu); Washington University (http://genome.wustl.edu/groups/informatics/software); and National Center for Biotechnology Information (http://www.ncbi.nlm.nih.gov/Sitemap/AlphaList.html).

Some of the software tools described in this chapter are highly specialized, easy to use, superbly perform a relatively small set of tasks (e.g., Sequencher), and new features are continuously added to address emerging needs. An example of such features would be the integrated primer design module, which can be as simple as user-defined regions where primers are needed or as complex as software-selected and user suggested primers to obtain full double-stranded coverage. The other feature would be an automated contig assembly of fragments with the same name. Though this feature is available in version 4.5 and it works relatively well with simple template names, but it is still cumbersome with complex template names that require defining not-so-straightforward regular expression rules. Often it is faster to highlight a set of chromatograms belonging to the same template and assemble them manually. It would also be helpful if the contig naming process was automated using "default names" dialog. Such requirements may be difficult to implement for any stand-alone program, and one possible solution can be through SOAP (simple object access protocol) access to other software tools (e.g., DNA LIMS that produces sequencing requests/orders and stores finished chromatograms).

On the other end of the spectrum are programs like Vector NTI or DNAstar, which contain many modules that can satisfy most computational needs of a typical molecular biology laboratory. Although they have an obvious advantage, because of integration, they too are limited in automation of repetitive tasks and are inherently more difficult to use.

We hope that the future will bring improved versions of current software tools as well as a new breed of highly integrated and flexible programs that will allow streamlining and automating many mundane tasks. Such developments will require closer and continuous collaboration of software developers with experienced bench scientists. Based on our observations, such collaboration is rather sporadic and typically involves scientists at the beta-stage testing level. This may already be too late in the development cycle and only cosmetic changes can be implemented without incurring huge redesign and delay costs. Table 17.1 lists major strengths and weaknesses of software packages described in this chapter.

**Table 17-1.** Strengths and weaknesses of the major software programs described in this chapter.

| Software Category and Short Description | Strengths | Weaknesses |
|---|---|---|
| **DNAStar**<br>Multipurpose package of computer programs for manipulating and analyzing DNA and protein sequence data | —Suite-like integration of all units<br>—Solid Web integration<br>—Over 23 year history on the market<br>—Ability to read Sequencher files | —Software does not address 21 CFR 11 guidelines<br>—Occasional unexpected crashes<br>—Lack of support for SNP analysis |
| **Ingenuity Pathways Analysis**<br>Comprehensive package to discover, visualize, and explore relevant networks among genes and proteins | —Extensive human-curated literature knowledge base<br>—Gene network visualization | —User interface is focused on gene-set centric analysis could be more flexible<br>—Software does not address 21 CFR 11 guidelines |
| **Mutation Surveyor**<br>DNA resequencing, sequence confirmation, mutation detection and analysis, SNP discovery, and validation | —Patented deconvoluted chromatogram profile displays<br>—Comprehensive and customizable reports<br>—Easy to use Auto Run and reference sequence database URL | —Software does not address 21 CFR 11 guidelines |

*(continues)*

349

**Table 17-1.** *Continued.*

| Software Category and Short Description | Strengths | Weaknesses |
|---|---|---|
| **SeqScape**<br>DNA resequencing, sequence confirmation, mutation detection and analysis, SNP discovery, and validation | —Integration with ABI equipment, software, and reagents (especially KB)<br>—Automatic identification of genes, exons, and introns with a GenBank file<br>—Comprehensive and customizable reports<br>—User log-in and project audit trailing to satisfy 21 CFR 11 requirements | —Not as straightforward as other desktop programs<br>—Slower project template set-up time<br>—Analysis is slow with large data sets even with a very fast CPU<br>—False positives/negatives called by KB require a fair amount of curation<br>—Only works with sequences from ABI instrumentation (.ab1 files only) |
| **Sequencher**<br>Multipurpose contig assembly and editing, de novo sequence assembly, comparative sequence analysis, heterozygote detection, restriction mapping, cDNA to gDNA mapping | —Simple and easy to use tool<br>—Multifunctional<br>—Very fast assembly algorithm with large data sets<br>—15 year history of continual and routine software additions and enhancements | —No reporting feature for identifying base edits—it is problematic for manual deletions and reviewing edits<br>—Software does not address 21 CFR 11<br>—Rudimentary variation reporting<br>—No primer design capability |

**Spotfire**

Comprehensive software for retrieval, visualization, and analysis of data from high-throughput screens, plate-based biological assays (e.g., siRNA screens, ELISA), gene expression data from microarray experiments, and single nucleotide polymorphism (SNP) studies

—Powerful dynamic visualization for multi-dimensional data exploration
—Extensive data integration capabilities

—The program is primarily a visualization tool. If complex analytical methods are required, it may require customization
—Software does not address 21 CFR 11

**Vector NTI**

Complete suite of DNA and protein sequence analysis, annotation, and database storage software

—Dynamic interactions between sequence panel and graphics panel as well as between alignments and original files
—Built-in database allows for sorting files and tracking lineages
—Reads and stores files in a Genbank format
—Web integration for search tools

—Software does not address 21 CFR 11 guidelines
—Steep learning curve resulting from relative complexity due to large number of functions

## Acknowledgments

Many thanks to Philippe Desjardins at NanoDrop Technologies for permission to use the NanoDrop® ND-1000 Spectrophotometer-related images. We would like to thank Todd Smith (Geospiza) and Jerry Tuneberg (DNAtools) for checking the accuracy of descriptions of Finch Server and DNAtools LIMS, respectively.

## References

1. ABI PRISM®. 2005. *SeqScapeR Software Version 2.5: User Guide: Rev A*. Foster City, CA: Applied Biosystems.
2. Alberts, B., Johnson, A., Lewis, J., et al. 2002. *Molecular Biology of the Cell*, 4th ed. New York: Garland Science.
3. Ausubel, F.M., Brent, R., Kingston, R.E., et al., eds. 1998. *Current Protocols in Molecular Biology*. New York: John Wiley.
4. Benson, D.A., Karsch-Mizrachi, I., Lipman, D.J., et al. 2004. GenBank: update. *Nucleic Acids Res* 32: D23–D26.
5. Brown, S.M. 2000. *Bioinformatics: A Biologist's Guide to Biocomputing and the Internet*. Natick, MA: Eaton Publishing.
6. Hannon, G.J., ed. 2003. RNAi. A Guide to Gene Silencing. Cold Spring Harbor, NY: Cold Spring Harbor Laboratory Press.
7. Innis, M.A., Gelfand, D.H., Sninsky, J.J., eds. 1999. *PCR Applications. Protocols for Functional Genomics*. San Diego: Academic Press.
8. International Human Genome Sequencing Consortium. 2001. Initial sequencing and analysis of the human genome. *Nature* 409: 860–921.
9. Kanz, C., Aldebert, P., Althorpe, N., et al. 2005. The EMBL Nucleotide Sequence Database. *Nucleic Acids Res* 33: D29–D33.
10. Koffman, D.M., and Sookdeo, H. 2005. DNA sequencing database. A flexible LIMS for DNA sequencing analysis. In: Kieleczawa, J., ed. *DNA Sequencing: Optimizing the Process and Analysis LIMS*. Sudbury, MA: Jones and Bartlett; 143–156.
11. Lehninger, A.L. 1970. *Biochemistry*. New York: Worth Publishers.
12. Lewin, B. 2003. *Genes VIII*. New York: Oxford University Press.
13. Maxam, A.M., and Gilbert, W. 1977. A new method of sequencing DNA. *Proc Natl Acad Sci U S A* 74: 560–564.
14. Misener, S., Krawetz, S.A. (eds). 2000. B*ioinformatics: Methods and Protocols*. Totowa, NJ: Humana Press.
15. Miyazaki, S., Sugawara, H., Ikeo, K., et al. 2004. DDBJ in the stream of various biological data. *Nucleic Acids Res* 32: D31–D34.
16. Mutation Surveyor™. 2005. *Operation Manual*. State College, PA: Softgenetics LLC.
17. Nelson, D.L., and Cox, M.M. 1999. *Lehninger Principles of Biochemistry*. New York: Worth Publishers.
18. Porter, S., Slagel, J., Smith, T. 2005. Geospiza's Finch-Server: a complete data management system for DNA sequencing. In: Kieleczawa, J., eds. *DNA*

*Sequencing: Optimizing the Process and Analysis.* Sudbury, MA: Jones and Bartlett; 131–142.

19. Rapley, R., ed. 1999. *The Nucleic Acid Protocols. Handbook.* Totowa, NJ: Humana Press.
20. Salisbury, M. 2005. At ASHG, Collins reports on sequencing, haplotypes, more. *Genome Technology* 49: 14.
21. Salzberg, S.L., Searls, D.B., Kasif, S., eds. 1998. *Computational Methods in Molecular Biology.* Amsterdam, Netherlands: Elsevier.
22. Sambrook, J., and Russell, D.W. 2001. *Molecular Cloning,* 3rd ed. Cold Spring Harbor, NY: Cold Spring Harbor Laboratory Press.
23. Sanger, F., Nicklen, S., and Coulson, A.R. 1977. DNA sequencing with chain-terminating inhibitors. *Proc Natl Acad Sci U S A* 74: 5463–5467.
24. Schena, M., ed. 2005. *Protein Microarrays.* Sudbury, MA: Jones and Bartlett Publishers.
25. *Sequencher™ Tour Guide for Windows.* 2005. Ann Arbor, MI: Gene Codes Corporation.
26. Venter, J.C., Adams, M.D., Myers, E.W., et al. 2001. The sequence of human genome. *Science* 291: 1304–1351.

# Index